Grand Challenges in the Diagnosis and Management of Cardiovascular and Cerebrovascular Diseases

Grand Challenges in the Diagnosis and Management of Cardiovascular and Cerebrovascular Diseases

Guest Editor

Sonu M. M. Bhaskar

MDPI

Basel • Beijing • Wuhan • Barcelona • Belgrade • Novi Sad • Cluj • Manchester

Guest Editor
Sonu M. M. Bhaskar
Department of Neurology
National Cerebral and
Cardiovascular Center
Osaka
Japan

Editorial Office
MDPI AG
Grosspeteranlage 5
4052 Basel, Switzerland

This is a reprint of the Special Issue, published open access by the journal *Diagnostics* (ISSN 2075-4418), freely accessible at: https://www.mdpi.com/journal/diagnostics/special_issues/1O8T2QFJ98.

For citation purposes, cite each article independently as indicated on the article page online and as indicated below:

Lastname, A.A.; Lastname, B.B. Article Title. *Journal Name* **Year**, *Volume Number*, Page Range.

ISBN 978-3-7258-5877-4 (Hbk)
ISBN 978-3-7258-5878-1 (PDF)
https://doi.org/10.3390/books978-3-7258-5878-1

Cover image courtesy of Sonu M. M. Bhaskar

Contents

About the Editor

Sonu M. M. Bhaskar

Sonu M. M. Bhaskar, MD, PhD, PHF, FANA, is an internationally recognized neurologist and physician-scientist specializing in vascular neurology, stroke, and neuroimaging. He is the Founding Director of the Global Health Neurology Lab and the NSW Brain Clot Bank in Sydney, Australia. Currently, he is on sabbatical as Visiting Director (2023–2025) at the National Cerebral and Cardiovascular Center in Osaka, Japan, supported by the Japan Society for the Promotion of Science (JSPS) and the Australian Academy of Science. Dr. Bhaskar's work lies at the intersection of cerebrovascular research, neurotechnology, health systems, and translational science. He leads multidisciplinary teams working across the areas of acute stroke mechanisms, health inequities, and biomarker discovery, with a strong emphasis on global health impact. His scholarly output includes over 170 peer-reviewed publications, with an H-index of 56 and over 38,000 citations—placing him among the top 2–3% of most-cited physicians and researchers in neurology worldwide. His work has been featured in journals such as *Lancet Neurology*, *Annals of Clinical and Translational Neurology*, and the *European Journal of Neurology*. Dr. Bhaskar has delivered more than 150 invited talks, chaired sessions at leading conferences, and contributed to training programs in neurology and public health leadership. His research and leadership have informed and shaped healthcare policy not only across the Asia-Pacific region but also globally. His numerous accolades include the European Academy of Neurology Investigator Award (2019, two-time recipient); the Rotary Vocational Excellence Award (2020); the Fellowship of the American Neurological Association (2022); the Australian Global Talent Award (2023); the Paul Harris Fellow citation (PHF, four-time recipient); and the TouchNEUROLOGY Future Leader Award (2025). He was elected as a full member of Sigma Xi, one of the world's oldest and most prestigious scientific research honor societies, in 2023.

Preface

The Reprint "Grand Challenges in the Diagnosis and Management of Cardiovascular and Cerebrovascular Diseases" brings together a body of work addressing some of the most urgent and complex problems in modern medicine. Cardiovascular and cerebrovascular conditions remain the foremost causes of disability and death worldwide, placing a heavy burden on patients, families, and healthcare systems. Despite advances in imaging, pharmacotherapy, and interventional approaches, significant gaps remain in timely diagnosis, optimal treatment, and equitable access to care.

The scope of this Reprint is deliberately broad, reflecting the multidisciplinary nature of the field. It spans innovations in diagnostic imaging and biomarker discovery, the application of artificial intelligence and machine learning, and the genetic underpinnings of cardiac and cerebrovascular disorders. It also highlights clinical investigations into heart failure, atrial fibrillation, and acute stroke, alongside critical appraisals of surgical and device-based interventions.

This Reprint presents a curated collection that stimulates further inquiry, fosters collaboration, and helps translate scientific advances into improved patient outcomes. It is motivated by the recognition that solving these challenges requires integrative approaches that unite laboratory discovery, clinical insight, and health system innovation.

This Reprint is intended for a wide audience of clinicians, researchers, and policy makers who are engaged in advancing cardiovascular and cerebrovascular health. It also serves as a resource for students and early-career professionals seeking an overview of current frontiers in the field.

Sonu M. M. Bhaskar
Guest Editor

diagnostics

MDPI

Editorial

Bridging the Heart and Brain—Grand Challenges in the Diagnosis and Management of Cardiovascular and Cerebrovascular Diseases

Sonu M. M. Bhaskar [1,2,3,4,5]

1 Global Health Neurology Lab., Sydney, NSW 2150, Australia; sonu.bhaskar@globalhealthneurolab.org
2 Clinical Sciences Stream, Ingham Institute for Applied Medical Research, Liverpool, NSW 2170, Australia
3 NSW Brain Clot Bank, NSW Health Pathology, Sydney, NSW 2170, Australia
4 Department of Neurology & Neurophysiology, Liverpool Hospital, South West Sydney Local Health District, Liverpool, NSW 2170, Australia
5 Department of Neurology, Division of Cerebrovascular Medicine and Neurology, National Cerebral and Cardiovascular Center (NCVC), Suita 564-8565, Osaka, Japan

Cardiovascular and cerebrovascular diseases (CVDs) remain the leading global causes of mortality and long-term disability [1,2]. Collectively responsible for nearly one-third of all deaths worldwide, they impose a rising economic and societal burden, particularly among aging populations [3] and in low- and middle-income settings [2]. Despite substantial progress in acute intervention, secondary prevention, and imaging, diagnostic and therapeutic gaps persist in early disease detection, risk stratification, and equitable access to advanced care [4,5].

This Special Issue of *Diagnostics*, entitled "*Grand Challenges in the Diagnosis and Management of Cardiovascular and Cerebrovascular Diseases*", brings together original research and comprehensive reviews that reflect the growing convergence between cardiovascular and cerebrovascular medicine. The eleven papers featured span neurovascular imaging, genetic and inflammatory biomarkers, artificial intelligence-driven prediction models, and early identification frameworks for heart failure. Together, these contributions highlight the need for multidisciplinary collaboration and global standardization in diagnostic practice [6].

Accurate diagnosis remains foundational to effective vascular management [7]. Several studies in this Special Issue demonstrate how advances in imaging continue to redefine both cardiac and cerebrovascular assessment [8]. The NEUROGEN-SVD Study (Jin et al.) analyzed 18,425 participants to examine MRI-visible markers of cerebral small vessel disease (white matter hyperintensities, lacunes, microbleeds) and the APOE ε4 allele in predicting cognitive outcomes [9]. The findings strengthen the concept of an integrated *vascular–neurodegenerative continuum* and support multimodal imaging-genetic models for risk assessment and early detection of cognitive impairment [10]. In coronary imaging, Singhal et al. reported that computed tomography coronary angiography (CTCA) performed at initial presentation significantly improved diagnostic accuracy in children with Kawasaki disease [11]. Conducted at a large tertiary care hospital in North India, the study demonstrated that CTCA identified distal aneurysms and coronary thrombi frequently missed on transthoracic echocardiography, thereby facilitating more precise treatment decisions and, in several cases, allowing for the appropriate de-escalation of therapy. Bonanni et al. evaluated global longitudinal strain (GLS) after surgical aortic valve replacement. GLS improved significantly after one year, particularly among patients with mild diastolic dysfunction, suggesting that strain metrics are sensitive indicators of myocardial remodeling and postoperative recovery beyond conventional ejection fraction [12]. Together, these studies demonstrate the increasing role of quantitative imaging parameters and how modern imaging enhances diagnostic accuracy, treatment planning, and longitudinal assessment in vascular disease.

The contribution of systemic inflammation and circulating biomarkers to vascular disease pathophysiology also features prominently in this Special Issue [13]. Memiş et al. examined the prognostic role of the neutrophil-to-lymphocyte ratio (NLR) in over 1000 patients with large-vessel occlusion stroke undergoing thrombectomy [14]. While NLR correlated with initial stroke severity, it did not independently predict 90-day outcomes after successful reperfusion, suggesting that its prognostic value may be limited to the acute inflammatory phase [15]. On a population scale, Sastre-Alzamora et al. analyzed data from 139,634 Spanish workers and demonstrated that atherogenic indices (total cholesterol/HDL, LDL/HDL, and triglyceride/HDL ratios) strongly correlated with elevated "*Heart Age*" [16]. These accessible and low-cost markers have potential utility in population-level screening and risk communication, bridging laboratory science and preventive cardiology.

Genomic and molecular studies continue to advance precision diagnostics in cardiology [17]. Jiang et al. identified two novel truncating mutations in SOX4 among patients with familial and idiopathic atrial fibrillation (AF) [18]. Functional assays confirmed impaired transactivation of GJA1 (connexin-43) and SCN5A, implicating SOX4 as a contributor to atrial electrical remodeling. These findings expand the molecular understanding of AF and highlight how genetic testing can inform individualized risk profiling and early detection strategies [19,20].

Artificial intelligence (AI) is increasingly being integrated into vascular diagnostics [21]. Goh and colleagues conducted a meta-analysis of machine learning (ML) models for stroke and AF across more than seven million patients [22]. The pooled area under the receiver operating curve (AUROC) ranged between 0.68 and 0.79, indicating moderate accuracy with superior performance observed in ensemble approaches such as Random Forest and Extreme Gradient Boosting. Despite these promising results, the analysis revealed major gaps in validation, calibration, and reporting standards. The study reinforces that AI tools must be transparent, reproducible, and population-representative to achieve reliable clinical translation. Standardized evaluation frameworks, including TRIPOD-AI guidelines, remain essential for benchmarking their real-world utility.

Interventional and critical care aspects of cardiovascular management are also represented. Ziayee et al. compared outcomes in patients treated with veno-arterial extracorporeal membrane oxygenation (va-ECMO) alone versus ECMO combined with intra-aortic balloon pump (IABP) [23]. No significant difference in procedural brain infarction rates was observed between groups, and thromboembolic strokes predominated. These results suggest that IABP use neither increases nor mitigates the neurological risk inherent in ECMO but underscores the importance of integrated neuro-monitoring and anticoagulation management in these high-risk populations.

Two comprehensive reviews in this Special Issue address heart failure (HF), emphasizing early detection and stratified care [24,25]. Jankajova et al. introduce a six-stage HF framework linking behavioral and biological risk factors with advanced imaging parameters, including speckle-tracking echocardiography and global longitudinal strain, to identify subclinical dysfunction before symptom onset [24]. Manzi et al. reviewed acute HF across non-ischemic cardiomyopathies, delineating significant heterogeneity in etiology, clinical presentation, and prognosis [25]. Both studies advocate for substrate-based diagnostic strategies and targeted therapies, key steps toward the realization of precision cardiology.

Beyond individual modalities, this Special Issue highlights an overarching imperative: the integration of diagnostic data into coherent, clinically actionable frameworks. Cardiovascular and cerebrovascular disorders are physiologically interconnected manifestations of systemic vascular dysfunction. Progress, therefore, depends on harmonizing molecular biomarkers, imaging metrics, and computational models to capture the full spectrum of vascular health and disease. The globally diverse studies represented here, spanning Asia, Europe, and Australia, underscore the shared commitment to addressing these challenges collaboratively. Yet, equitable access remains central; diagnostic innovation must be accompanied by implementation strategies that ensure applicability across varied healthcare systems and populations.

This Special Issue of *Diagnostics* highlights current progress and ongoing challenges in the diagnosis and management of cardiovascular and cerebrovascular diseases. Advances in imaging, molecular profiling, genetics, and AI are reshaping clinical practice by improving diagnostic precision and guiding individualized management [17,21]. However, further work is needed to standardize methodologies, validate tools across populations, and embed innovations within clinical workflows or everyday clinical practice. Multidisciplinary collaboration and cross-regional partnerships will be essential to translating technological advances into tangible improvements in patient outcomes. The integration of cardiovascular and cerebrovascular diagnostics represents an important step toward a unified approach to vascular health, one that aligns technological innovation with patient-centered, equitable care.

Funding: No funding was received for the writing of this manuscript. S.M.M.B. is supported through the JSPS International Fellowship, funded by the Japan Society for the Promotion of Science (JSPS) and the Ministry of Education, Culture, Sports, Science and Technology (MEXT), Japan (Grant ID: P23712), and the Australian Academy of Science. A separate project is supported by KAKENHI (Grant ID: 23KF0126).

Institutional Review Board Statement: This article does not contain any studies with human participants or animals performed by any of the authors.

Informed Consent Statement: For this type of study, informed consent is not required.

Acknowledgments: The author thanks all contributing authors, reviewers, and the *Diagnostics* editorial staff for their support in the editorial process and commitment to advancing cardiovascular and cerebrovascular diagnostics and care.

Conflicts of Interest: The author reports leadership or fiduciary role in other board, society, committee, or advocacy group, paid or unpaid, with the National Cerebral and Cardiovascular Center (Osaka, Japan) as Visiting Director (2023–2025); *Rotary District 9675* (Sydney, Australia) as District Chair for Diversity, Equity, and Inclusion; the Global Health and Migration Hub Community, *Global Health Hub Germany* (Berlin, Germany) as Chair, Founding Member, and Manager; and Editorial Board Memberships at *PLoS ONE, BMC Neurology, Frontiers in Neurology, Frontiers in Stroke, Frontiers in Public Health, Journal of Aging Research, Neurology International, VasCog, Diagnostics, and BMC Medical Research Methodology*. Additionally, S.M.M.B. serves as a Member of the *College of Reviewers* for the Canadian Institutes of Health Research (CIHR), Government of Canada; Director of Research for the *World Headache Society* (Bengaluru, India); a member of the Scientific Review Committee at *Cardiff University Biobank* (Cardiff, UK); Chair of the *Rotary Reconciliation Action Plan* (RAP), Rotary District 9675 (NSW, Australia), Expert Adviser/Reviewer for the *Cariplo Foundation* (Milan, Italy) and as Healthcare and Medical Adviser for the *Japan Connect* (Osaka, Japan). These roles are unrelated to the submitted work. The funding body has no role in the study design, data collection, analysis, interpretation of findings, or manuscript preparation. The content is solely the responsibility of the authors and does not necessarily represent the official views of the affiliated/funding organization/s.

References

1. Global Burden of Cardiovascular Diseases and Risks 2023 Collaborators. Global, Regional, and National Burden of Cardiovascular Diseases and Risk Factors in 204 Countries and Territories, 1990–2023. *J. Am. Coll. Cardiol.* **2025**, *ahead of print*. [CrossRef]
2. Bhaskar, S. Advancing Understanding and Addressing Disparities in Cardiomyopathy Care in Southern Africa. *JACC Adv.* **2024**, *3*, 100957. [CrossRef]
3. Bhaskar, S. Population aging in Japan offers a warning and a template for action. *Nat. Aging* **2025**, *5*, 1921–1923. [CrossRef]
4. Feigin, V.L.; Owolabi, M.O.; Feigin, V.L.; Abd-Allah, F.; Akinyemi, R.O.; Bhattacharjee, N.V.; Brainin, M.; Cao, J.; Caso, V.; Dalton, B.; et al. Pragmatic solutions to reduce the global burden of stroke: A World Stroke Organization-Lancet Neurology Commission. *Lancet Neurol.* **2023**, *22*, 1160–1206. [CrossRef] [PubMed]
5. Salahuddin, H.; Nguyen, T.N.; Masoud, H.E.; Martins, S.O.; Mansour, O.Y.; Al-Mufti, F.; Zaidi, S.F. Bridging the Gap: Training and Infrastructure Solutions for Mechanical Thrombectomy in Low- and Middle-Income Countries. *Stroke* **2025**, *56*, e157–e160. [CrossRef]
6. Batchelor, W.B.; Anwaruddin, S.; Wang, D.D.; Perpetua, E.M.; Krishnaswami, A.; Velagapudi, P.; Wyman, J.F.; Fullerton, D.; Keegan, P.; Phillips, A.; et al. The Multidisciplinary Heart Team in Cardiovascular Medicine: Current Role and Future Challenges. *JACC Adv.* **2023**, *2*, 100160. [CrossRef] [PubMed]
7. Isselbacher, E.M.; Preventza, O.; Black, J.H., III; Augoustides, J.G.; Beck, A.W.; Bolen, M.A.; Braverman, A.C.; Bray, B.E.; Brown-Zimmerman, M.M.; Chen, E.P.; et al. 2022 ACC/AHA Guideline for the Diagnosis and Management of Aortic Disease: A Report of the American Heart Association/American College of Cardiology Joint Committee on Clinical Practice Guidelines. *Circulation* **2022**, *146*, e334–e482. [CrossRef]
8. West, H.W.; Dangas, K.; Antoniades, C. Advances in Clinical Imaging of Vascular Inflammation: A State-of-the-Art Review. *JACC Basic Transl. Sci.* **2024**, *9*, 710–732. [CrossRef]
9. Jin, C.; Beran, R.G.; Bhaskar, S.M.M. Neuroimaging and Genetic Markers of Cerebral Small Vessel Disease and Cognitive Outcomes: A Systematic Review and Meta-Analysis (NEUROGEN-SVD Study). *Diagnostics* **2025**, *15*, 2585. [CrossRef]
10. Jin, C.; Bhaskar, S. Unifying Vascular Injury and Neurodegeneration: A Mechanistic Continuum in Cerebral Small Vessel Disease and Dementia. *Eur. J. Neurosci.* **2025**, *62*, e70246. [CrossRef]
11. Singhal, M.; Pilania, R.K.; Basu, S.; Desai, D.; Thangaraj, A.; Singh, R.; Semwal, R.; Kaur, T.; Sri, G.; Sudhakar, M.; et al. Role of CT Coronary Angiography at Initial Presentation in Kawasaki Disease—Insights from a Tertiary Care Center in North India. *Diagnostics* **2025**, *15*, 1806. [CrossRef]
12. Bonanni, F.; Caciolli, S.; Berteotti, M.; Grasso Granchietti, A.; Tozzetti, V.; Cenni, N.; Servoli, C.; Bandini, M.; Marchi, E.; Del Pace, S.; et al. Left Ventricular Diastolic Dysfunction Predicts Global Longitudinal Strain Recovery after Surgical Aortic Valve Replacement. *Diagnostics* **2024**, *14*, 2176. [CrossRef]
13. Sharma, D.; Spring, K.J.; Bhaskar, S.M.M. Neutrophil-lymphocyte ratio in acute ischemic stroke: Immunopathology, management, and prognosis. *Acta Neurol. Scand.* **2021**, *144*, 486–499. [CrossRef]
14. Memiş, Z.; Gürkaş, E.; Özdemir, A.Ö.; Acar, B.A.; Ögün, M.N.; Aytaç, E.; Akpınar, Ç.K.; Akıl, E.; Çabalar, M.; Özkul, A.; et al. Impact of Neutrophil-to-Lymphocyte Ratio on Stroke Severity and Clinical Outcome in Anterior Circulation Large Vessel Occlusion Stroke. *Diagnostics* **2024**, *14*, 2880. [CrossRef] [PubMed]
15. Sharma, D.; Spring, K.J.; Bhaskar, S.M.M. Role of Neutrophil-Lymphocyte Ratio in the Prognosis of Acute Ischaemic Stroke After Reperfusion Therapy: A Systematic Review and Meta-analysis. *J. Cent. Nerv. Syst. Dis.* **2022**, *14*, 11795735221092518. [CrossRef] [PubMed]
16. Sastre-Alzamora, T.; Tárraga López, P.J.; López-González, Á.A.; Vallejos, D.; Paublini, H.; Ramírez Manent, J.I. Usefulness of Atherogenic Indices for Predicting High Values of Avoidable Lost Life Years Heart Age in 139,634 Spanish Workers. *Diagnostics* **2024**, *14*, 2388. [CrossRef]
17. Dainis, A.M.; Ashley, E.A. Cardiovascular Precision Medicine in the Genomics Era. *JACC Basic Transl. Sci.* **2018**, *3*, 313–326. [CrossRef]
18. Jiang, W.-F.; Sun, Y.-M.; Qiu, X.-B.; Wu, S.-H.; Ding, Y.-Y.; Li, N.; Yang, C.-X.; Xu, Y.-J.; Jiang, T.-B.; Yang, Y.-Q. Identification and Functional Investigation of SOX4 as a Novel Gene Underpinning Familial Atrial Fibrillation. *Diagnostics* **2024**, *14*, 2376. [CrossRef]
19. Patel, J.; Bhaskar, S.M.M. Diagnosis and Management of Atrial Fibrillation in Acute Ischemic Stroke in the Setting of Reperfusion Therapy: Insights and Strategies for Optimized Care. *J. Cardiovasc. Dev. Dis.* **2023**, *10*, 458. [CrossRef] [PubMed]
20. Karakasis, P.; Vlachakis, P.K.; Theofilis, P.; Ktenopoulos, N.; Patoulias, D.; Fyntanidou, B.; Antoniadis, A.P.; Fragakis, N. Atrial Cardiomyopathy in Atrial Fibrillation: A Multimodal Diagnostic Framework. *Diagnostics* **2025**, *15*, 1207. [CrossRef]
21. Goh, B.; Bhaskar, S.M.M. The role of artificial intelligence in optimizing management of atrial fibrillation in acute ischemic stroke. *Ann. N. Y. Acad. Sci.* **2024**, *1541*, 24–36. [CrossRef]
22. Goh, B.; Bhaskar, S.M.M. Evaluating Machine Learning Models for Stroke Prognosis and Prediction in Atrial Fibrillation Patients: A Comprehensive Meta-Analysis. *Diagnostics* **2024**, *14*, 2391. [CrossRef]

23. Ziayee, F.; Dalyanoglu, H.; Schnitzler, C.; Jannusch, K.; Boschheidgen, M.; Boeven, J.; Aubin, H.; Turowski, B.; Kaschner, M.G.; Mathys, C. A Retrospective Analysis of the Effects of Concomitant Use of Intra-Aortic Balloon Pump (IABP) and Veno-Arterial Extracorporeal Membrane Oxygenation (va-ECMO) Therapy on Procedural Brain Infarction. *Diagnostics* **2025**, *15*, 699. [CrossRef] [PubMed]
24. Jankajova, M.; Singh, R.B.; Hristova, K.; Elkilany, G.; Fatima, G.; Singh, J.; Fedacko, J. Identification of Pre-Heart Failure in Early Stages: The Role of Six Stages of Heart Failure. *Diagnostics* **2024**, *14*, 2618. [CrossRef] [PubMed]
25. Manzi, L.; Buongiorno, F.; Narciso, V.; Florimonte, D.; Forzano, I.; Castiello, D.S.; Sperandeo, L.; Paolillo, R.; Verde, N.; Spinelli, A.; et al. Acute Heart Failure and Non-Ischemic Cardiomyopathies: A Comprehensive Review and Critical Appraisal. *Diagnostics* **2025**, *15*, 540. [CrossRef] [PubMed]

diagnostics MDPI

Article

Role of CT Coronary Angiography at Initial Presentation in Kawasaki Disease—Insights from a Tertiary Care Center in North India

Manphool Singhal *, Rakesh Kumar Pilania *, Suprit Basu, Dev Desai, Abarna Thangaraj, Ripudaman Singh, Radhika Semwal, Taranpreet Kaur, Gopika Sri, Murugan Sudhakar, Arun Sharma, Pandiarajan Vignesh, Deepti Suri and Surjit Singh

Postgraduate Institute of Medical Education and Research, Chandigarh 160012, India; basusuprit@gmail.com (S.B.); devcdesai@gmail.com (D.D.); dr.abarna.t@gmail.com (A.T.); singhripudaman57@gmail.com (R.S.); radhikasemwal23@gmail.com (R.S.); taranpreetkaur2196@gmail.com (T.K.); gopikasri1711@gmail.com (G.S.); drmurugansudhakar20@gmail.com (M.S.); drarungautam@gmail.com (A.S.); vigimmc@gmail.com (P.V.); surideepti@gmail.com (D.S.); surjitsinghapc@gmail.com (S.S.)
* Correspondence: drmsinghal74@gmail.com (M.S.); kumarpilania007@gmail.com (R.K.P.)

Abstract: Background: Kawasaki disease (KD) is a systemic vasculitis and the leading cause of acquired heart disease in children. Early identification of coronary artery abnormalities (CAAs) is crucial to guide treatment and improve outcomes. While transthoracic 2D echocardiography (TTE) remains the first-line imaging modality, it has limitations, particularly in visualizing distal coronary artery segments and detecting thrombi. Computed tomography coronary angiography (CTCA) offers enhanced visualization, but its role at initial presentation of KD remains underexplored. **Methods**: We reviewed the records of 71 children with KD who underwent CTCA at their initial presentation at a tertiary center between November 2013 and December 2024. The CTCA findings were compared with those of TTE. CTCA was performed after stabilization using radiation-minimized protocols. **Results**: Of 71 patients, 62 had CAAs on baseline TTE. CTCA confirmed CAAs in 39 patients, identified additional lesions in 23, and detected distal aneurysms and coronary branch involvement missed by TTE. In 20 patients with initially abnormal TTE, CTCA demonstrated normal coronaries, facilitating treatment de-escalation. CTCA identified coronary thrombi missed on TTE in two patients and congenital coronary anomalies in three patients. CTCA findings led to modification of therapy in multiple cases. **Conclusions**: CTCA is a valuable adjunct to TTE in evaluating coronary artery involvement at the time of initial presentation of children with KD. Given its superior visualization of the entire length of coronary arteries, CTCA has a vital role in therapeutic decision-making in KD.

Keywords: CT coronary angiography; coronary artery abnormality; Kawasaki disease; acute phase; missed aneurysm

1. Introduction

Kawasaki disease (KD) is an acute systemic vasculitis of unknown etiology that predominantly affects children below 5 years. It is the most common cause of acquired heart disease in children [1–3]. Without treatment, one out of four children with KD can develop coronary artery abnormalities (CAAs), underscoring the critical need for early diagnosis and treatment to reduce this risk [3–5].

Cardiac imaging is crucial for assessing coronary involvement in KD. Transthoracic 2D echocardiogram (TTE) is the primary imaging modality for diagnosing coronary arteries in KD [6,7]. However, it has several limitations. These include observer dependency and inability to assess CAAs in distal segments of coronary arteries and the left circumflex coronary artery (LCx). TTE is also not the best imaging modality for detecting coronary thrombosis. Computed tomography coronary angiography (CTCA) has the potential to address many of these limitations of TTE [8,9].

The 2024 American Heart Association (AHA) scientific statement on diagnosing and managing KD recommends CTCA on follow-up for evaluating patients with KD who develop CAAs. It has been suggested that CTCA may be performed 1 year after diagnosis, and subsequently every 3–5 years [2]. However, there are currently no guidelines on the role of CTCA as an imaging modality at the time of initial presentation. There is a paucity of literature on this aspect of KD. Herein, we report our experience on the role of CTCA in the evaluation of coronary arteries in children with KD at initial presentation.

2. Methodology

We reviewed records of children with KD who underwent CTCA for coronary artery assessment at our center. Our institute is a federally funded, not-for-profit, tertiary care teaching institute in North India. The diagnosis of KD was based on the American Heart Association (AHA) scientific statement [2,3,10]. CTCA was performed on 376 patients between November 2013 and December 2024. Amongst these, patients' records of those who underwent CTCA at the initial presentation were analyzed in detail. Patient demographics (age at diagnosis, gender, date of TTE, and date of CTCA) and findings of TTE and CTCA were entered on a predesigned proforma. CTCA was performed at the time of initial presentation after the child had stabilized. Informed written consent was obtained from all parents/guardians before subjecting the children to CTCA. Findings on CTCA were compared with TTE performed around that time and with baseline TTE. The cardiac radiologist (MS) was, however, blinded to the findings of TTE. The manuscript was approved by the Departmental Publication Review Board (No. RDG/EC/Pub/27). No funding was involved in this study.

CTCAs were performed on a 128-slice scanner (Siemens Somatom Definition Flash, Erlangen, Germany) during the period January 2013 to February 2022, and thereafter on a 192-detector dual-source CT scanner (Siemens Somatom Force, Erlangen, Germany). No sedation was necessary in children above the age of 5 years. For children below 5, oral triclofos (50 mg/kg/dose) was used as a single dose. In addition, some children required intravenous midazolam (0.1–0.2 mg/kg/dose).

Radiation exposure was optimized using an adaptive prospective electrocardiography-gated sequence, automated tube current modulation (Care Dose 4D; Siemens Healthineers, Erlangen, Germany), lower kilovoltage settings (fixed at 80 kVp), and iterative image reconstruction algorithms (Safire/Admire; Siemens Healthineers, Erlangen, Germany).

Echocardiography assessment and interpretation of CAAs: TTE was performed at admission, and repeated during the hospital course as mandated by clinical requirements. In most cases, it was also performed at discharge and, thereafter, 2–6 weeks on follow-up. TTE was carried out by postdoctoral fellows trained in cardiovascular assessment of KD under the supervision of faculty members. Before 2019, TTE was performed using an Esaote MyLab30Gold ultrasound machine (Esaote S.p.A., Genowa, Italy), and since then, it has been conducted on the Philips EPIQ 7G ultrasound system (US 51881625) (Philips Ultrasound, Inc., Bothell, WA, USA). In 2013, we began assessing coronary arteries in children with KD using Z-scores [11]. Before this, we evaluated CAAs using absolute diameters.

CTCA assessment and interpretation of CAAs: Post-processing analysis was carried out on a dedicated Syngovia (Siemens Healthineers) workstation to reconstruct and evaluate coronary arteries. All CTCAs were reviewed by a single, experienced cardiac radiologist

(M.S.). Coronary arteries and side branches were included in the analysis wherever these were visualized. CAAs were evaluated in terms of location, number, and morphology (i.e., aneurysms—saccular/fusiform; and dilatation). The term 'aneurysm' was used if the internal diameter of the coronary artery was $\geq$1.5 times that of an adjacent segment. Coronary arteries were said to be 'dilated' if the internal diameter was increased, but was <1.5 times that of an adjacent segment [3,10]. CAAs on CTCA were recorded, and the results were compared with the findings on TTE, including both the baseline TTE and the one performed immediately prior to CTCA.

3. Results

A total of 71 patients with KD who underwent both TTE and CTCA at the time of initial presentation were identified through a review of records.

(a) Patient characteristics:

Among the 71 patients, 50 (70.4%) were male. Median age was 39 months (2–360 months), and 22 (30.1%) were infants. All patients had fever at presentation, with a median duration of fever of 11 days (range 3–60 days). Median time to KD diagnosis was 15 days (4–60 days) from the onset of symptoms. Complete KD was diagnosed in 49 patients, and 22 had incomplete KD. KD shock syndrome was seen in eight (11.3%) patients, and macrophage activation syndrome in two (2.8%) patients, respectively. Laboratory investigations revealed a median hemoglobin value of 92.5 g/L (range 58–138), total leukocyte count 16.02 $\times$ 10^9/L (range 3.22–53.20), platelet count 6.94 $\times$ 10^9/L (range 230–1635), serum albumin 3.15 g/dL (range 1.2–6.4), and C-reactive protein 44.5 mg/L (range 0.1–279). All patients received treatment with intravenous immunoglobulin (IVIg) and aspirin. Treatment augmentation was required in 54 patients. This was carried out with infliximab in 46 patients (64.8%) and steroids in 31 (43.7%) patients, respectively (Table 1).

Table 1. Clinical details, laboratory profile, and treatment of the patients in our cohort.

Characteristic	*N* = 71
Demographics	
Age at presentation (median, range)	39 months (2–360)
Male sex	50 (70.4%)
Clinical characteristics	
Duration of fever (median, range)	11 days (3–60)
Delay in diagnosis (median, range)	15 days (4–60)
Interval between onset of disease and CTCA (median, range)	23 days (9–85)
Fever	71 (100%)
Rash	57 (80.3%)
Mucosal changes	57 (80.3%)
Extremity changes	27 (38%)
Eye changes	39 (54.9%)
Cervical lymphadenopathy	21 (29.6%)
Complete KD	49 (69%)
Peeling	9 (12.7%)
Beau's lines	13 (18.3%)

Table 1. *Cont.*

Characteristic	N = 71
Laboratory features	
Hemoglobin, g/L (median, range)	92.5 (58–138)
Total leukocyte count, cells/mm^3 (median, range)	16,015 (3220–53,020)
Neutrophil, % (median, range)	57% (13–98)
Platelet count, cells/mm^3 (median, range)	694,000 (230,000–1,635,000)
AST, U/L (median, range)	41.0 (12–1236)
ALT, U/L (median, range)	30.5 (6–428)
Albumin, g/dL (median, range)	3.15 (1.2–6.4)
Serum sodium, mmol/L (median, range)	134.0 (123–140)
ESR, mm in 1st hr (median, range)	42.0 (2–115)
CRP, mg/L (median, range)	44.5 (0.1–279)
Treatment	
Standard of care	
IVIg	71 (100%)
Aspirin	71 (100%)
Treatment intensification	
Infliximab	46 (64.8%)
Steroids	31 (43.7%)
Repeat IVIg	7 (9.9%)
Cyclosporine	21 (29.6%)
Anti-coagulation	
LMWH	28 (39.4%)
Warfarin	4 (5.6%)

(b) Indications of CTCA:

Median interval between disease onset and the performance of CTCA was 23 days (range: 9–85 days) (Figure 1). Indications for CTCA were: (i.) CAAs (aneurysms/dilatations/non-tapering of coronary arteries) on TTE at presentation (n = 62); (ii.) non-visualization of one or more coronaries on TTE (n = 8); and (iii.) non-availability of an appropriate window to carry out TTE (n = 1) (Figure 2).

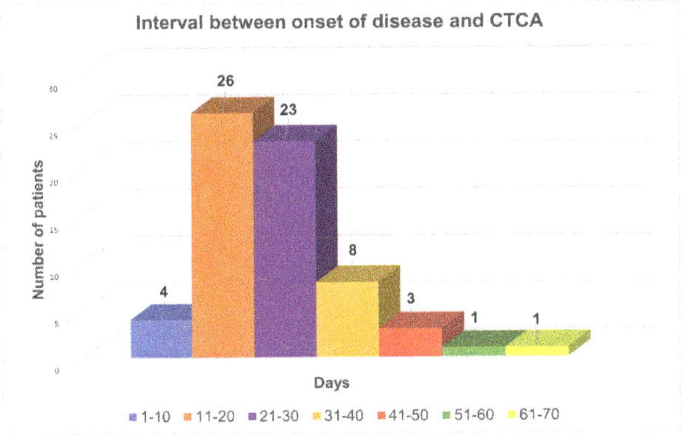

Figure 1. Distribution of patients by interval (in days) between the onset of disease and CTCA.

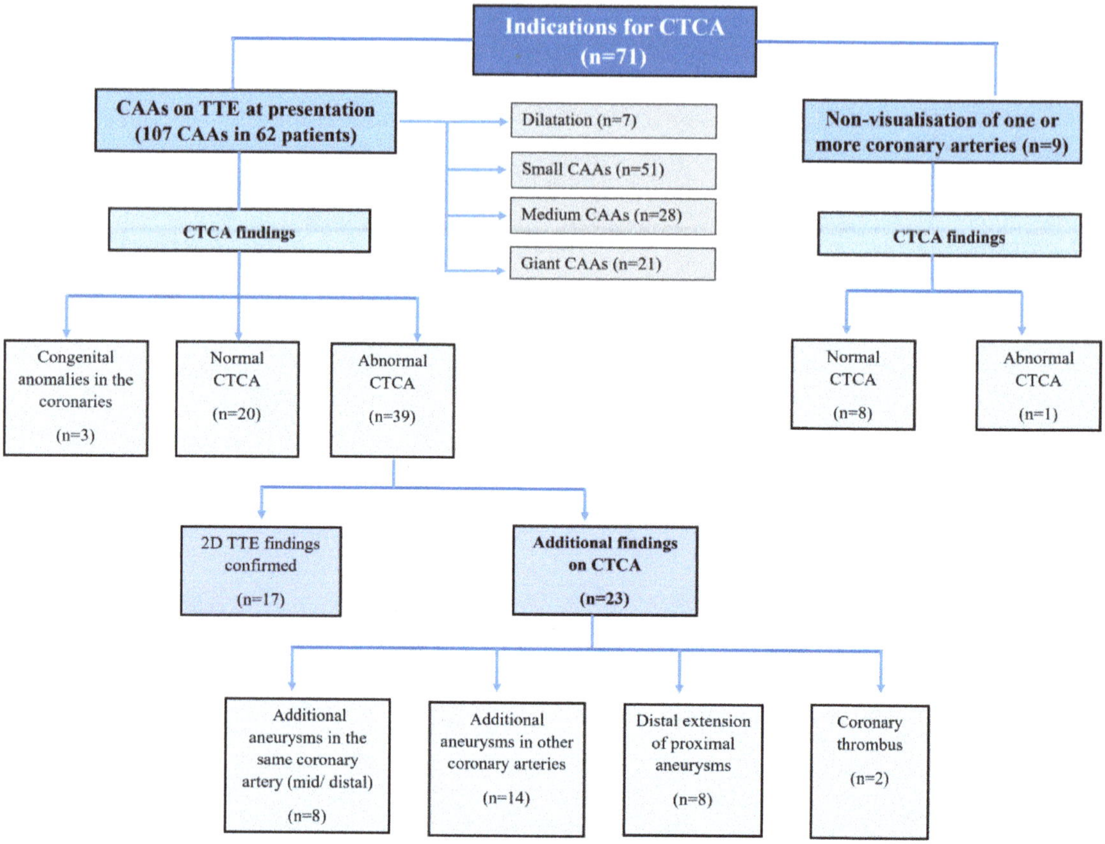

Figure 2. Indications for CTCA in our cohort and correlation between CTCA and TTE findings.

(c) Findings on TTE and CTCA

i. Imaging by TTE:

TTE identified 107 CAAs in 62 patients (Figure 3). Amongst these, 42 patients had CAAs detected on both baseline TTE and the TTE performed at the time of CTCA. In the remaining 20 patients, CAAs were observed on baseline TTE, but subsequent TTE at the time of CTCA showed normalization of coronary artery findings. In eight patients, one or more coronaries could not be assessed on TTE. In one patient, TTE could not be performed because of an infectious skin lesion over the anterior chest wall. On TTE, the left anterior descending (LAD) artery was the most commonly affected (n = 39), followed by the left main coronary artery (LMCA) (n = 35), right coronary artery (RCA) (n = 26), and left circumflex artery (LCx) (n = 7) (Figure 2).

ii. Imaging by CTCA:

Of the 62 patients with CAAs detected on TTE, CTCA confirmed the presence of CAAs in 39 patients, while 20 had normal coronary arteries. These 20 patients also showed normal findings on TTE performed at the time of CTCA. In three patients, congenital coronary anomalies were later identified, which had been misinterpreted as CAAs on the initial TTE [12].

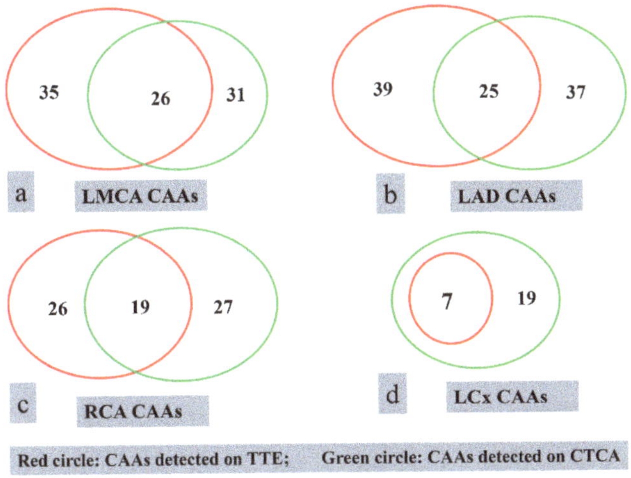

Figure 3. Comparison of CAAs detected on TTE (red circle) and CTCA (green circle).

iii. CAAs detected, or confirmed, on CTCA

CTCA detected a total of 114 CAAs in 39 patients. These were seen in LAD (n = 37), LMCA (n = 31), RCA (n = 27), and LCx (n = 19) (Figure 3). CTCA detected six aneurysms in the distal segments of the coronary arteries—these had been missed on TTE. Among the 39 patients with CAAs detected on CTCA, 23 had additional findings missed on TTE (Figure 2). CTCA also identified additional aneurysms in the same coronary artery (n = 8) and new aneurysms in another coronary artery (n = 14) that had been missed on TTE (Figure 4). There was distal extension of the proximal CAAs in 12 patients. At the same time, involvement of coronary artery branches was noted in nine (Figures 4 and 5)—these findings had been missed on TTE and demonstrated on CTCA (Table 2).

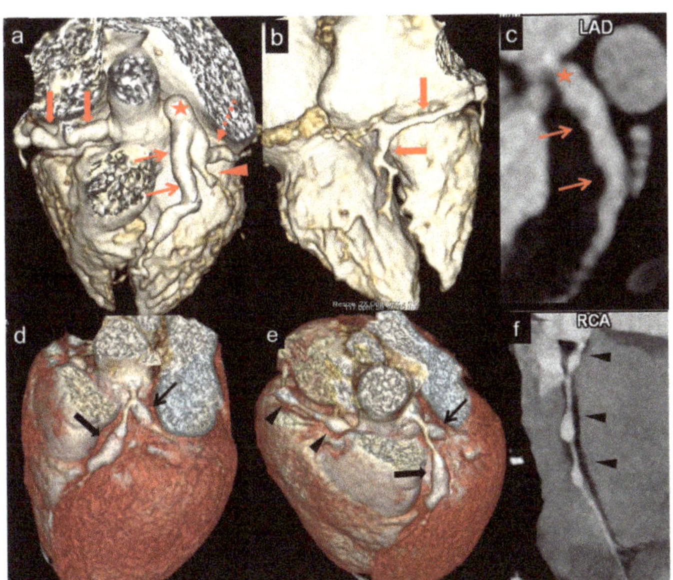

Figure 4. (**a–c**) (Patient No.3 in Table 2): CT coronary angiography (CTCA) volume rendered (**a**,**b**) and curved reformatted of left anterior descending (LAD) (**c**) images in a 7 months male infant at presentation

show fusiform aneurysm of right coronary artery (RCA) along its entire course with extension into posterior descending branch (PDA) (thick arrows in **a**,**b**). The left main coronary (LMCA) (asterisk in **a**,**c**), proximal and mid anterior descending (LAD) (thin arrows in **a**,**c**), and proximal left circumflex (LCX) (interrupted arrows in **a**) show fusiform aneurysms. Fusiform aneurysm is extending to involve the proximal segment of the obtuse marginal (OM) branch of LCX (arrowhead in **a**). Although TTE showed aneurysms in LMCA, proximal LAD, LCX, and RCA, it failed to demonstrate the distal extension of the aneurysm of RCA and involvement of the OM branch of LCX. (**d**–**f**) (Patient No.15 in Table 2): CTCA volume rendered (**a**,**b**) and curved reformatted images of RCA (**c**) in a 76 months boy at presentation show fusiform aneurysms of proximal left anterior descending (LAD) (thick arrows in **a**,**b**) and left circumflex (LCX) (thin arrows in **d**,**e**). RCA shows three discontinuous fusiform aneurysms in the proximal and mid segments (arrowheads in **e**,**f**). Although TTE showed aneurysms in proximal LAD, LCX, and RCA, it failed to demonstrate aneurysms in mid RCA.

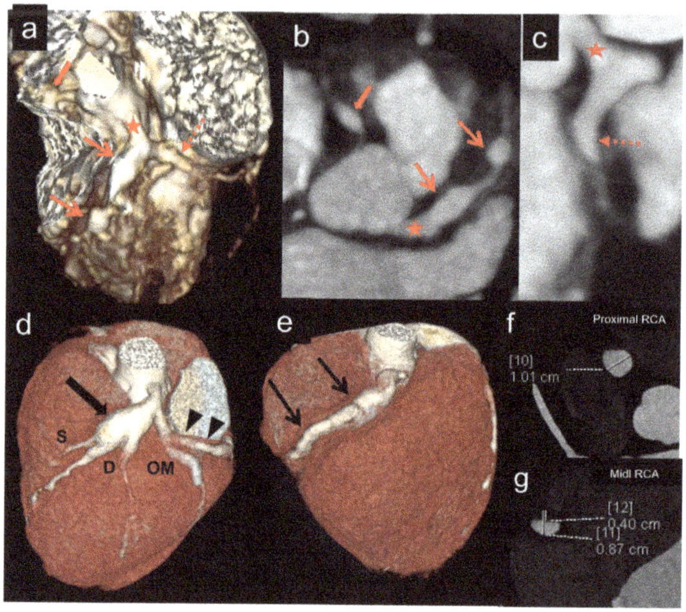

Figure 5. (**a**–**c**): (Patient No.12 in Table 2): CT coronary angiography (CTCA) volume rendered (**a**), axial (**b**) and curved reformatted images of left circumflex (LCX) (**c**) in a 2 months female infant at presentation show fusiform aneurysm of left main coronary artery (LMCA) (asterisk in **a**–**c**) with extension into osteo-proximal segments of LAD (thin arrow in **a**,**b**) and LCX (interrupted arrow in **a**,**c**). Another non-contiguous aneurysm is seen in mid LAD (second thin arrow in **a**,**b**). Although TTE showed aneurysms in LMCA, proximal LAD, and RCA, it failed to demonstrate a second non-contiguous aneurysm in mid LAD and a fusiform aneurysm in LCX. (**d**–**g**): (Patient No.22 in Table 2): CTCA volume rendered (**d**,**e**), axial proximal RCA (**f**), and distal RCA (**g**) images in a 96 months boy at presentation show fusiform aneurysm of LMCA with extension into osteo-proximal segments of LAD (thick arrow in d) and LCX (arrow heads in **d**). Note the extension/involvement of septal (S) and diagonal (D) branches of LAD and the obtuse marginal branch (OM) of LCX (**a**). RCA shows a fusiform aneurysm in its entire course (thin arrows in **e**). Axial images of RCA (**f**,**g**) show dilated RCA with eccentric thrombus in mid RCA (**g**). TTE showed aneurysms in LMCA, LAD, and RCA; however, it failed to demonstrate LCX aneurysm, distal extension of RCA aneurysm, and extension/involvement of septal (S) and diagonal (D) branches of LAD and obtuse marginal branch (OM) of LCX. The thrombus in the mid RCA aneurysm was also missed.

Table 2. Details of coronary artery abnormalities on TTE and CTCA in the subgroup of patients who had additional findings on CTCA.

S.no.	Age (Months)/Sex	TTE Findings (Measurement in mm/Z-Score)				Interval Between Diagnosis and CTCA (Days)	CTCA Findings				Detailed Description of Findings on CTCA
		LMCA	LAD	RCA	LCx		LMCA (mm)	LAD (mm)	RCA (mm)	LCx (mm)	
1	8/M	4.8 (8.36)	7 (26.24)	8.3 (23.52)	NV	15	8	7	6.4		Fusiform and saccular aneurysms in proximal and distal RCA, respectively. Saccular aneurysm involving distal LCA, incorporating/involving the origin of proximal LAD and LCX. No e/o thrombosis
2	55/F	2.9 (1.8)	2.5 (2.2)	2.6 (1.7)	NV	18	Normal	2.4	Normal	Normal	Segmental areas of dilatations in the proximal and mid LAD
3	7/M	5 (10.3)	4.7 (11.3)	4.2 (8.50)	1.7 (1.7)	19	7.9	3.5	4.6	3.2	Fusiform aneurysm involving all the coronary arteries; entire course of RCA with extension into PDA; LMCA, proximal and mid LAD, and proximal LCX with involvement of OM1 branch
4	39/M	3.3 (2.9)	5.1 (10.5)	0.8 (−2.83)	0.8 (−2.11)	16	5.2	5.4	3		Saccular aneurysm in mid LAD; areas of dilatations in RCA, LCA, proximal LAD
5	20/M	2.4 (2.03)	2.9 (4.98)	1.5 (−0.49)	1.2 (−0.12)	12	2.7	2.7	1.9	1.5	Dilated distal LCA and proximal LAD
6	64/M	3.1 (1.6)	1.7 −0.5	1.4 (−1.80)	1 (−1.98)	24	2.6	1.8	1.4	1.5	Dilated mild LAD
7	156/M	6.17 (7.42)	4.68 (6)	7.5 (10.63)	NV	41	6.3	5	6.6	4.6	Fusiform aneurysms in all coronaries: LMCA, proximal and mid RCA, LAD, and LCX.
8	132/M	2.89 (−0.37)	3.9 (3.62)	2.14 (−1.26)	NV	38	3.8	2.7	1.8	2.7	Dilated all coronary arteries: Diffuse LMCA and LAD; proximal and mid segments of RCA and proximal LCX
9	42/M	3.1 (3.85)	2.7 (2.5)	3.85 (6.43)	NV	14	3	3.2	4.4	2.7	Fusiform aneurysms in all coronaries: LMCA, entire RCA, including PDA and PLV branches, and proximal LAD and LCX.
10	43/M	5.2 (7.6)	3.2 (4.1)	3.2 (3.00)	NV	10	7	3.5	2.4	2.4	Fusiform aneurysm in LMCA with extension into origins of LAD and LCX. Dilated the entire RCA and LCX.
11	14/F	5.2 (9.93)	3.6 (4.03)	3.3 (2.76)	1.6 (2.76)	26	6.4	3.6	1	2	LMCA saccular aneurysm, which is extending into the proximal LAD and proximal LCX
12	2/F	2.6 (4.41)	3.1 (7.53)	2.8 (5.75)	NV	24	2.4	3.8	2.7	2.5	Fusiform aneurysm of LMCA with extension into osteo-proximal segments of LAD and LCX; another non-contiguous aneurysm was seen in mid-LAD LAD

Table 2. *Cont.*

S.no.	Age (Months)/Sex	TTE Findings (Measurement in mm/Z-Score)				Interval Between Diagnosis and CTCA (Days)	CTCA Findings				Detailed Description of Findings on CTCA
		LMCA	LAD	RCA	LCx		LMCA (mm)	LAD (mm)	RCA (mm)	LCx (mm)	
13	204/M	3.2 (−0.33)	2.96 (0.84)	2.7 (−0.62)	NV	28	3.2	3.1	2.3	Normal	Dilated LMCA and proximal LAD
14	7/M	2.97 (5.33)	3.2 (5.91)	4.9 (6.88)	NV	19	2.7	4.2	5.8	3.3	Fusiform aneurysms in RCA (2 in number). Fusiform aneurysms in LCx (2 in number)
15	76/M	3 (1.13)	6.68 (12.33)	6.5 (9.89)	2 (0.22)	58	1.8	6.4	5.1	12.6 × 4.1	(1) Fusiform aneurysm 6.4 mm W ∗ 18 mm L in proximal LAD (2) Three discontinuous fusiform aneurysms in proximal and mid RCA 6.5 mm W ∗ 5.1 mm L, 10.6 mm W ∗ 6.3 mm L, 9.2 mm W ∗ 3.4 mm L (3) Fusiform aneurysms in LCx just beyond origin, 12.6 mm L × 4.1 mm W
16	44/M	4.5 (6.4)	4.2 (7.35)	4.2 (6.06)	2.3 (1.97)	33	Normal	5	8.2	4	Fusiform aneurysm in proximal RCA (8.2 mm W ∗ 31 mm L giant aneurysm), LAD (5 mm W ∗ 25 mm L), LCx (4 mm W ∗ 25 mm L). No e/o thrombus
17	132/M	4.2 (3.15)	3.6 (3.5)	2 (−1.13)	1.5 (−1.36)	32	5.8	9.1	12.6	6.1	Multiple aneurysms in the entire course of RCA, with a giant aneurysm in its proximal segment; fusiform giant aneurysm in proximal LAD; fusiform aneurysm in proximal LCx; and a dilated LMCA
18	60/M	3.5 (2.23)	6.8 (16.05)	5.7 (10.04)	NV	26	3.6	7.7	6	3.1	LMCA normal; proximal LAD: fusiform aneurysm (7.7 mm W ∗ 50 mm L), with involvement of diagonal branch (3.6 mm), mid-segment: skip areas of dilatation. Distal segment: dilated (3.6 mm). RCA proximal segment dilated (6.6 mm for a length of 40 mm). AM branch: multiple skip aneurysms. LCx: proximal segment fusiform aneurysm (3.6 mm) with extension into OM branch (3.9 mm)
19	7/F	2.1 (2.06)	1.8 (0.65)	1.8 (3.30)	1.5 (0.16)	27	2.3	1.6	2.2	1.5	Dilated proximal and distal LAD with normal intervening segment
20	27/M	3.2 (2.95)	3.5 (5.42)	2.6 (1.98)		13	3.5	5.1		3.6	Fusiform aneurysm of LMCA extending to proximal LAD and LCx. No e/o thrombus
21	6/M	3.1 (4.6)	6.7 (16.82)	9.4 (24.00)	NV	35	3.8	6.6	8	4.1	Fusiform aneurysmal dilatation of all coronary arteries, no thrombus

Table 2. *Cont.*

S.no.	Age (Months)/Sex	TTE Findings (Measurement in mm/Z-Score)				Interval Between Diagnosis and CTCA (Days)	CTCA Findings				Detailed Description of Findings on CTCA
		LMCA	LAD	RCA	LCx		LMCA (mm)	LAD (mm)	RCA (mm)	LCx (mm)	
22	96/M	4.85 (0.63)	8 (34)	7.5 (14.3)	NV	85	4.5	13	5.1	10	Fusiform aneurysm LMCA with extension into osteo-proximal segments of LAD and LCX and involving septal and diagonal branches of LAD and obtuse marginal branch of LCX; fusiform aneurysm in the entire course of RCA with an eccentric thrombus in mid RCA
23	7/M	2.47 (3.46)	14.9 (47.2)	4.17 (9.50)	4.39 (11.48)	51	2	18.3	3.5	7.2	Giant aneurysm in proximal LAD with partial thrombus (detected on TTE), proximal RCA saccular aneurysm with thrombus (not detected on TTE), LCx fusiform aneurysm (not detected on TTE)

TTE—Transthoracic 2D echocardiogram; CTCA—Computed tomography coronary angiography; LMCA—left main coronary artery, RCA—right coronary artery; LAD—left anterior descending artery; LCX—left circumflex coronary artery; OM—obtuse marginal branch; NV—Not visualized.

In one patient suspected of having a thrombus on TTE, imaging by CTCA revealed normal findings. This finding facilitated the discontinuation of anticoagulation therapy. In two patients, three thrombi in CAAs were diagnosed on CTCA. Of these three thrombi, one had been detected on TTE, while two had been missed. Two thrombi were in RCA (Figure 5), whereas one was in LAD. Both the thrombi missed on TTE were seen in the RCA.

iv. Patients with congenital coronary artery anomalies on CTCA

CTCA identified incidental congenital coronary artery anomalies in three children, but on TTE, these had been misinterpreted as CAAs. One patient had an anomalous left coronary artery from the pulmonary artery (ALCAPA), and TTE showed a dilated RCA. In the remaining two patients, one had separate origins of the left LAD and LCx from the left sinus with an absent left main trunk, while the other had a single coronary artery. Both had been misinterpreted as dilated LMCA on TTE [12].

v. Patients with no CAAs on CTCA

CTCA demonstrated normal coronary arteries in 20 patients. In all of these cases, baseline TTE had shown CAAs; however, TTE performed on the day of CTCA revealed normal coronary findings. The median interval between the initial TTE and CTCA was 72 h (24 h to 50 days) (Figure 1). Among these 20 patients, 15 had CAAs involving a single coronary artery, while 5 had involvement of two coronary arteries. All aneurysms detected in the LMCA (n = 10) and RCA (n = 7) were small. In the LAD, eleven aneurysms were small, while three were medium-sized. The time intervals between the first TTE and CTCA in patients with medium-sized LAD aneurysms were 21 days, 40 days, and 50 days, respectively.

(d) Impact of CTCA on management:

Aspirin treatment was discontinued in three patients with congenital coronary anomalies once CTCA ruled out the presence of CAAs. In one patient with an infective skin lesion over the anterior chest wall, TTE could not be performed, and CTCA was carried out. As this showed no abnormality, aspirin could be discontinued. Moreover, CTCA provided critical additional information in two patients where TTE had failed to detect coronary thrombi. CTCA revealed previously unrecognized intraluminal thrombi within the coronary arteries in these patients. This led to a change in management strategy—low molecular weight heparin (LMWH) was introduced as an anticoagulant. In both cases, this was accompanied by intensified immunomodulatory therapy, which included IVIg, infliximab, corticosteroids, and cyclosporine.

4. Discussion

The most important complication of KD is the development of CAAs [9]. Although CAAs have been reported to occur in the proximal segments of coronary arteries, these can also occur in distal segments and branches of coronary arteries [13,14]. This study compares the diagnostic performance of TTE and CTCA in children with KD at the time of initial presentation. Our findings suggest that CTCA provides explicit visualization of coronary arteries along their entire length. This helps detect CAAs in distal segments of coronary arteries that may be missed on TTE. The role of CTCA at the time of initial presentation of KD is particularly valuable, as early identification of coronary involvement enables appropriate risk stratification and timely intervention. As the highest risk of developing coronary aneurysms occurs within the first two weeks of illness, a state-of-the-art imaging modality like CTCA can significantly impact treatment decisions. Unlike TTE, which has limited visualization of distal coronary artery segments, CTCA facilitates comprehensive 3D imaging. This provides a more detailed assessment of aneurysm morphology, distal extension of proximal aneurysms, non-contiguous aneurysms in distal segments, involvement of coronary branch arteries, and coronary thrombus [8,9,15].

In 20 patients, where baseline TTE was abnormal (small aneurysms (n = 17); medium aneurysms (n = 3)), CTCA showed normal coronaries. A normal CTCA study facilitated the discontinuation of aspirin at 6 weeks. Furthermore, a regular CTCA examination helped de-escalate the therapeutic regimens, reducing unnecessary medication exposure and obviating the need for follow-up CTCA [16]. Additionally, normal CTCA findings can mitigate unnecessary anxiety and concern for patients and families. All patients in this cohort with a normal CTCA have remained well on follow-up. In our study, CTCA identified aneurysms in 23 patients that had been missed or underestimated by TTE. In nine patients, CTCA was performed due to the inability of TTE to visualize the coronary arteries adequately. CTCA helped optimize therapeutic decisions in these patients, such as stopping anticoagulation therapy in patients where CTCA ruled out a suspected thrombus on TTE. CTCA also helped initiate anticoagulation therapy in two patients, wherein a thrombus had been missed on TTE. These findings highlight the potential of CTCA to guide both treatment escalation (e.g., immunomodulatory therapy, anticoagulation) and de-escalation decisions in children with KD based on a more accurate assessment of coronary artery involvement [2].

The superiority of CTCA over TTE in detecting CAAs in KD has been well documented [14,17–19]. We have previously shown that CTCA is particularly useful in evaluating distal coronary artery segments, which are often difficult to visualize on TTE because of inappropriate acoustic windows [14]. In the present study, CTCA revealed six additional distal aneurysms missed on TTE. Similarly, CTCA identified more aneurysms in the LCx artery (n = 19) than TTE (n = 7). In patients with suspected coronary thrombi, CTCA helped clarify the diagnoses and facilitated therapeutic decision-making. These findings reinforce that CTCA enhances diagnostic accuracy, especially in patients with CAAs involving distal coronary arteries, LCx, or branches of coronary arteries, and also in patients with coronary thrombi. Additionally, TTE may miss congenital coronary anomalies, especially when these are not suspected clinically before TTE. Further, these are liable to be misdiagnosed as CAAs [12]. We contend that CTCA is a useful adjunct to TTE at the initial presentation of KD. While TTE is widely accessible and cost-effective, it may not be sufficient as the sole imaging modality in KD [14,17,20].

The 2024 AHA scientific statement recommended serial CTCAs on follow-up in KD patients with CAAs at presentation. However, its use during initial presentation has not been explicitly addressed [2]. Based on our findings and experience, we propose that CTCA be integrated into the standard evaluation protocol of coronary arteries in children with KD at the time of initial presentation, in selected cases. These include cases where CAAs have been detected on 2DE, older children and adolescents with poor acoustic windows for 2DE, children with a severe disease course (e.g., KD shock syndrome, KD with macrophage activation syndrome, or symptomatic myocarditis), infants under 6 months of age, and children with equivocal 2DE findings [9]. By providing comprehensive coronary artery visualization, CTCA offers critical information for risk stratification, treatment planning, and long-term cardiovascular monitoring.

The strengths of our study include our center's more than 30 years of clinical experience in managing children with KD. The diagnosis and treatment of KD were based on standard guidelines, with the senior author (SS) personally managing this cohort. Additionally, the study utilized a standardized imaging protocol and evaluated CTCA in a large cohort of KD patients at the time of initial presentation. All CTCAs were performed under the direct supervision of a senior cardiac radiologist (MS). To the best of our knowledge, this is the first study to examine the role of CTCA in children with KD at initial presentation. The retrospective nature of the research and a single-center setting are limitations that may affect generalizability. Additionally, while radiation exposure from CTCA was minimized using radiation-optimized protocols, it remains a consideration in children. Importantly, no z-score systems are currently available for CTCA in children; hence, absolute coronary dimensions were used. Applying TTE-based z-scores to CTCA coronary diameters may not be appropriate, especially for the mid/distal segments of coronary arteries. Furthermore, our cohort's median time to diagnosis was 11 days (range: 3–60 days), reflecting potential diagnostic delays commonly encountered in low- and middle-income countries. Larger and multicentric prospective studies are needed to validate our findings and assess the long-term impact of early CTCA assessment on clinical outcomes of KD. Additionally, advancements in CT imaging, such as ultra-low-dose protocols and artificial intelligence-assisted analysis, may further enhance its safety and diagnostic yield in children with KD. It is important to note that CTCA was primarily performed based on abnormal or inconclusive echocardiographic findings. In cases with incomplete visualization on echocardiography, CTCA identified significant abnormalities, demonstrating its added diagnostic value. We propose that evolving clinical indications may lead to broader use of CTCA, particularly in selected cases during the acute phase of KD.

5. Conclusions

We have shown the utility of CTCA as a valuable adjunct to TTE in evaluating coronary artery involvement at the time of initial presentation of children with KD. Given its superior visualization of the entire length of coronary arteries, CTCA has a vital role in therapeutic decision-making in KD. This can facilitate earlier risk stratification and adoption of more precise treatment strategies. Future updates to KD clinical guidelines should consider incorporating CTCA during initial presentation in the management algorithm, especially in cases where TTE findings are inconclusive.

Author Contributions: M.S. (Manphool Singhal) and R.K.P.: Conceptualization, study design, data curation, formal analysis, investigation, methodology, supervision, validation, figures, writing—original draft, and writing–review and editing. M.S. (Manphool Singhal) and R.K.P. contributed equally, and share the first authorship and corresponding authorship. S.B. and D.D.: data curation, formal analysis, investigation, methodology, supervision, validation, figures, writing—original draft, and writing. A.T., R.S. (Ripudaman Singh), R.S. (Radhika Semwal), T.K., M.S. (Murugan Sudhakar), P.V. and D.S.: Literature search, data collection, and manuscript editing. G.S. and A.S.: Figures, radiology data interpretation, literature search, and manuscript editing. S.S.: conceptualization, study design, overall supervision, and critical editing of the manuscript at each step. All authors have read and agreed to the published version of the manuscript.

Funding: This research received no external funding.

Institutional Review Board Statement: The manuscript has been approved by the publication review board (Department Ethics Committee, Department of Radiodiagnosis and Imaging, PGIMER, Chandigarh No. RDG/EC/Pub/27; dated 20 April 2025).

Informed Consent Statement: Informed consent was obtained from all subjects involved in the study.

Data Availability Statement: This article's data are available in the manuscript and in our records. These data can be shared on request.

Acknowledgments: The authors gratefully acknowledge the role of all faculty members, clinical fellows, nursing officers, research, and technical staff who have been involved in the care of these patients.

Conflicts of Interest: The authors declare no conflicts of interest.

References

1. Pilania, R.K.; Tremoulet, A.H.; Prinja, S.; Dahdah, N.; Singh, S. Kawasaki disease: The most common cause of acquired heart disease among children globally. *Cardiol. Young* **2025**, *35*, 441–443. [CrossRef] [PubMed]
2. Jone, P.N.; Tremoulet, A.; Choueiter, N.; Dominguez, S.R.; Harahsheh, A.S.; Mitani, Y.; Zimmerman, M.; Lin, M.-T.; Friedman, K.G. Update on Diagnosis and Management of Kawasaki Disease: A Scientific Statement From the American Heart Association. *Circulation* **2024**, *150*, e481–e500. [CrossRef] [PubMed]
3. McCrindle, B.W.; Rowley, A.H.; Newburger, J.W.; Burns, J.C.; Bolger, A.F.; Gewitz, M.; Baker, A.L.; Jackson, M.A.; Takahashi, M.; Shah, P.B.; et al. Diagnosis, Treatment, and Long-Term Management of Kawasaki Disease: A Scientific Statement for Health Professionals From the American Heart Association. *Circulation* **2017**, *135*, e927–e999. [CrossRef] [PubMed]
4. Pilania, R.K.; Jindal, A.K.; Bhattarai, D.; Naganur, S.H.; Singh, S. Cardiovascular Involvement in Kawasaki Disease Is Much More Than Mere Coronary Arteritis. *Front. Pediatr.* **2020**, *8*, 526969. [CrossRef] [PubMed]
5. Pilania, R.K.; Basu, S.; Dixit, J.; Kumrah, R.; Jindal, A.K.; Thangaraj, A.; Nimesh, R.; Kaur, T.; Vignesh, P.; Suri, D.; et al. Incidence of Kawasaki disease among children in Chandigarh, India during 2015–2019: A trend analysis. *Lancet Reg. Health—Southeast Asia* **2024**, *29*, 100474. Available online: https://www.thelancet.com/journals/lansea/article/PIIS2772-3682(24)00124-0/fulltext (accessed on 12 September 2024). [CrossRef] [PubMed]
6. Tsuda, E.; Singhal, M. Role of imaging studies in Kawasaki disease. *Int. J. Rheum. Dis.* **2018**, *21*, 56–63. [CrossRef] [PubMed]
7. Singhal, M.; Singh, S.; Gupta, P.; Sharma, A.; Khandelwal, N.; Burns, J.C. Computed Tomography Coronary Angiography for Evaluation of Children With Kawasaki Disease. *Curr. Probl. Diagn. Radiol.* **2018**, *47*, 238–244. [CrossRef] [PubMed]
8. Pilania, R.K.; Basu, S.; Singh, S.; Sabui, T.; Singhal, M. Computed tomography coronary angiography for imaging in children with Kawasaki disease: An update. *Int. J. Rheum. Dis.* **2024**, *27*, e15331. [CrossRef] [PubMed]
9. Singhal, M.; Pilania, R.K.; Basu, S.; Sharma, A.; Singh, S. CT coronary angiography in Kawasaki disease: Current perspectives. *Arch. Dis. Childhood* **2025**, *110*, 499–500. Available online: https://adc.bmj.com/content/early/2025/03/27/archdischild-2024-327190 (accessed on 18 April 2025). [CrossRef] [PubMed]
10. Newburger, J.W.; Takahashi, M.; Gerber, M.A.; Gewitz, M.H.; Tani, L.Y.; Burns, J.C.; Shulman, S.T.; Bolger, A.F.; Ferrieri, P.; Baltimore, R.S.; et al. Diagnosis, Treatment, and Long-Term Management of Kawasaki Disease. *Circulation* **2004**, *110*, 2747–2771. [CrossRef] [PubMed]
11. Dallaire, F.; Dahdah, N. New equations and a critical appraisal of coronary artery Z scores in healthy children. *J. Am. Soc. Echocardiogr.* **2011**, *24*, 60–74. [CrossRef] [PubMed]
12. Pilania, R.K.; Nadig, P.L.; Basu, S.; Tyagi, R.; Thangaraj, A.; Aggarwal, R.; Arora, M.; Sharma, A.; Singh, S.; Singhal, M. Congenital anomalies of coronary artery misdiagnosed as coronary dilatations in Kawasaki disease: A clinical predicament. *World J. Clin. Pediatr.* **2025**, *14*, 99177. [CrossRef] [PubMed]
13. Thangathurai, J.; Kalashnikova, M.; Takahashi, M.; Shinbane, J.S. Coronary Artery Aneurysm in Kawasaki Disease: Coronary CT Angiography through the Lens of Pathophysiology and Differential Diagnosis. *Radiol. Cardiothorac. Imaging* **2021**, *3*, e200550. [CrossRef] [PubMed]
14. Singhal, M.; Pilania, R.K.; Jindal, A.K.; Gupta, A.; Sharma, A.; Guleria, S.; Johnson, N.; Maralakunte, M.; Vignesh, P.; Suri, D.; et al. Distal coronary artery abnormalities in Kawasaki disease: Experience on CT coronary angiography in 176 children. *Rheumatology* **2023**, *62*, 815–823. [CrossRef] [PubMed]
15. Borhanuddin, B.K.; Abdul Latiff, H.; Mohamed Yusof, A.K. CT coronary angiogram in children with Kawasaki disease: Experience in 52 patients. *Cardiol. Young* **2022**, *32*, 1994–1998. [CrossRef] [PubMed]
16. Barman, P.; Pilania, R.K.; Cv, G.; Thangaraj, A.; Arora, M.; Singh, S. Treatment intensification in Kawasaki disease—Current perspectives. *Expert. Rev. Clin. Immunol.* **2024**, *20*, 1179–1191. [CrossRef] [PubMed]
17. Singhal, M.; Pilania, R.K.; Thangaraj, A.; Chaudhary, H.; Gummadi, A.; Soundararajan, R.; Loganathan, S.; Sharma, A.; Gupta, A.; Bhattad, S.; et al. The value of CT coronary angiography for a comprehensive assessment of left circumflex artery in Kawasaki disease: 9 years of experience from a tertiary center. *Lancet Reg. Health—Southeast Asia* **2024**, *29*, 100471. [CrossRef] [PubMed]
18. Carbone, I.; Cannata, D.; Algeri, E.; Galea, N.; Napoli, A.; De Zorzi, A.; Bosco, G.; D'Agostino, R.; Menezes, L.; Catalano, C.; et al. Adolescent Kawasaki disease: Usefulness of 64-slice CT coronary angiography for follow-up investigation. *Pediatr. Radiol.* **2011**, *41*, 1165–1173. [CrossRef] [PubMed]
19. Chakraborty, R.; Singhal, M.; Pandiarajan, V.; Sharma, A.; Pilania, R.K.; Singh, S. Coronary arterial abnormalities detected in children over 10 years following initial Kawasaki disease using cardiac computed tomography. *Cardiol. Young* **2021**, *31*, 998–1002. [CrossRef] [PubMed]
20. Dietz SM: Tacke, C.E.; Kuipers, I.M.; Wiegman, A.; de Winter, R.J.; Burns, J.C.; Gordon, J.B.; Groenink, M.; Kuijpers, T.W. Cardiovascular imaging in children and adults following Kawasaki disease. *Insights Imaging* **2015**, *6*, 697–705. [CrossRef] [PubMed]

diagnostics

MDPI

Article

A Retrospective Analysis of the Effects of Concomitant Use of Intra-Aortic Balloon Pump (IABP) and Veno-Arterial Extracorporeal Membrane Oxygenation (va-ECMO) Therapy on Procedural Brain Infarction

Farid Ziayee [1], Hannan Dalyanoglu [2,*], Christian Schnitzler [1], Kai Jannusch [1], Matthias Boschheidgen [1], Judith Boeven [1], Hug Aubin [2], Bernd Turowski [1], Marius Georg Kaschner [1,*] and Christian Mathys [1,3]

1 Department of Diagnostic and Interventional Radiology, Medical Faculty, University Dusseldorf, 40225 Dusseldorf, Germany; farid.ziayee@med.uni-duesseldorf.de (F.Z.); christian.schnitzler@med.uni-duesseldorf.de (C.S.); kai.jannusch@med.uni-duesseldorf.de (K.J.)
2 Department of Cardiovascular Surgery, Medical Faculty, University Dusseldorf, 40225 Dusseldorf, Germany
3 Institute of Radiology and Neuroradiology, Evangelisches Krankenhaus Oldenburg, Universitätsmedizin Oldenburg, Steinweg 13–17, 26122 Oldenburg, Germany
* Correspondence: hannan.dalyanoglu@med.uni-duesseldorf.de (H.D.); marius.kaschner@med.uni-duesseldorf.de (M.G.K.); Tel.: +49-(0)211-8107643 (H.D.); +49-(0)211-8108553 (M.G.K.)

Abstract: Background/Objectives: Brain ischemia is a frequent complication in patients undergoing veno-arterial extracorporeal membrane oxygenation (va-ECMO) therapy due to hypoperfusion, low oxygenation, and thromboembolism. While concomitant intra-aortic balloon pump (IABP) therapy may improve the perfusion of the supra-aortic branches, it may also favor thromboembolism. This retrospective study aimed to evaluate the effects of combined va-ECMO and IABP therapy on procedural brain infarction compared to va-ECMO therapy alone, with a specific focus on analyzing the types of infarctions. **Methods:** Cranial computed tomography (CCT) scans of consecutive patients receiving va-ECMO therapy were analyzed retrospectively. Subgroups were formed for patients with combined therapy (ECMO and IABP) and va-ECMO therapy only. The types of infarctions and the potential impacts of va-ECMO vs. combined therapy with IABP on stroke were investigated. **Results:** Overall, 146 patients (36 female, 110 male, mean age 61 $\pm$ 13.3 years) were included, with 69 undergoing combined therapy and 77 patients receiving va-ECMO therapy alone. In total, 14 stroke events occurred in 11 patients in the ECMO-only group and there were 12 events in 12 patients in the ECMO + IABP-group, showing no significant difference ($p = 0.61$). The majority of infarctions were of thromboembolic ($n = 23$; 88%) origin, with 14 stroke-events in 12 patients in the ECMO + IABP-group and 9 stroke events in the ECMO-only group. The survival rate within 30 days of treatment was 29% in the ECMO-only group and 32% in the ECMO + IABP group. **Conclusions:** The results of this retrospective study show that concomitant IABP therapy appears to be neither protective nor more hazardous in relation to ECMO-related stroke. Thus, the indication for additional IABP therapy should be assessed independently from the procedural risk of brain ischemia. Thromboembolic infarctions seem to represent the most common type of infarction in ECMO, especially within the first 48 h of treatment.

Keywords: thromboembolism; watershed; stroke; intra-aortic balloon pump; extracorporeal membrane oxygenation

 https://doi.org/10.3390/diagnostics15060699

1. Introduction

Extracorporeal membrane oxygenation (ECMO) has emerged as a pivotal component of modern extracorporeal life support, serving as a vital temporary measure for patients requiring cardiopulmonary support in critical situations [1]. It provides a crucial bridge to more definitive interventions, such as heart transplantation or recovery from severe organ failure [2]. Despite its life-saving potential, ECMO therapy is associated with a range of severe and often life-threatening complications that can significantly impact patient outcomes.

In the case of veno-arterial ECMO (va-ECMO), venous blood is extracted from the inferior vena cava and returned to the arterial system. This technique is predominantly employed as a bridge therapy, aiming to stabilize patients with acute cardiogenic shock or severe cardiac failure. However, va-ECMO carries considerable risks, including the potential for brain ischemia, which can result in devastating neurological outcomes [3,4].

Despite significant advancements in ECMO management over the past four decades, serious complications remain common, including bleeding, vascular injury, and ischemia affecting the brain, visceral organs, and limbs. Diagnosing brain ischemia in ECMO patients poses unique challenges due to their often altered consciousness and complex clinical presentations, which can lead to the potential underreporting of neurological complications [5,6]. Current estimates of neurological complications among ECMO patients vary widely, ranging from 1–78% [7]. The most prevalent cerebral complications associated with ECMO therapy include intracranial hemorrhage, stroke, and intracranial edema [5–7].

Brain infarction during va-ECMO can arise from several mechanisms. One potential cause is the formation of a thrombus within the ECMO system that may enter the cerebral circulation through either central or peripheral cannulation. This can result in territorial or small lacunar infarctions. Hemodynamic alterations, especially those associated with peripheral va-ECMO, can reduce blood flow in supraaortic vessels, resulting in hemodynamic infarctions. Furthermore, a watershed phenomenon in the aortic arc and device dysfunction may lead to hepoxemic infarctions in the basal ganglia or cortex [6–9].

The intra-aortic balloon pump (IABP) has been introduced as an adjunctive therapy to enhance the efficacy of peripheral va-ECMO. The IABP operates by inflating during the diastole and deflating during the systole, which enhances diastolic pressure and is thought to reduce afterload. This mechanism can improve cardiac output by up to 20% and enhance perfusion to the coronary arteries, supra-aortic branches, and visceral vessels [10]. However, the combined use of IABP with va-ECMO remains a topic of debate among clinicians [3,11]. Despite its potential cerebral benefits, the IABP is a thrombogenic device, raising concerns about the risk of thromboembolic complications.

Currently, there is a notable gap in the literature regarding the incidence of brain infarctions in patients receiving combined ECMO and IABP therapy. The simultaneous implantation of IABP during peripheral va-ECMO is not considered standard practice, and its effect on stroke risk remains largely unclear. We hypothesize that the addition of IABP therapy may reduce the risk of hemodynamic intracranial infarctions by improving cerebral perfusion [12]. Conversely, an embolus originating from the ECMO system during peripheral cannulation may be more likely to reach the cerebral circulation in the presence of IABP therapy. Therefore, unless the watershed area is distal to the supra-aortic vessels, the use of IABP may inadvertently increase the risk of thromboembolic infarctions in this scenario. To our knowledge, no studies have specifically reported the incidence rates of brain infarctions or differentiated between the types of infarctions (e.g., hypoxic, thromboembolic, or hemodynamic) in patients undergoing combined va-ECMO and IABP therapy.

The primary objective of this study is to assess the impact of adjunctive IABP therapy on the risk of brain ischemia in patients receiving va-ECMO therapy and to classify the types of infarctions observed. By enhancing our understanding of these relationships, we aim to improve patient management strategies and clinical outcomes in this vulnerable population.

2. Materials and Methods

This retrospective study analyzed data from consecutive patients who underwent va-ECMO therapy between January 2010 and December 2015 at the Department of Cardiovascular Surgery, University Hospital of Düsseldorf, Germany. This study, approved by the local institutional review board, aimed to assess the impact of va-ECMO with or without adjunct intra-aortic balloon pump (IABP) therapy on the occurrence of brain infarctions. Patients were divided into two groups based on their treatment modality: the ECMO-only group and the ECMO + IABP group. The choice to employ va-ECMO alone or alongside IABP for patients in cardiogenic shock depends on a range of clinical parameters and individual patient conditions. Va-ECMO is often preferred initially for its robust circulatory and respiratory support capabilities. The addition of IABP was performed in cases of severe left ventricular dysfunction or when afterload reduction was essential to optimize left ventricular unloading and enhance coronary perfusion. In instances where va-ECMO alone led to increased afterload and inadequate left ventricular unloading, IABP was integrated to mitigate complications such as pulmonary edema and left ventricular thrombus formation.

2.1. Inclusion and Exclusion Criteria

2.1.1. Inclusion Criteria Included the Following

Patients treated with va-ECMO.
A minimum age of 18 years.
Cranial computed tomography (CCT) performed during ECMO therapy.

2.1.2. Exclusion Criteria Included the Following

Patients who received additional left or right ventricular assist devices (LVADs/RVADs).
The incorrect positioning of the IABP tip or any ECMO component.
Patients with hypoxic infarctions not associated with ECMO therapy.

The proper placement of the IABP tip was defined as being distal to the left subclavian artery within the thoracic aorta. For peripheral cannulation, the arterial ECMO cannula was inserted into either the external or common iliac artery. For central cannulation, it was positioned in the ascending aorta. The venous cannula was inserted into the inferior vena cava, right atrium, or superior vena cava. Cannulation positions and IABP placements were verified through thoracic X-ray analysis and CT imaging.

2.1.3. Data Collection

Patient baseline characteristics were gathered, including details on cannulation sites, the duration of ECMO therapy, indications for device therapy, occurrences of cardiopulmonary resuscitation (CPR), 30-day survival rates post-device implantation, and neuron-specific enolase (NSE) levels. This information was extracted from the hospital's electronic health records and digital medical information system.

2.1.4. Imaging Analysis

Cranial computed tomography (CCT) scans were performed using four different CT systems from Siemens Healthcare GmbH (Erlangen, Germany). All CCT scans followed a standardized protocol with supine patient positioning to ensure consistency across images. Detailed parameters for each CT system are outlined in Table 1.

Table 1. Four different CT models employed in this study.

Model	Siemens Definition AS Plus	Siemens Definition AS with Sliding Gantry	Siemens Definition Flash	Siemens Emotion
Slices	128	64	2 × 128	16
Reconstruction slice thickness	2 mm	2 mm	2 mm	3 mm infratentorial/ 5 mm supratentorial
Tube voltage	100 kV	100 kV	120 kV	130 kV
Field of view	213 mm	220 mm	220 mm	220 mm
Automatic dose exposure control (Siemens care dose 4 D)	Yes	Yes	Yes	No

The CCT images were retrospectively reviewed by a team of board-certified radiologists and neuroradiologists, working based on consensus to ensure the accuracy of findings. Brain infarctions were identified and classified based on lesion age, using established radiological criteria to differentiate lesions as acute, subacute, or chronic. This classification enabled the detailed analysis of the temporal relationship between infarctions and ECMO therapy. All CCTs during device therapy were included, so that, eventually, early infarctions missed on the first CCT could potentially be detected in the follow-up scan. Ischemic strokes in the ECMO and IABP group were only taken into account after the implantation of both devices. Each infarction was further categorized as a thromboembolic, hemodynamic, or hypoxemic type based on established criteria from previous studies (Figure 1). Infarctions were assigned to vascular territories: anterior, medial and posterior cerebral arteries (ACAs, MCAs, and PCAs), superior, posterior inferior and anterior inferior cerebellar arteries (SCAs, PICAs, and AICAs), and basilar arteries (BAs). Infarctions were categorized as thromboembolic, hemodynamic, or global cerebral hypoxemia.

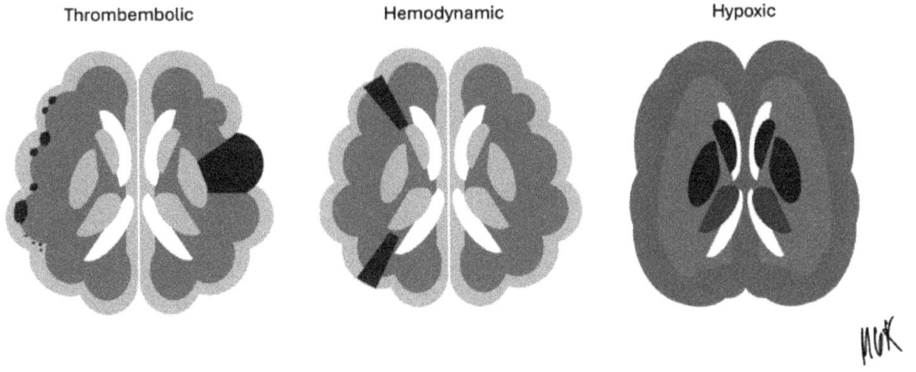

Figure 1. Thromboembolic, hemodynamic, and hypoxic stroke pattern for the classification of infarction.

The Alberta Stroke Program Early CT Score (ASPECTS) was used to quantify parenchymal involvement in the MCA territory of each side [13]. A score of 10 points represents healthy MCA territory, with one point deducted for each affected region (the nucleus caudatus, putamen, capsula interna, insular cortex, frontal operculum, anterior temporal lobe, and posterior temporal lobe, as well as the cortex regions immediately superior to the frontal operculum and the anterior and posterior lobe). A modified ASPECTS was introduced for other territories, with a maximum score of 19 points per side, accounting for ACA, PCA, cerebellar arteries, and BA territories. In addition to the ASPECTS, 2 more points were deducted for ACA territory (1 point for less than and 1 point for more than 50% of the territory), 3 points were deducted for PCA territory (1 point for the medial temporal lobe, 1 point for the thalamus, and 1 point for the occipital lobe), 1 point for SCA, AICA, and PICA each and 1 point for BA. Patients with multiple stroke events had each event recorded and classified to provide a comprehensive view of complications.

2.1.5. Statistical Analysis

Statistical analysis was carried out using SPSS Statistics 24 software (IBM, Armonk, NY, USA). The chi-square test was applied to dichotomous data with a sample size of five or greater. In cases with smaller sample sizes, Fisher's exact test was used to ensure precision. The Kolmogorov–Smirnov test assessed the normal distribution of metric data. For non-normally distributed data, the Mann–Whitney U test was applied, while Student's *t*-test was used for normally distributed data. A p-value of < 0.05 was considered statistically significant.

2.1.6. Additional Parameters and Considerations

Neuron-Specific Enolase (NSE): NSE levels, a biomarker for neurological damage, were monitored and cross-referenced with imaging findings to validate the presence and extent of brain injury.

Device therapy duration: The duration of ECMO and IABP support was recorded and analyzed to explore any correlation between the length of therapy and the risk of ischemic events. This analysis aimed to identify critical time points where the risk of complications might increase.

Patient outcomes: A detailed survival analysis was conducted, extending beyond the initial 30-day period to capture long-term outcomes associated with brain ischemia in the study cohort. This included functional outcomes and quality-of-life assessments when such data were available.

This comprehensive methodology offers a thorough evaluation of the effects of va-ECMO and IABP on brain ischemia, contributing valuable information on ECMO-associated complications and helping to guide future clinical practices.

It must always be noted that retrospective data collection is not feasible due to the absence of consistent documentation. Therefore, the combined use of mechanical circulatory support systems should be guided by standardized protocols that capture cerebral perfusion pressure, cerebral saturation, and hemodynamic parameters at the time of cerebral events. This will facilitate a better causal explanation of stroke rates whenever possible.

3. Results

A total of 146 patients (36 females and 110 males) with an average age of 61.3 years (range: 21–85 years) were included in this study, all of whom underwent va-ECMO therapy. Out of this cohort, 69 patients received simultaneous intra-aortic balloon pump (IABP) treatment. Within this subgroup, 44 patients underwent peripheral cannulation, and 25 patients underwent central cannulation. In contrast, among the 77 patients treated with va-ECMO alone, 66 received peripheral cannulation, while 11 received central cannulation. This detailed distribution of cannulation types provides insight into the management strategies and potential differences in clinical outcomes between the groups.

The baseline characteristics of the two groups (ECMO-only and ECMO + IABP) are presented comprehensively in Table 2. The table includes demographic information, such as age and sex distribution, along with details on underlying conditions that led to the initiation of ECMO therapy. These conditions include cardiogenic shock, myocardial infarction, and cardiac arrest, highlighting the critical status of patients who typically receive this intervention. The table also lists pre-existing risk factors, such as hypertension, diabetes mellitus, and a history of stroke, which are essential for understanding how comorbidities might influence clinical outcomes, complication rates, and overall survival in ECMO therapy.

Additionally, the table provides information on the duration of device therapy for each patient. These data enable an analysis of the relationship between ECMO duration and clinical complications or mortality. By examining the duration of therapy, the study aims to identify any patterns that might suggest optimal or critical periods for intervention or adjustment in patient management. This approach offers nuanced insights into balancing life-saving therapy by minimizing risks such as thrombosis, bleeding, and ischemic events.

Incorporating details on cannulation site specifics also enables a more refined analysis of the impact of peripheral versus central cannulation on patient outcomes. This differentiation is particularly important when comparing the ECMO-only group to those receiving additional IABP therapy, as it may highlight variations in perfusion and complication rates based on the cannulation strategy used. Peripheral and central cannulations present distinct challenges and risks, and understanding their implications is essential for optimizing patient management protocols.

Table 2 offers a detailed comparison of the patient groups, including therapy duration and baseline characteristics, which are instrumental in evaluating the efficacy and safety of va-ECMO with or without concomitant IABP. By providing a comprehensive overview of demographics, pre-existing conditions, and procedural details, the table supports a deeper exploration into how these factors influence patient outcomes and may guide future improvements in ECMO management practices.

Table 2. Baseline characteristics, duration of device therapy, cannulation site, and indication for ECMO-therapy.

	ECMO Only (*n* = 77)	**ECMO + IABP (*n* = 69)**	***p*-Values**
Age (years)	58.5 (±13.0)	64.5 (±13.0)	0.98
Sex	21 females, 27%	15 females, 22%	0.60
Duration of device therapy in days mean (±SD)	7.1 (±5.9)	ECMO: 9.7 (±6.1) IABP: 8.2 (±4.6) Simultaneous: 6.8 (±4.2)	0.60
Duration of ECMO without IABP in days	N/A	2.9 (±4.8)	N/A
Duration of IABP without ECMO (days)	N/A	1.3 (±1.9)	N/A
Cannulation site	66 peripheral (85.7%) 11 central (14.3%)	44 peripheral (60.9%) 25 central (39.1%)	N/A
Indication	Coronary artery disease and myocardial infarction: 44% Cardiac decompensation: 9% Cardiac arrhythmia: 4% Heart transplantation: 3% Aortic dissection: 3% Other: 7% Unknown: 30%	Coronary artery disease and myocardial infarction: 47% Artificial heart valve: 22% Cardiac decompensation: 17% Aortic dissection: 7% Heart transplantation: 1% Cardiac tamponade: 1% Unknown: 5%	N/A

SD = standard deviation, IABP = intra-aortic balloon pump, ECMO = extracorporeal membrane oxygenation.

The distribution of peripheral versus central cannulation was meticulously documented, as the choice of cannulation site can significantly influence perfusion outcomes and the risk of ischemic events in ECMO patients. Differentiating between these two types of cannulation allows for a more nuanced analysis of how each impacts brain perfusion, particularly when additional intra-aortic balloon pump (IABP) therapy is administered. Understanding these distinctions is crucial, as the study aims to evaluate the impact of combined ECMO and IABP therapy on brain ischemia compared to ECMO-only therapy, thereby providing insights that could guide future clinical decisions.

During the study, 23 out of 146 patients (16%) experienced a total of 26 ischemic stroke events, which were identified via CCT during their treatment. Of these, 23 events (88%) were classified as thromboembolic infarctions, highlighting the predominance of this type of ischemic event in the ECMO population. In the ECMO-only group, 11 out of 77 patients (14%) experienced stroke events, while in the ECMO + IABP group, 12 out of 69 patients (17%) had brain infarctions. Additionally, within the ECMO-only group, three patients experienced a second stroke event during the course of their treatment, emphasizing the recurrent risk of cerebrovascular events in this critically ill population.

Among the 26 total stroke events, 3 were identified as hemodynamic infarctions (12%), and notably, all of these cases occurred within the ECMO + IABP group. The occurrence of hemodynamic infarctions solely in this group suggests that the combination of ECMO and IABP therapy might influence the hemodynamic environment in a way that predisposes patients to this type of infarction. However, despite this observation, statistical analysis did not show a significant difference between the ECMO-only and ECMO + IABP groups regarding the overall infarction rate, infarction type, or survival rate. This lack of significant difference indicates that while there may be trends in the data, larger studies or additional parameters may be needed to confirm the influence of combined therapy on stroke outcomes.

Neuron-specific enolase (NSE), a marker of neuronal injury, was found to be elevated in all patients who experienced infarctions. The levels were higher in the ECMO + IABP group (160 µg/dL) compared to the ECMO-only group (120 µg/d, $p = 0.72$).

Further detailed information, including the distribution of stroke events between the groups, the types of strokes identified, and the 30-day survival rate following ECMO explantation, as well as the involvement of vascular territory in cases of infarction, assessed using the ASPECTS and the modified ASPECTS, is presented in Table 3. Often, several vascular territories were affected during one single stroke event. There were no significant differences regarding involvement of vascular territories or parenchymal involvement according to ASPECTS and modified ASPECTS between the two groups. All patients received conservative treatment because the infarction was already demarked on CCT, so no intravenous lysis or thrombectomy were performed in case of large vessel occlusion.

Table 3. Stroke rate, type of infarction, localization, ASPECTS, and modified ASPECTS, and survival rate in ECMO-only- and ECMO + IABP-therapy.

	ECMO Only (N = 77)	ECMO + IABP (N = 69)	*p*-Values
Patients with cerebral infarctions	11	12	0.61
Total no. of events with cerebral infarction	14	12	0.44
Hemodynamic cerebral infarctions (no. of events)	0	3	0.1
Thromboembolic cerebral infarctions (no. of events)	14	9	0.61
Vascular territories involved in thromboembolic infarctions			
ACA	3	1	
ACM	10	72	
ACP	8	6	

Table 3. *Cont.*

	ECMO Only (N = 77)	ECMO + IABP (N = 69)	*p*-Values
SCA	3	3	
AICA	3	2	
PICA	4	4	
BA	1	1	
Hypoxic cerebral infarctions (no. of events)	0	0	N/A
ASPECTS right MCA territory (median)	9	10	N/A
ASPECTS left MCA territory (median)	9	10	N/A
Modified ASPECTS right (median)	17	17	N/A
Modified ASPECTS left (median)	17	18	N/A
Survival rate within 30 days of treatment (%)	29	32	0.66
Cause of death during ECMO therapy	Multi organ dysfunction syndrome (49%) Cardiac failure (16%) Bleeding (5%) Neurological complication (11%) Other (19%)	Multi organ dysfunction syndrome (57%) Cardiac failure (19%) Bleeding (4%) Neurological complication (8%) Other (12%)	
CPR-rate (%)	68	36	<0.01

SD = standard deviation, IABP= intra-aortic balloon pump, ECMO = extracorporeal membrane oxygenation, CPR = cardiopulmonary resuscitation, ASPECTS = Alberta Stroke Program Early CT Score, ACA = anterior cerebral artery, ACM = middle cerebral artery, ACP = posterior cerebral artery, SCA = superior cerebellar artery, AICA = anterior inferior cerebellar artery, PICA = posterior inferior cerebellar artery, BA = basilar artery.

The amount of infarction per day after implantation is shown in Figure 2.

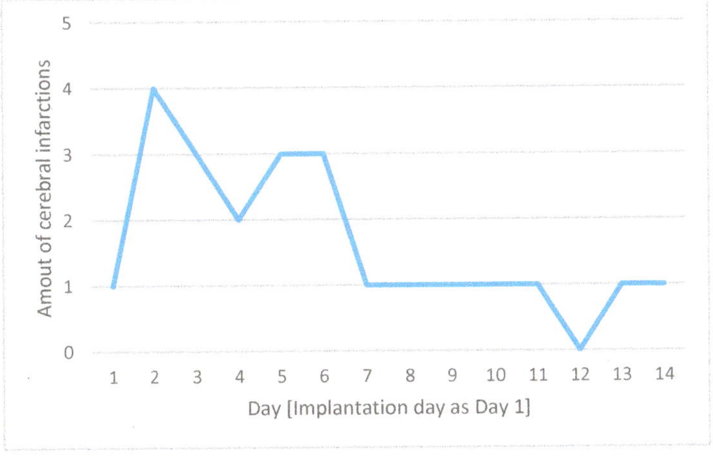

Figure 2. The number of cerebral complications for each day following implantation, with day 1 defined as the day of device implantation on the x-axis and the number of cerebral infarctions on the y-axis.

4. Discussion

Our retrospective study emphasizes that ischemic stroke remains a notable complication in va-ECMO therapy, with the majority of events being thromboembolic in nature. These events pose a critical risk that necessitates the further development of strategies for understanding and prevention. The findings suggest that patients receiving combined therapy involving ECMO and intra-aortic balloon pump (IABP) do not experience an increased risk of thromboembolic brain infarctions compared to those treated with ECMO alone. Furthermore, the addition of IABP therapy does not appear to mitigate the risk of hemodynamic infarctions in these patients.

The survival rate in our study's cohort was notably lower than the survival rates reported by the Extracorporeal Life Support Organization (ELSO) in its 2022 and 2014 data, which indicated survival rates of 59% and 40%, respectively, for va-ECMO patients [14]. Several factors may explain this discrepancy. First, our inclusion criteria were specific, excluding ECMO patients who did not undergo CCT. This selection could have inadvertently omitted patients with potentially better outcomes, thereby skewing our survival data toward a worse outcome. Additionally, a higher proportion of patients in our ECMO-only group presented with cardiopulmonary resuscitation (CPR), accounting for 68% of this group. In contrast, the combination of ECMO with an IABP may have been used less frequently in patients with less severe illnesses, potentially contributing to lower resuscitation rates in this group. Furthermore, research has demonstrated that ECMO is often used more frequently in patients with severe cardiogenic shock, increasing the need for resuscitation, while IABP is more commonly employed in patients with less severe conditions [15]. Differing therapeutic practices may have contributed to variations in outcomes and resuscitation rates [16]. These potential explanations are likely attributable to the retrospective data collection process and the inconsistent documentation of patient characteristics and treatment strategies, leading to biases and random discrepancies. This high resuscitation rate also may reflect the severity of illness in our patients, as our center specializes in the care of highly critical cases, which can negatively impact survival outcomes [17]. This level of severity could also be a contributing factor, as critically ill patients generally have poorer prognoses, regardless of therapeutic interventions, weakening the comparability of the two groups.

The absence of a significant difference in 30-day survival between the ECMO-only and ECMO + IABP groups is consistent with previous findings, such as those of Lin et al., who reported nearly identical two-week mortality rates of 48.5% in the IABP group versus 47% in the ECMO-alone group [11]. Gass et al. also found an in-hospital survival rate of 57.8% for va-ECMO and IABP patients ($p = 0.053$) [18]. Collectively, these data suggest that using IABP does not provide a survival advantage. This lack of clear benefit should be carefully weighed against the risks associated with IABP use in clinical decision-making.

The brain infarction rate in our study was 16% overall, with rates of 14% in the ECMO-only group and 17% in the ECMO + IABP group. These rates are in line with the wide range reported in the literature, where the brain infarction incidence in va-ECMO therapy varies from 3% to 33% [7,19,20]. This variability may stem from differences in diagnostic criteria, patient populations, and ECMO management strategies across studies. It is crucial to distinguish between different ECMO types when discussing stroke risk; for instance, venovenous ECMO involves the lungs, which act as a filter for emboli, offering a degree of protection to the brain. The protective role of the lungs in venovenous ECMO highlights the importance of carefully selecting the appropriate ECMO modality based on the patient's risk profile. Moreover, some studies diagnose ischemic events based solely on written radiological reports, while our study utilized second image readings, which may have enhanced our detection accuracy [7,21,22]. To our knowledge, no previous studies have specifically reported on the stroke incidence and infarction types in patients undergoing combined va-ECMO and IABP therapy, making our findings an important contribution to the field.

Thromboembolic events were identified as the predominant cause of infarctions in our cohort. These emboli may originate from multiple sources, including the ECMO circuit itself, the pulmonary veins, or the heart. In the case of cardiac thrombus formation, slow blood flow or reduced cardiac wall motion could lead to the development of thrombus, which can eventually detach and enter cerebral circulation [23]. In scenarios involving peripheral cannulation, thrombotic material might bypass the IABP during the diastole and enter the aortic arch. On the other hand, thrombi from the left ventricle or central ECMO cannula could reach the supra-aortic vessels through the antegrade flow [12,23]. The IABP's role in enhancing perfusion to the supra-aortic branches might alter the watershed location, thereby raising the risk of thromboembolic events from the ECMO system especially in cases of peripheral cannulation [24–27]. Despite our inability to demonstrate statistically significant differences in thromboembolic infarctions between groups, the higher CPR rate in the ECMO-only group may have skewed these results, as CPR itself is a recognized risk factor for thromboembolism. Similarly, more extensive neuronal damage was detected in patients receiving the combined therapy. However, the statistical analysis of NSE levels between the two groups did not demonstrate a significant difference, indicating that while there is an observable difference in NSE levels, it may not be clinically meaningful, or a larger sample size may be required to determine its significance.

Embolism and hypoperfusion seem to play synergetic roles [28,29]. In cases of lower levels of perfusion in delivering vessels, a thrombus that would typically pass the capillary system under normal hemodynamic circumstances is more likely to obstruct a terminal vessel and cause infarction [30]. Hemodynamic infarction was less common, with only three cases observed, all occurring within the ECMO + IABP group. This low incidence suggests that hemodynamic infarction is relatively rare in ECMO therapy, and the addition of IABP did not appear to provide any protective effect against these events. The absence of ECMO-related hypoxic infarctions in our cohort indicates that this type of infarction may be of limited relevance in the context of ECMO therapy. However, other authors have reported hypoxemic infarctions [31] and our data show a trend of increased risk of hemodynamic infarction in the ECMO + IABP group, which is perhaps not statistically significant due to the sample size. Our data further indicate that the risk of thromboembolic events is highest within the first 48 h following ECMO implantation, emphasizing the importance of vigilant monitoring during this critical period to promptly identify and manage these events. The early occurrence of strokes in patients undergoing ECMO can be attributed to a complex interplay of mechanisms and patient-specific factors. The first days of ECMO support are critical due to the combination of the patient's underlying critical condition and the physiological changes induced by the ECMO system. Patients requiring ECMO often have severe statuses that predispose them to thrombotic or embolic events. These pre-existing conditions, combined with the inherent risks of ECMO, significantly increase the likelihood of stroke early in the course of support, particularly before anticoagulation therapy is fully optimized [32].

The retrospective design of our study presented limitations, especially in assessing all relevant comorbidities or different coagulation protocols that could contribute to the risk of intracranial infarction. While CCT is an effective tool for detecting brain infarction, magnetic resonance imaging (MRI) could have offered superior resolution and potentially higher detection rates. Conventional MRI is not feasible in this arrangement, and so portable bedside MRI has been employed in a few studies [33]. The absence of a true control group limits the comparability of either intervention. Differences in CPR rates, cannulation sites, and the timing of CCT acquisition between the two groups might also have introduced biases that influenced our results. The absence of multivariate modeling is a significant limitation, as it hinders the ability to account for potential confounders such as age, gender, and comorbidities. For outcomes like ischemic stroke, logistic regression, or others, multivariable techniques should be utilized to adjust for these confounders and enhance the validity of the findings. Additionally, the limited sample size and retrospective nature of the study further restrict its reproducibility and external validity, compromising the generalizability of the results and undermining confidence in their clinical implications. The study's 5-year duration may have also confounded the results, as advancements in ECMO or IABP therapy during this period could have led to varying outcomes. Standardizing these factors in future studies could provide a more accurate understanding of the risks associated with ECMO and IABP therapy, ensuring more robust and generalizable findings. Low statistical power may reduce the likelihood of identifying significant associations or differences, especially when small sample sizes are involved. Due to these limitations, our study had the potential for Type II errors, where true effects may be present but are not detected due to an insufficient sample size or variability in patient characteristics.

The primary goal of our study was to compare ECMO-only therapy with combined ECMO and IABP therapy, offering initial insights into the types of brain infarctions in va-ECMO patients. In summary, this study provides a detailed account of the stroke events in ECMO therapy, emphasizing the predominance of thromboembolic events and the complexity of managing such complications in patients receiving combined ECMO and IABP therapy. Our findings highlight that brain infarctions, which are predominantly thromboembolic, remain a critical and frequent complication in ECMO therapy. The addition of IABP does not appear to reduce the risk of ischemic stroke, including hemodynamic strokes, which were infrequent in our cohort. Hypoxemic infarctions were also rare, regardless of the cannulation site, suggesting that they are less relevant in this context. To deepen the understanding of the hemodynamic mechanisms and the potential sources of thromboembolic events in ECMO and IABP therapy, further studies should include prospective data collection, including echocardiographic evaluations, to assess cardiac and aortic flow dynamics more accurately.

In conclusion, the decision to use additional IABP therapy in ECMO patients should be guided by considerations other than the risk of brain ischemia, as our retrospective study demonstrates that it does not significantly alter the likelihood of ischemic stroke. It is essential to interpret these findings with caution, considering the study's design, sample size, and limitations specific to the population. Given the declining use of IABP in favor of more frequently deployed devices such as the Impella, further prospective studies are essential to evaluate the impact of these newer mechanical circulatory support strategies on cerebrovascular outcomes in ECMO patients. The focus in future studies must be on multivariate modeling and standardized protocols to ensure transparency and improve interpretability.

Author Contributions: Conceptualization, F.Z. and C.M.; methodology, F.Z. and C.M.; software, C.M.; validation, M.G.K., C.M., H.A. and H.D.; formal analysis C.S. and C.M.; investigation, C.M. and F.Z.; data curation, C.S. and C.M.; writing—original draft preparation, F.Z.; writing—review and editing, B.T., C.M., J.B., K.J., M.B. and M.G.K.; visualization C.S. All authors have read and agreed to the published version of the manuscript.

Funding: This research received no external funding.

Institutional Review Board Statement: The study was conducted in accordance with the Declaration of Helsinki, and approved by the Ethics Committee of the Heinrich-Heine University (protocol code 5102, approved 27 May 2015).

Informed Consent Statement: Patient consent was waived due to the retrospective nature of this study.

Data Availability Statement: The data presented in this study are available on reasonable request from the corresponding author due to patient privacy.

Conflicts of Interest: The authors declare no conflicts of interest.

Abbreviations

CCT	cranial computed tomography
CPR	cardiopulmonary resuscitation
IABP	intra-aortic balloon pump
va-ECMO	veno-arterial extracorporeal membrane oxygenation
ASPECT	Alberta Stroke Program Early CT Score

References

1. Schaheen, B.W.; Thiele, R.H.; Isbell, J.M. Extracorporeal life support for adult cardiopulmonary failure. *Best. Pr. Res. Clin. Anaesthesiol.* **2015**, *29*, 229–239. [CrossRef] [PubMed]
2. Gajkowski, E.F.; Herrera, G.; Hatton, L.; Antonini, M.V.; Vercaemst, L.; Cooley, E. ELSO Guidelines for Adult and Pediatric Extracorporeal Membrane Oxygenation Circuits. *ASAIO J.* **2022**, *68*, 133–152. [CrossRef] [PubMed]
3. Ma, P.; Zhang, Z.; Song, T.; Yang, Y.; Meng, G.; Zhao, J.; Wang, C.; Gu, K.; Peng, J.; Jiang, B.; et al. Combining ECMO with IABP for the treatment of critically Ill adult heart failure patients. *Heart Lung Circ.* **2014**, *23*, 363–368. [CrossRef]
4. Hoeper, M.M.; Tudorache, I.; Kühn, C.; Marsch, G.; Hartung, D.; Wiesner, O.; Boenisch, O.; Haverich, A.; Hinrichs, J. Extracorporeal membrane oxygenation watershed. *Circulation* **2014**, *130*, 864–865. [CrossRef]
5. Thomas, J.; Kostousov, V.; Teruja, J. Bleeding and Thrombotic complications in the Use of Extracorporeal Membrane Oxygenation. *Semin. Thromb. Hemost.* **2018**, *44*, 20–29. [CrossRef] [PubMed]
6. Zangrillo, A.; Landoni, G.; Biondi-Zoccai, G.; Greco, M.; Greco, T.; Frati, G.; Patroniti, N.; Antonelli, M.; Pesenti, A.; Pappalardo, F. A meta-analysis of complications and mortality of extracorporeal membrane oxygenation. *Crit. Care Resusc.* **1013**, *15*, 172–178. [CrossRef]
7. Sutter, R.; Tisljar, K.; Marsch, S. Acute Neurologic Complications During Extracorporeal Membrane Osygenation: A Systematic Review. *Crit. Care Med.* **2018**, *46*, 1506–1513. [CrossRef]
8. Jonas, R.A. The effect of extracorporeal life support on the brain: Cardiopulmonary bypass. *Semin. Perinatol.* **2005**, *29*, 51–57. [CrossRef]
9. Osborn, A.G. Osborn's brain. In *Imaging, Pathology, and Anatomy*, 2nd ed.; Elsevier: Salt Lake City, UT, USA, 2018; pp. 208–232.
10. Krishna, M.; Zacharowski, K. Principles of intra-aortic balloon pump counterpulsation. *Contin. Educ. Anaesth. Crit. Care Pain.* **2009**, *9*, 24–28. [CrossRef]
11. Lin, L.Y.; Liao, C.W.; Wang, C.H.; Chi, N.H.; Yu, H.Y.; Chou, N.K.; Hwang, J.J.; Lin, J.L.; Chiang, F.T.; Chen, Y.S. Effects of Additional Intra-aortic Balloon Counter-Pulsation Therapy to Cardiogenic Shock Patients Supported by Extra-corporeal Membranous Oxygenation. *Sci. Rep.* **2016**, *6*, 23838. [CrossRef]
12. Themas, K.; Zisis, M.; Kourek, C.; Konstantinou, G.; D'anna, L.; Papanagiotou, P.; Ntaios, G.; Dimopoulos, S.; Korompoki, E. Acute Ischemic Stroke during Extracorporeal Membrane Oxygenation (ECMO): A Narrative Review of the Literature. *J. Clin. Med.* **2024**, *13*, 6014. [CrossRef] [PubMed]
13. Aviv, R.I.; Mandelcron, J.; Chakraborty, S.; Gladstone, D.; Malham, S.; Tomlinson, G.; Fox, A.J.; Symons, S. Alberta Stroke Program Early CT Scoring of CT Perfusion in Early Stroke Visualization and Assessment. *AJNR* **2007**, *28*, 1075–1980. [CrossRef]
14. Rycus, P.; Stead, C. Extracorporeal Life Support Organization Registry Report 2022. *J Card. Crit. Care* **2022**, *6*, 100–102. [CrossRef]
15. Vallabhajosyula, S.; O'Horo, J.C.; Antharam, P.; Ananthaneni, S.; Vallabhajosyula, S.; Stulak, J.M.; Eleid, M.F.; Dunlay, S.M.; Gersh, B.J.; Rihal, C.S.; et al. Concomitant Intra-Aortic Balloon Pump Use in Cardiogenic Shock Requiring Veno-Arterial Extracorporeal Membrane Oxygenation. *Circ. Cardiovasc. Interv.* **2018**, *11*, e006930. [CrossRef] [PubMed]
16. May, T.L.; Lary, C.W.; Riker, R.R.; Friberg, H.; Patel, N.; Søreide, E.; McPherson, J.A.; Undén, J.; Hand, R.; Sunde, K.; et al. Correction to: Variability in functional outcome and treatment practices by treatment center after out-of-hospital cardiac arrest: Analysis of International Cardiac Arrest Registry. *Intensive Care Med.* **2019**, *45*, 1176. [CrossRef]
17. Aubin, H.; Petrov, G.; Dalyanoglu, H.; Saeed, D.; Akhyari, P.; Paprotny, G.; Richter, M.; Westenfeld, R.; Schelzig, H.; Kelm, M.; et al. A Suprainstitutional Network for Remote Extracorporeal Life Support: A Retrospective Cohort Study. *JACC Heart Fail.* **2016**, *4*, 698–708. [CrossRef] [PubMed]

18. Gass, A.; Palaniswamy, C.; Aronow, W.S.; Kolte, D.; Khera, S.; Ahmad, H.; Cuomo, L.J.; Timmermans, R.; Cohen, M.; Tang, G.H.; et al. Peripheral venoarterial extracorporeal membrane oxygenation in combination with intra-aortic balloon counterpulsation in patients with cardiovascular compromise. *Cardiology* **2014**, *129*, 137–143. [CrossRef]
19. Kelly, J.; Malloy, R.; Knowles, D. Comparison of anticoagulated versus non-anticoagulated patients with intra-aortic balloon pumps. *Thromb. J.* **2021**, *19*, 46. [CrossRef]
20. Kietaibl, C.; Horvat Menih, I.; Engel, A.; Ullrich, R.; Klein, K.U.; Erdoes, G. Cerebral microemboli during extracorporeal life support: A single-centre cohort study. *Eur. J. Cardiothorac. Surg.* **2021**, *61*, 172–179. [CrossRef]
21. Cheng, R.; Hachamovitch, R.; Kittleson, M.; Patel, J.; Arabia, F.; Moriguchi, J.; Esmailian, F.; Azarbal, B. Complications of extracorporeal membrane oxygenation for treatment of cardiogenic shock and cardiac arrest: A meta-analysis of 1, 866 adult patients. *Amm. Thorac. Surg.* **2014**, *97*, 610–616. [CrossRef]
22. Graber, L.C.; Quillinan, N.; Marrotte, E.J.; McDonagh, D.L.; Bartels, K. Neurocognitive outcomes after extracorporeal membrane oxygenation. *Best. Pr. Res. Clin. Anaesthesiol.* **2015**, *29*, 125–135. [CrossRef] [PubMed]
23. Shigematsu, Y.; Hamada, M.; Mukai, M.; Matsuoka, H.; Sumimoto, T.; Hiwada, K. Mechanism of Atrial Fibrillation and Increased Incidence of Thromboembolism in Patients with Hypertrophic Cardiomyopathy. *Jpn. Circ. J.* **2005**, *59*, 329–336. [CrossRef]
24. Pucher, P.H.; Cummings, I.G.; Shipolini, A.R.; McCormack, D.J. Is heparin needed for patients with an intra-aortic balloon pump? *Interact. Cardiovasc. Thorac. Surg.* **2012**, *15*, 136–139. [CrossRef] [PubMed]
25. Du, G.; Zhang, J.; Liu, J.; Fan, L. Case Report: Two Cases of Watershed Phenomenon in Mechanical Circulatory Support Devices: Computed Tomography Angiography Imaging and Literature Review. *Front. Cardiovasc. Med.* **2022**, *9*, 893355. [CrossRef]
26. Gehron, J.; Schuster, M.; Rindler, F.; Bongert, M.; Böning, A.; Krombach, G.; Fiebich, M.; Grieshaber, P. Watershed phenomena during extracorporeal life support and their clinical impact: A systematic in vitro investigation. *ESC Heart Fail.* **2020**, *7*, 1850–1861. [CrossRef] [PubMed]
27. Xenitopoulou, M.P.; Ziampa, K.; Evangeliou, A.P.; Tzikas, S.; Vassilikos, V. Percutaneous Mechanical Circulatory Support in Acute Heart Failure Complicated with Cardiogenic Shock. *J. Clin. Med.* **2024**, *13*, 2642. [CrossRef]
28. Carturegli, G.; Shrey, K.; Ponomarev, V.; Kim, B.S.; Whitman, G.J.; Ziai, W.; Cho, S.M.; Zhang, L.Q.; Mayasi, Y.; Gusdon, A.; et al. Transcranial Doppler microemboli and acute brain injury in extracorporeal membrane oxygenation: A prospective observational study. *JTCVS Tech.* **2022**, *15*, 111–122. [CrossRef]
29. De Georgia, M.; Bowen, T.; Duncan, K.R.; Chebl, A.B. Blood pressure management in ischemic stroke patients undergoing mechanical thrombectomy. *Neurol. Res. Pract.* **2023**, *5*, 12. [CrossRef]
30. Moustafa, R.R.; Momjian-Mayor, I.; Jones, P.S.; Morbelli, S.; Day, D.J.; Aigbirhio, F.I.; Fryer, T.D.; Warburton, E.A.; Baron, J.C. Microembolism versus hemodynamic impairment in rosary-like deep watershed infarcts: A combined positron emission tomography and transcranial Doppler study. *Stroke* **2011**, *42*, 3138–3143. [CrossRef]
31. Khanduja, S.; Kim, J.; Kang, J.K.; Morbelli, S.; Day, D.J.; Aigbirhio, F.I.; Fryer, T.D.; Warburton, E.A.; Baron, J.C. Hypoxic-Ischemic Brain Injury in ECMO: Pathophysiology, Neuromonitoring, and Therapeutic Opportunities. *Cells.* **2023**, *12*, 1546. [CrossRef]
32. Zhang, H.; Xu, J.; Yang, X.; Zou, X.; Shu, H.; Liu, Z.; Shang, Y. Narrative Review of Neurologic Complications in Adults on ECMO: Prevalence, Risks, Outcomes, and Prevention strategies. *Front. Med.* **2021**, *8*, 713333. [CrossRef] [PubMed]
33. Wilcox, C.; Acton, M.; Rando, H.; Keller, S.; Sair, H.I.; Chinedozi, I.; Pitts, J.; Kim, B.S.; Whitman, G.; Cho, S.M. Safety of Bedside Portable Low-Field Brain MRI in ECMO Patients Supported on Intra-Aortic Balloon Pump. *Diagnostics* **2022**, *12*, 2871. [CrossRef] [PubMed]

diagnostics

MDPI

Article

Impact of Neutrophil-to-Lymphocyte Ratio on Stroke Severity and Clinical Outcome in Anterior Circulation Large Vessel Occlusion Stroke

Zülfikar Memiş [1], Erdem Gürkaş [2], Atilla Özcan Özdemir [3], Bilgehan Atılgan Acar [4,*], Muhammed Nur Ögün [5], Emrah Aytaç [6], Çetin Kürşad Akpınar [7], Eşref Akıl [8], Murat Çabalar [9], Ayça Özkul [9], Ümit Görgülü [10], Hasan Bayındır [10], Zaur Mehdiyev [11], Şennur Delibaş Katı [12], Recep Baydemir [13], Ahmet Yabalak [14], Ayşenur Önalan [2], Türkan Acar [4], Özlem Aykaç [3], Zehra Uysal Kocabaş [3], Serhan Yıldırım [15], Hasan Doğan [7], Mehmet Semih Arı [9], Mustafa Çetiner [16], Ferhat Balgetir [6], Fettah Eren [17], Alper Eren [18], Nazım Kızıldağ [18], Utku Cenikli [19], Aysel Büşra Şişman Bayar [20], Ebru Temel [1], Alihan Abdullah Akbaş [21], Emine Saygın Uysal [16], Hamza Gültekin [8], Cebrail Durmaz [8], Sena Boncuk Ulaş [22] and Talip Asil [23]

1. Department of Neurology, Haseki Training and Research Hospital, University of Health Sciences, Istanbul 34130, Turkey; zulfikarmemis1@gmail.com (Z.M.); tmlebru@outlook.com (E.T.)
2. Department of Neurology, Kartal Dr. Lütfi Kırdar City Hospital, University of Health Sciences, Istanbul 34130, Turkey; erdemgurkas@yahoo.com (E.G.); draysenurkaymaz@gmail.com (A.Ö.)
3. Department of Neurology, Faculty of Medicine, Eskişehir Osmangazi University, Eskişehir 26040, Turkey; atillaozcanozdemir@gmail.com (A.Ö.Ö.); drzlm@yahoo.com (Ö.A.); drzuysal@hotmail.com (Z.U.K.)
4. Department of Neurology, Faculty of Medicine, Sakarya University, Sakarya 54100, Turkey; tdeniz38@hotmail.com
5. Department of Neurology, Faculty of Medicine, Bolu Abant İzzet Baysal University, Bolu 14030, Turkey; dr.mogun@gmail.com
6. Department of Neurology, Faculty of Medicine, Fırat University, Elazig 23119, Turkey; dremrah_aytac@hotmail.com (E.A.); ferhatbalgetir@hotmail.com (F.B.)
7. Department of Neurology, Faculty of Medicine, Samsun University, Samsun 55080, Turkey; dr_ckakpinar@hotmail.com (Ç.K.A.); dr.hasandogan@outlook.com (H.D.)
8. Department of Neurology, Faculty of Medicine, Dicle University, Diyarbakir 21280, Turkey; esrefakil@gmail.com (E.A.); hamzagultekin72@hotmail.com (H.G.); dr.cdurmaz@gmail.com (C.D.)
9. Department of Neurology, İstanbul Başakşehir Çam Sakura City Hospital, Istanbul 34480, Turkey; mcabalar@gmail.com (M.Ç.); ozkulayca@hotmail.com (A.Ö.); semih-ari@hotmail.com (M.S.A.)
10. Department of Neurology, Ankara Bilkent City Hospital, Ankara 06800, Turkey; drumitgorgulu@hotmail.com (Ü.G.); h.bayindir@hotmail.com (H.B.)
11. Department of Neurology, Ankara Etlik City Hospital, Ankara 06710, Turkey; dr_mehdiyev@yahoo.com
12. Department of Neurology, Antalya Training and Research Hospital, Antalya 07070, Turkey; sennurdelibas@yahoo.com
13. Department of Neurology, Faculty of Medicine, Erciyes University, Kayseri 38030, Turkey; recep.baydemir@gmail.com
14. Department of Neurology, Faculty of Medicine, Düzce University, Duzce 81620, Turkey; yabalakahmet@gmail.com
15. Department of Neurology, Kocaeli City Hospital, Kocaeli 41060, Turkey; serhan_yildirim@yahoo.com
16. Department of Neurology, Faculty of Medicine, Kutahya Health Sciences University, Kütahya 43020, Turkey; drcetiner76@gmail.com (M.Ç.); sayginemine43@gmail.com (E.S.U.)
17. Department of Neurology, Faculty of Medicine, Selçuk University, Konya 42130, Turkey; dreren42@hotmail.com
18. Department of Neurology, Faculty of Medicine, Atatürk University, Erzurum 25240, Turkey; alpereren25@gmail.com (A.E.); kizildag.noro@gmail.com (N.K.)
19. Department of Neurology, Faculty of Medicine, Muğla Sıtkı Koçman University, Muğla 48000, Turkey; utkucenikli@yahoo.com
20. Silvan Dr. Yusuf Azizoğlu State Hospital, Diyarbakır 21640, Turkey; busrasisman@gmail.com
21. Department of Neurology, Sakarya University Training and Research Hospital, Sakarya 54100, Turkey; alihanakbas97@gmail.com
22. Independent Researcher, Sakarya 54100, Turkey; senaboncuk@gmail.com
23. Department of Neurology, Faculty of Medicine, Biruni University, Istanbul 34015, Turkey; talip_asil@yahoo.com

* Correspondence: bilgehanacar@sakarya.edu.tr

Abstract: Background: The prognostic value of the neutrophil–lymphocyte ratio (NLR) in ischemic stroke remains debated due to cohort variability and treatment heterogeneity across studies. This study evaluates the relationship between admission NLR, stroke severity and 90-day outcomes in patients with anterior circulation large vessel occlusion (LVO) undergoing early, successful revascularization. **Methods:** A retrospective multicenter study was conducted with 1082 patients treated with mechanical thrombectomy for acute ischemic stroke. The relationship between admission NLR, baseline National Institutes of Health Stroke Scale (NIHSS), 24 h NIHSS and 90-day modified Rankin Scale (mRS) outcomes was analyzed using logistic regression. **Results:** Admission NLR correlated weakly but significantly with both baseline ($p = 0.018$) and 24 h ($p = 0.005$) NIHSS scores, reflecting stroke severity. However, multivariate analysis showed that higher 24 h NIHSS scores (OR 0.831, $p = 0.000$) and prolonged puncture-to-recanalization times (OR 0.981, $p = 0.000$) were independent predictors of poor 90-day outcomes, whereas NLR was not ($p = 0.557$). **Conclusions:** Admission NLR is associated with stroke severity but does not independently predict clinical outcomes at 90 days in patients achieving early and successful revascularization. These findings underscore the critical role of inflammation in the acute phase of stroke but suggest that its prognostic value for long-term outcomes is limited in this context.

Keywords: stroke; neutrophil–lymphocyte ratio; inflammation; prognostic biomarkers; mechanical thrombectomy

1. Introduction

Ischemic stroke is considered to be the second most common cause of death and the most common cause of morbidity. Therefore, its treatment and long-term follow-up are of critical importance for the patient. Although various treatment options exist, the primary goal remains the restoration of reperfusion. The treatments are thrombolytic therapy and thrombectomy for suitable patients. Despite appropriate treatment, more than half of patients remain seriously disabled or die [1,2]. According to a 2023 study, the incidence of ischemic stroke in Turkey is 93.2–108.6 per 100,000 and the total stroke incidence is 141.7–158 per 100,000 [3]. According to 2019 data, the most common cause of death in Turkey is cardiovascular diseases (36.8%); stroke ranks second in this subgroup, after coronary syndromes, and stroke-related deaths have increased by 56% since 2002 [3]. The etiology of ischemic stroke in 40% of patients occurs secondary to large vessel occlusion [4]. There are no clear data on the cause of this occlusion and the content of the thrombus that forms. In 2017, Brinjikji et al. examined the histopathological features of the thrombi and looked at the platelet, leukocyte and erythrocyte ratios and thought that the ratios varied depending on the source of the thrombus [5]. Recent studies suggest that inflammation plays an important role in thrombus formation and that neutrophils are the most responsible for this [6]. Neutrophil extracellular traps (NETs) are released by neutrophils, initiate pro-inflammatory processes and cause arterial and venous thrombogenesis [7,8].

Although it has been stated that inflammation is important in thrombus formation, anti-inflammatory therapies remain ineffective for ischemic stroke. This has led researchers to question whether inflammatory markers can be used in the prognostic evaluation of patients. Neutrophil–lymphocyte ratio (NLR), which has recently been frequently used as an inflammatory biomarker, combines the destructive effect of neutrophils and the anti-inflammatory effect of lymphocytes and reflects the interplay between neutrophilic inflammation and lymphocyte-mediated immunity, offering insights into systemic inflammation [9].

After the appearance of acute ischemic stroke (AIS), neutrophils and lymphocytes are involved in the pathophysiological process. Neutrophils migrate to vascular lesions, release proteases and free oxygen radicals, and directly or indirectly aggravate brain damage [10]. Neutrophils can also increase the expression of matrix metalloproteinase-9 and directly destroy the blood–brain barrier (BBB), resulting in secondary brain damage or hemorrhagic transformation (HT) [11].

On the other hand, lymphocytes also play an important role in the development of inflammation. The accumulation of lymphocytes in the damaged vascular endothelium causes the progression of the disease. In the advanced stages of the disease, lymphocytes play a protective role, mainly as anti-inflammatory actors [11]. In addition, neutrophils are the first cells to cross the BBB and initiate damage within the first hour after stroke. Other lymphocytes then follow, increasing the damage to the BBB. In this context, NLR gains importance both as an easily accessible blood-based biomarker and because it is involved in the pathogenesis of stroke from the first stages of damage to the BBB [12].

NLR reflects the combined roles of neutrophils and lymphocytes, indicating the body's inflammatory and stress status [13]. NLR, which shows the balance between innate immunity and acquired immunity, has been evaluated in terms of prognosis prediction in sepsis, cancer patients and cardiac events, and significant results have been obtained [14–16].

Although there are studies showing that NLR can be used in short-term prognosis prediction of ischemic stroke, there are still no clear data [17–19]. Our aim in this study is to investigate the effect of NLR on stroke severity and prognosis in AIS patients who underwent successful revascularization in the early period.

2. Materials and Methods

2.1. Study Population

This is a multicenter, retrospective study conducted with the participation of 19 stroke centers in Turkey, including the prospective and consecutively collected data of 1289 adult patients with anterior circulation AIS between January 2021 and December 2022. After applying the exclusion criteria and using consecutive patient sampling to reduce selection bias, 1082 patients who achieved successful revascularization were included in the study.

All patients in the study population met standard acute stroke treatment criteria and underwent mechanical thrombectomy in accordance with current stroke guidelines. The decision to perform thrombectomy was based on the identification of large vessel occlusion via computed tomography angiography in patients presenting within 6 h of estimated onset.

Exclusion criteria were as follows: age < 18 or >80 years, modified Rankin Scale (mRS) > 2, brain hemorrhage, intracerebral mass or computed tomography diagnosis of cerebrovascular injury due to trauma, systemic diseases that may cause changes in complete blood count, autoimmune diseases, hematological diseases, chronic infections or using immunosuppressive therapy, posterior circulation large vessel occlusion or internal carotid artery (ICA) tandem occlusion. Patients over 80 years of age and those with a baseline mRS > 2 were excluded to minimize potential confounding factors. Advanced age and significant pre-stroke disability can independently influence clinical outcomes, frailty and systemic inflammation, thereby introducing bias in the analysis of NLR and stroke outcomes. Patients were categorized based on their mRS score at 90 days (mRS 0,1,2; good clinical outcome, mRS 3,4,5,6; poor clinical outcome) (Figure 1). Stroke severity at baseline was determined using the admission National Institutes of Health Stroke Scale (NIHSS) score.

A total of 1289 potentially eligible participants were recruited from 19 comprehensive stroke centers in Turkey. After excluding 207 patients (malignancy: n = 126, chronic infections: n = 26 and hematological diseases: n = 55), 1082 patients were included in the study. The included patients were categorized based on their modified Rankin Scale (mRS) scores on the 90th day. Patients with mRS scores of 0-1-2 (favorable outcomes) accounted for 636 participants, while those with mRS scores of 3-4-5-6 (unfavorable outcomes) totaled 446 participants.

The primary outcome was poor functional outcome at 90 days reflected by an mRS score of 3 to 6. Successful revascularization was defined as a post-treatment modified treatment in cerebral infarction (mTICI) score of 2b to 3.

Written consent was obtained from patients or their relatives (for aphasic or unconscious patients) for participation in the study, and the study was approved by the ethics committee of Haseki Training and Research Hospital.

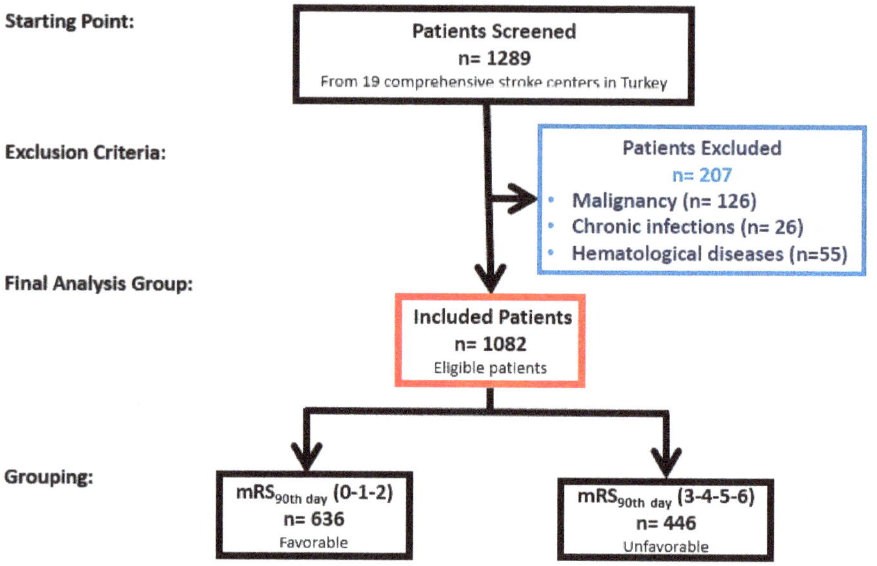

Figure 1. Patient selection flowchart.

2.2. Patient Data Evaluation and Analysis

We collected patient data including demographic characteristics, clinical data and laboratory parameters. Patient characteristics included age, gender, smoking status, alcohol consumption, hypertension, diabetes, atrial fibrillation (AF), asthma, hyperlipidemia (HL) and coronary heart disease (CHD). Clinical data included admission systolic blood pressure (BP), admission diastolic BP, admission NIHSS score, symptom–puncture time (time from onset to inguinal puncture), puncture–recanalization time (time from inguinal puncture to reperfusion) and Alberta Stroke Program Early Computed Tomography score (ASPECTS).

Laboratory parameters were blood glucose (BG), lymphocyte count, neutrophil count, red blood cell distribution width (RDW), platelet (PLT) count, NLR, platelet–lymphocyte ratio (PLR). All peripheral venous blood tests were performed before thrombectomy. Laboratory parameters were determined by optical laser scatter and fluorescent staining methods using Mindray BC6200 (Shenzhen Mindray Bio-Medical Electronics Co., Ltd., Shenzhen, China), Sysmex XN-1000 (Sysmex Corporation, Kobe, Japan) and/or Celltac G MEK-9100 (Nihon Kohden Corporation, Tokyo, Japan) devices. NLR was calculated by dividing the neutrophil count by the lymphocyte count. PLR was calculated by dividing the platelet count (10^9/L) by the lymphocyte count.

2.3. Statistics

The Kolmogorov–Simirnov test was used to examine whether continuous quantitative variables showed normal distribution, and, at the same time, the skewness and kurtosis coefficients of the variables were examined. While *n* and percentage were reported for categorical variables, minimum, maximum, median and mean ± standard deviation (SD) were reported for continuous variables. When examining the relationships between variables, the correlations for variables that did not show normal distribution (neutrophil, lymphocyte, platelets, NLR, PLR) were calculated using Spearman's rho; Pearson's product-moment correlation coefficient was calculated for normally distributed variables (24th hour NIHSS score, admission NIHSS score).

The skewness coefficient of variables with non-normal distribution ranged between 2.09 and 32.10, while the skewness coefficient of variables with normal distribution was in the range of 0.12–0.92. Relationships between categorical variables were examined using chi-square analysis. Fisher's exact test was used for those with an expected value below 5. Logistic regression analysis was performed to estimate the dichotomized mRS score (mRS scores of 0, 1 and 2 as zero and those of 3, 4, 5 and 6 as one). As a result of both single and multiple logistic regression analysis, 95% confidence intervals and likelihood ratios were calculated. NLR, 24th hour NIHSS score and Neutrophil variables were not considered together in multiple logistic regression because the variables could affect each other. All analyses were performed using SPSS 26 software with a significance level of $p < 0.05$.

3. Results

Of the 1082 patients, 545 (50.4%) were female and 537 (49.6%) were male, with a mean age of 65.1 ± 12.2 years. The mean NLR was 5.3 ± 6.3. The time between the onset of symptoms and inguinal puncture was 193 ± 85.7 min (range 9–360). The mean NIHSS score on admission was 14.81 ± 4.8, with a range of 6–32 and a median of 15. A total of 636 (58.7%) patients had good functional outcomes 3 months after acute ischemic stroke. Of the total cohort, 914 patients (80%) had M1 occlusion, M2 occlusion occurred in 164 patients (14.3%), carotid T or L occlusion in 63 patients (5.5%), and anterior cerebral artery A1 occlusion in 2 patients (0.2%). Table 1 summarizes the basic demographic, laboratory, and clinical data of the patients.

Table 1. Patient characteristics, clinical findings and inflammatory parameters.

		Min–Max	Mean ± SD	n-%
Ages		19–80	65.1 ± 12.2	1082
Gender	Female			545–50.4%
	Male			537–49.6%
Body-Mass Index	Normal			784–72.5%
	Obesity			267–24.7%
	Cachectic			31–2.9%
Clinical Characteristics				
Hypertension	Yes			426–39.4%
	No			656–60.6%
Diabetes	No			725–67.0%
	Yes			357–33.0%
Coronary Artery Disease	No			692–64.0%
	Yes			390–36.0%
Congestive Heart Failure	No			905–83.6%
	Yes			177–16.4%
Asthma and/or COPD	No			964–89.1%
	Yes			118–10.9%
Atrial Fibrillation	No			640–59.1%
	Yes			442–40.9%
Alcohol Consumption	No			996–92.1%
	Yes			86–7.9%
Smoker	No			788–72.8%
	Yes			294–27.2%
Previous Stroke History	No			934–86.3%
	Yes			148–13.7%
Hyperlipidemia	No			515–47.6%
	Yes			567–52.4%
Inflammatory Parameters				
Neutrophil (10^9/L)		0.1–112.18	8.05 ± 8.6	1082 (%100)

Table 1. *Cont.*

		Min–Max	Mean ± SD	n-%
Lymphocyte (10^9/L)		0.2–51.7	2.2 ± 2.6	1082 (%100)
PLT (10^9/L)		10–996	236.6 ± 88.2	1082 (%100)
RDW (%)		9.5–172	14.8 ± 5.8	1082 (%100)
NLR		0.02–124.45	5.3 ± 6.3	1082 (%100)
PLR		3.33–193.0	155.9 ± 123.5	1082 (%100)
Admission NIHSS		4–32	14.81 ± 4.8	1082 (%100)
Procedural Data				
Symptom-to-Inguinal Puncture Time (min)		9–360	193 ± 85.7	1082 (%100)
Puncture-to-Recanalization Time (min)		6–182	46.13 ± 23.6	1082 (%100)
Number of Passes		1–15	2 ± 1.2	1082 (%100)
First Pass Recanalization	No			596–55.1%
	Yes			486–44.9%
24th Hour NIHSS		0–32	9.6 ± 7	1082 (%100)
ASPECTS Score		6–10	8.4 ± 1.2	1082 (%100)
Occlusion Location	MCA M1			914–80%
	MCA M2			164–14.3%
	ICA T,L			63–5.5%
	ACA A1			2–0.2%

Min: minimum; Max: maximum; SD: standard deviation; COPD: chronic obstructive pulmonary disease; PLT: platelet; RDW: red cell distribution width; NLR: neutrophil-to-lymphocyte ratio; PLR: platelet-to-lymphocyte ratio; NIHSS: National Institutes of Health Stroke Scale; min: minutes; ASPECTS: Alberta Stroke Program Early CT Score; MCA: middle cerebral artery; ICA: internal carotid artery; ACA: anterior cerebral artery.

In the correlation analysis of inflammatory parameters and admission NIHSS, neutrophil values and NLR values showed a statistically significant, weak positive correlation with admission NIHSS score. (p = 0.003, rho = 0.131, p = 0.018, rho = 0.143, respectively).

Similarly, in the correlation analysis of inflammatory parameters with 24th hour NIHSS score, neutrophil value and NLR values showed a statistically significant, weak positive correlation with 24th hour NIHSS score (p = 0.000, rho = 0.152, p = 0.005, rho = 0.186, respectively) (Table 2).

Table 2. Correlation of inflammatory parameters with admission NIHSS score and 24th hour NIHSS score.

			Neutrophil	Lymphocyte	PLT	NLR	PLR	Admission NIHSS	24th Hour NIHSS
Spearman's rho	Neutrophil	Correlation Coefficient	1.000	−0.217 **	0.245 **	0.716 **	0.327 **	0.131 **	0.152 **
		p		0.000	0.000	0.000	0.000	0.003 *	0.000 *
		n	1082	1082	1082	1082	1082	1082	1082
	Lymphocyte	Correlation Coefficient	−0.217 **	1.000	0.162 **	−0.801 **	−0.833 **	−0.061	−0.095
		p	0.000		0.000	0.000	0.000	0.482	0.627
		n	1082	1082	1082	1082	1082	1082	1082
	PLT	Correlation Coefficient	0.245 **	0.162 **	1.000	0.043	0.353 **	0.004	0.042
		p	0.000	0.000		0.159	0.000	0.891	0.170
		n	1082	1082	1082	1082	1082	1082	1082
	NLR	Correlation Coefficient	0.716 **	−0.801 **	0.043	1.000	0.766 **	0.143 **	0.186 **
		p	0.000	0.000	0.159		0.000	0.018 *	0.005 *
		n	1082	1082	1082	1082	1082	1082	1082
	PLR	Correlation Coefficient	0.327 **	−0.833 **	0.353 **	0.766 **	1.000	0.021	0.037
		p	0.000	0.000	0.000	0.000		0.498	0.218
		n	1082	1082	1082	1082	1082	1082	1082

Correlation coefficients (rho): negligible: 0.00–0.10; weak: 0.10–0.39; moderate: 0.40–0.69; strong: 0.70–0.89. PLT: platelet; NLR: neutrophil-to-lymphocyte ratio; PLR: platelet-to-lymphocyte ratio; NIHSS: National Institutes of Health Stroke Scale. * correlation is significant at the 0.05 level (2-tailed). ** correlation is significant at the 0.01 level (2-tailed).

Of the 1082 patients, 446 (41.3%) had unfavorable 90-day functional outcomes. The favorable group had a lower proportion of males and younger individuals than the unfavorable group. The differences between the two groups are detailed in Table 3. Hypertension and diabetes were more prevalent in the unfavorable outcome group (p value 0.003 and 0.001, respectively). Patients in the unfavorable outcome group also had higher admission NIHSS scores ($p = 0.001$), higher neutrophil counts ($p = 0.001$), higher NLR values ($p = 0.013$), longer puncture–revascularization time ($p = 0.001$) and lower first pass revascularization rates ($p = 0.002$) (Table 3, Supplementary Table S1, Supplementary Figure S1).

Table 3. Association of demographic, clinical and laboratory data with 90th day clinical outcome.

		mRS 0-1-2			**mRS 3-4-5-6**			
		Min–Max	**Mean ± SD**	***n*-%**	**Min–Max**	**Mean ± SD**	***n*-%**	***p* Value**
Risk Factors								
Age		19–80	63.4 ± 12.8	636–58.7%	31–80	67.6 ± 10.7	446–41.3%	0.538 ^{t}
Gender	Female			321–50.5%			224–50.2%	0.936 ^{x2}
	Male			315–49.5%			222–49.8%	
Hypertension	No			274–43.1%			152–34.1%	0.003 * ^{x2}
	Yes			362–56.9%			294–65.9%	
Diabetes Mellitus	No			468–73.6%			257–57.6%	0.001 * ^{x2}
	Yes			168–26.4%			189–42.4%	
Coronary Artery Disease	No			426–67%			266–59.6%	0.013 * ^{x2}
	Yes			210–33%			180–40.4%	
Asthma and/or COPD	No			582–91.5%			382–85.7%	0.002 * ^{x2}
	Yes			54–8.5%			64–14.3%	
Atrial Fibrillation	No			382–60.1%			258–57.8%	0.466 ^{x2}
	Yes			254–39.9%			188–42.2%	
Alcohol Consumption	No			588–92.5%			408–91.5%	0.56 ^{x2}
	Yes			48–7.5%			38–8.5%	
Smoker	No			453–71.2%			335–75.1%	0.157 ^{x2}
	Yes			183–28.8%			111–24.9%	
Previous Stroke History	No			561–88.2%			373–83.6%	0.031 * ^{x2}
	Yes			75–11.8%			73–16.4%	
Hyperlipidemia	No			278–43.7%			237–53.1%	0.002 * ^{x2}
	Yes			358–56.3%			209–46.9%	
Inflammatory Parameters								
Neutrophil (10^9/L)		0.1–90.8	7 ± 5.08	636	0.6–112.18	9.5 ± 11.8	446	0.001 * ^{t}
Lymphocyte (10^9/L)		0.2–51.7	2.19 ± 2.7	636	0.36–20.5	2.3 ± 2.5	446	0.956 ^{t}
PLT (10^9/L)		10–996	236.6 ± 88.2	636	10–996	243.1 ± 97.8	446	0.104 ^{t}
RDW (%)		9.5–172	14.8 ± 5.8	636	9.9–53.9	14.9 ± 3.89	446	0.03 * ^{t}
NLR		0.02–124.45	4.9 ± 6.3	636	0.13–56.37	5.98 ± 6.2	446	0.013 * ^{t}
PLR		4.74–1930	152.4 ± 123.2	636	3.33–1350	160.9 ± 123.9	446	0.318 ^{t}
Admission NIHSS		2–27	13.5 ± 4.5	636	5–32	16.5 ± 4.6	446	0.001 * ^{t}
Symptom-to-Inguinal Puncture Time (min)		15–360	185.5 ± 83.8	636	9–360	205.4 ± 87	446	0.001 * ^{t}
Puncture-to-Recanalization Time (min)		6–172	43.2 ± 21.8	636	7–182	50.1 ± 25.4	446	0.001 * ^{t}
Number of Passes		1–15	1.9 ± 1.2	636	1–8	2.1 ± 1.2	446	0.001 * ^{t}
First Pass Recanalization	No			325–51.1%			271–60.8%	0.002 * ^{x2}
	Yes			311–48.9%			175–39.2%	
24th Hour NIHSS		0–32	6.4 ± 5.3	636	0–32	14.3 ± 6.6	446	0.001 * ^{t}

* p-values based on chi-square test for categorical variables and t-test for continuous variables. mRS: modified Rankin Scale; Min: minimum; Max: maximum; SD: standard deviation; COPD: chronic obstructive pulmonary disease; PLT: platelet; RDW: red cell distribution width; NLR: neutrophil-to-lymphocyte ratio; PLR: platelet-to-lymphocyte ratio; min: minutes; NIHSS: National Institutes of Health Stroke Scale; t: t-test; ^{x2}: chi-square test.

Univariate logistic regression analyses and multivariate logistic regression analyses performed after adjusting for confounding variables revealed no independent association between any of the inflammatory parameters and 90th day functional outcomes. It was observed that the high 24th hour NIHSS score (OR, 0.831; 95% CI, 0.803–0.859; $p = 0.000$) and the length of puncture–revascularization time (OR, 0.981; 95% CI, 0.973–0.989; $p = 0.000$)

were independent predictors of poor clinical outcome on the 90th day. In univariate logistic regression analysis, it was seen that high 24th hour NIHSS score (OR, 1.242; 95% CI, 1.209–1.277; $p = 0.000$), long puncture–revascularization time (OR, 1.013; 95% CI, 1.007–1.018; $p = 0.000$) and first pass recanalization (OR, 0.675; 95% CI, 0.528–0.863; $p = 0.591$) were independent predictors in favor of poor clinical outcomes on the 90th day (Table 4, Supplementary Table S2).

Table 4. Univariate and multivariate logistic regression analysis.

	Univariate Analysis Results		Multivariate Analysis Results	
	OR (95%CI)	*p*	aOR (95%CI)a	aP
Neutrophil, $\times 10^9$/L	1.000 (0.999–1.001)	0.649	0.988 (0.962–1.014)	0.630
Lymphocyte, $\times 10^9$/L	1.032 (0.990–1.075)	0.134	1.000 (0.959–1.044)	0.096
NLR	1.000 (0.998–1.001)	0.594	1.027 (0.989–1.066)	0.557
PLR	1.001 (1.000–1.002)	0.273	1.002 (0.998–1.006)	0.268
RDW	1.007 (0.985–1.028)	0.539	0.996 (0.974–1.020)	0.527
24th Hour NIHSS	1.242 (1.209–1.277)	0.000	0.831 (0.803–0.859)	0.000
Puncture–Recanalization Time	1.013 (1.007–1.018)	0.000	0.981 (0.973–0.989	0.000
First Pass Recanalization	0.675 (0.528–0.863	0.002	1.130 (0.723–1.766)	0.591

OR: odds ratio; aOR: adjusted odds ratio; aP: adjusted *p* value; CI: confidence interval, adjusted for age, sex, diabetes mellitus, coronary heart disease, prior stroke; NLR: neutrophil-to-lymphocyte ratio; PLR: platelet-to-lymphocyte ratio; RDW: red cell distribution width; NIHSS: National Institutes of Health Stroke Scale.

Receiver Operating Characteristic (ROC) curve analysis was performed for the significant variables. Variables with an area under curve (AUC) value above 0.60 were analyzed. The ROC curve for 24th hour NIHSS score was found to be highly associated with predicting poor 90-day outcomes, with an AUC of 0.84 (Figure 2, Table 5).

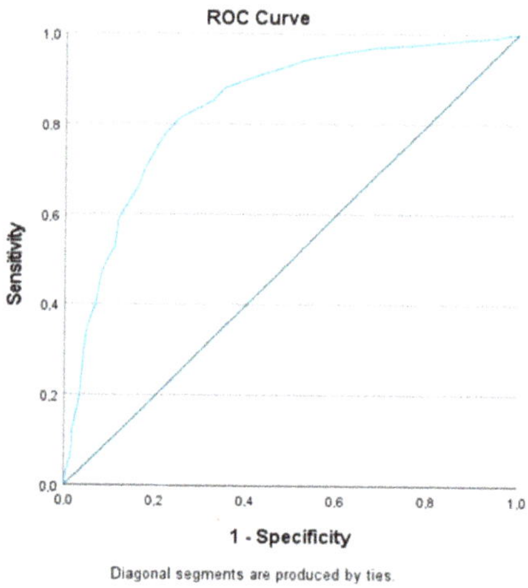

Figure 2. ROC curve for 24th hour NIHSS score predicting poor 90-day outcomes (AUC = 0.84, $p < 0.001$).

Table 5. ROC analysis for 24th hour NIHSS score and puncture-to-recanalization time predicting poor 90-day outcomes.

Variables	AUC	95%CI	*p*	Cut-Off	Sensitivity	Specificity
24th hour NIHSS	0.839	0.815–0.863	0.00	8.50	0.812	0.753
Puncture-to-recanalization time (min)	0.578	0.543–0.613	0.00	43.5	0.543	0.583

AUC > 0.60 is weak; AUC > 0.70 is acceptable; 0.80–0.90 is excellent; and greater than 0.90 is outstanding. Therefore, variables with an AUC value above 0.60 were analyzed. AUC: area under curve; CI: confidence interval; NIHSS: National Institutes of Health Stroke Scale; min: minutes.

4. Discussion

In this study, firstly, the predictive value of NLR levels at admission in AIS patients for stroke severity was investigated. The results showed that high NLR levels were blunted and statistically associated with the elevation of admission NIHSS, which indicates stroke severity. In addition, we observed that a high 24th hour NIHSS score, which can be considered as an indicator of early neurological deterioration, and high admission-NLR levels were blunted and statistically associated. Many studies show that NLR is closely related to cerebrovascular diseases. A large meta-analysis of data from 9563 patients by Wang et al. showed that high NLR predicts short-term adverse outcomes in AIS patients [20]. Another review study showed that NLRs were higher in patients with unfavorable prognosis in hemorrhagic strokes and that this could be a predictive marker in the short term [21]. In a multicenter study by Piri Cinar B et al., which included 737 patients, high NLR levels were associated with high admission-NIHSS [22]. Aytaç E et al., in their study with 209 patients, showed that high admission-NLR values were associated with high admission-NIHS scores [23]. In a recent meta-analysis of nine studies, high NLR values were found to be associated with high NIHSS scores and early neurological deterioration [24]. Our study's data, which demonstrate the relationship between baseline NLR values and NIHSS scores at admission and 24 h, align with findings in the literature.

To our knowledge, this is the first study to evaluate the relationship between NLR levels and 90-day functional outcome in patients with a symptom–puncture time within the first 6 h and successful anterior circulation revascularization. We focused on the NLR values in this patient group for several important reasons. First, most of the previous studies have assessed inflammatory responses in all ischemic stroke patients, regardless of revascularization success [25,26]. Second, considering that mechanical thrombectomy (MT) can be performed up to 24 h, symptom–puncture times vary greatly across studies in the literature [22–24,27,28]. Finally, most studies evaluated ischemic stroke patients in the same pool without making any distinction between the anterior and posterior systems [24,29,30]. We thought that this study, which was conducted in a specific patient group with a higher number of patients, could provide more valuable data. At the end of our study, the admission NLR value was not associated with the 3rd month mRS score. In contrast to our study, a recent prospective study of 341 ischemic stroke patients found that high NLR levels were associated with poor clinical outcome in the 3rd month. However, only those over 50 years of age and those with a baseline NIHSS score below 10 were included in that study [26]. In a meta-analysis conducted in 2022 evaluating PLR in AIS patients treated with reperfusion therapy, data of 4788 patients were analyzed and it was shown that low admission-PLR values were associated with early neurologic recovery and good functional outcome on day 90. This meta-analysis emphasizes the importance of inflammatory parameters other than NLR in AIS and is inconsistent with our results in terms of the association with 90th day clinical outcome [31]. In another meta-analysis study of 10,308 patients examining the role of NLR in AIS patients undergoing reperfusion therapy, higher admission or delayed NLR are significantly associated with worse morbidity, mortality and safety outcomes in AIS patients [32]. Unlike these meta-analyses, the lack of association between outcome and NLR in our study may have resulted from methodological differences such as the inclusion of successful recanalization patients and the analysis of pre-procedural NLR values.

In another study conducted by Chen et al. on AIS patients receiving MT and/or intravenous thrombolytic therapy (IVTPA), it was shown that both baseline NLR and first-day NLR values were high, reducing functional independence. However, it has been emphasized that the first-day NLR value has a stronger relationship [30]. In studies evaluating patients who underwent successful revascularization, the focus has generally been on the NLR values studied after revascularization rather than the admission NLR values. Li S et al. studied 237 cases who were admitted within the first 12 h and had successful revascularization; NLR was calculated for each day in the first 3 days of treatment and its relationship with 1st month mortality was examined. All three days' NLR values were found to be associated with 1st month mortality [33]. In a study performed by Zou F et al., which included 160 ischemic stroke patients who underwent successful revascularization, high NLR values in the early period after revascularization were associated with poor prognosis at 3 months [34].

In another study of patients with anterior system stroke who were successfully revascularized, the admission NLR value was not associated with the 3rd month mRS score, but a significant relationship was found between the NLR value studied on the 1st day after revascularization and the 3rd month mRS score [28]. Similar to our study, in this study where patients who had successful revascularization were included in the study, the average symptom–recanalization time was calculated as 438 min. In our study, the average symptom–recanalization time was 239 min. Considering this situation, it is thought that the majority of the patients in this study were revascularized late, that the inflammatory process may have already started in these patients and that this situation may have affected the study results in terms of recanalization benefit.

In a study by Li SJ et al., there was no relationship found between the admission NLR value and neurological outcomes, but the post-MT NLR value was found to be predictive of neurological outcomes. Subgroup analyses of NLR assessment showed that patients with MT, successful revascularization, and higher inflammatory biomarkers did not have an increased risk of poor 3-month functional outcome; this result is consistent with the results in our study [25]. The relationship between inflammation and acute ischemic stroke is currently a focus of attention. In our study, we focused on one of the most common markers of inflammation, NLR, and ultimately demonstrated that NLR is not associated with poor prognosis after stroke in the setting of successful and timely revascularization. Given that NLR is increased by many factors, such as cardiovascular disease [35], chronic inflammatory diseases [36], systemic infections [37], and post-stroke pneumonia [38], this value may not be an independent predictor of clinical outcome when beneficial reperfusion is achieved, which is the primary goal in ischemic stroke.

This study has several limitations but also notable strengths, such as including a large cohort of patients who underwent early and successful revascularization. First, NLR values were studied only from venous blood taken at presentation. Considering the variability of inflammation over time, studying NLR values after MT or in the following days may yield different results. Secondly, although we listed the patient group that we selected as an advantage (early and successful revascularization) and included a more specific patient group, an analysis was made that still included too many variables affecting the mRS result. It is possible to obtain more valuable results by reducing the variables of the study group. Another limitation of our study is that infections or other systemic inflammatory responses after stroke were not analyzed. These data may provide important information about stroke when monitored over time. In addition, successful revascularization may reduce inflammation and affect NLR and other inflammatory parameter outcomes. However, we analyzed the NLR values before the procedure, when the patient was first admitted to the hospital, and the aim of our study was to examine the association of these values with stroke severity and clinical outcome. Variability in systemic inflammatory responses (e.g., pneumonia after stroke) may affect the prognostic value of NLR. Future studies may plan study designs with repeated NLR measurements after MT to address this limitation.

5. Conclusions

This study highlights a significant relationship between admission NLR and stroke severity in patients with anterior circulation large vessel occlusion undergoing early, successful revascularization. However, NLR was not found to be an independent predictor of long-term functional outcomes at 90 days. These results suggest that inflammation, as measured by NLR, is a determinant of acute stroke severity but not of long-term recovery in the context of effective and timely reperfusion therapy.

The findings of this study should be interpreted in the context of its design, which included only patients with early and successful revascularization and analyzed pre-procedural NLR values. Post-stroke inflammatory changes and delayed NLR measurements were not evaluated, which may limit the generalizability of the results. Future research should consider incorporating serial NLR measurements, broader patient cohorts, and additional inflammatory biomarkers to comprehensively assess their prognostic potential.

Supplementary Materials: The following supporting information can be downloaded at: https://www.mdpi.com/article/10.3390/diagnostics14242880/s1, Figure S1: NLR box Plot in terms of mRS; Table S1: Association of Demographic, Clinical and Laboratory data with 90th Day Clinical Outcome; Table S2: Multivariate Regression Analysis Results.

Author Contributions: Conceptualization, Z.M. (Zülfikar Memiş) and T.A. (Talip Asil); methodology, Z.M. (Zülfikar Memiş), E.G., A.Ö.Ö., B.A.A., M.N.Ö., E.A. (Emrah Aytaç), Ç.K.A. and T.A. (Talip Asil); software, M.Ç. (Murat Çabalar), A.Ö. (Ayça Özkul) and H.B.; validation, E.A. (Eşref Akıl), Ü.G., R.B. and A.Ö. (Ayşenur Önalan); formal analysis, Z.M. (Zülfikar Memiş), Z.M. (Zaur Mehdiyev), A.Y. and T.A. (Türkan Acar); investigation, Ş.D.K., Ö.A., Z.U.K., H.D. and F.B.; resources, E.G., A.Ö.Ö., B.A.A., M.N.Ö., E.A. (Emrah Aytaç), Ç.K.A. and T.A. (Talip Asil); data curation, S.Y., M.S.A., M.Ç. (Mustafa Çetiner), F.E., A.E., N.K., U.C., A.B.Ş.B., E.T., A.A.A., E.S.U., H.G., C.D. and S.B.U.; writing—original draft preparation, Z.M. (Zülfikar Memiş); writing—review and editing, B.A.A., A.E., A.B.Ş.B. and S.B.U.; visualization, Ö.A., Z.U.K. and M.S.A.; supervision, T.A. (Talip Asil); project administration, Z.M. (Zülfikar Memiş). All authors have read and agreed to the published version of the manuscript.

Funding: This research received no external funding.

Institutional Review Board Statement: This study was reviewed and approved by the Clinical Research Ethics Committee of Haseki Training and Research Hospital, with the approval number: 23–2024.

Informed Consent Statement: Written informed consent was obtained from all subjects involved in the study.

Data Availability Statement: Due to the nature of the research, for ethical reasons supporting data is not available.

Conflicts of Interest: The authors declare no conflicts of interest.

Abbreviations

NLR	Neutrophil-to-Lymphocyte Ratio
mRS	Modified Rankin Scale
NIHSS	National Institutes of Health Stroke Scale
OR	Odds ratio
CI	Confidence Interval
NETs	Neutrophil extracellular traps
AIS	Acute Ischemic Stroke
BBB	Blood Brain Barrier
HT	Hemorrhagic Transformation
ICA	Internal Carotid Artery
mTICI	Modified treatment in cerebral infarction
AF	Atrial Fibrilation
HL	Hyperlipidemia
CHD	Coronary Heart Disease
BP	Blood Pressure
ASPECTS	Alberta Stroke Program Early Computed Tomography score
BG	Blood Glucose
RDW	Red Blood Cell Distribution Width
PLT	Platelet
PLR	Platelet/Lymphocyte Ratio
SD	Standard Deviation
COPD	Chronic Obstructive Pulmonary Disease
Min	Minimum
Max	Maximum

min	Minutes
MCA	Middle Cerebral Artery
ACA	Anterior Cerebral Artery
aOR	Adjusted Odds Ratio
aP	Adjusted P Value
ROC	Receiver Operating Characteristic
AUC	Area Under curve
MT	Mechanical Thrombectomy

References

1. Goyal, M.; Yu, A.Y.; Menon, B.K.; Dippel, D.W.; Hacke, W.; Davis, S.M.; Fisher, M.; Yavagal, D.R.; Turjman, F.; Ross, J.; et al. Endovascular therapy in acute ischemic stroke: Challenges and transition from trials to bedside. *Stroke* **2016**, *47*, 548–553. [CrossRef] [PubMed]
2. Hankey, G.J. Stroke. *Lancet.* **2017**, *389*, 641–654. [CrossRef] [PubMed]
3. Topcuoglu, M.A.; Ozdemir, A.O. Acute stroke management in Turkey: Current situation and future projection. *Eur Stroke J.* **2023**, *8*, 16–20. [CrossRef]
4. Rennert, R.C.; Wali, A.R.; A Steinberg, J.; Santiago-Dieppa, D.R.; E Olson, S.; Pannell, J.S.; A Khalessi, A. Epidemiology, natural history, and clinical presentation of large vessel ischemic stroke. *Neurosurgery* **2019**, *85*, S4–S8. [CrossRef]
5. Brinjikji, W.; Duffy, S.; Burrows, A.; Hacke, W.; Liebeskind, D.; Majoie, C.B.L.M.; Dippel, D.W.J.; Siddiqui, A.H.; Khatri, P.; Baxter, B.; et al. Correlation of imaging and histopathology of thrombi in acute ischemic stroke with etiology and outcome: A systematic review. *J. Neurointerventional Surg.* **2017**, *9*, 529–534. [CrossRef]
6. Tirandi, A.; Sgura, C.; Carbone, F.; Montecucco, F.; Liberale, L. Inflammatory biomarkers of ischemic stroke. *Intern. Emerg. Med.* **2023**, *18*, 723–732. [CrossRef]
7. Kimball, A.S.; Obi, A.T.; Diaz, J.A.; Henke, P.K. The emerging role of NETs in venous thrombosis and immunothrombosis. *Front. Immunol.* **2016**, *7*, 236. [CrossRef] [PubMed]
8. Laridan, E.; Martinod, K.; De Meyer, S.F. Neutrophil extracellular traps in arterial and venous thrombosis. *Semin. Thromb. Hemost.* **2019**, *45*, 86–93. [CrossRef] [PubMed]
9. Song, S.Y.; Zhao, X.X.; Rajah, G.; Hua, C.; Kang, R.J.; Han, Y.P.; Ding, Y.C.; Meng, R. Clinical significance of baseline Neutrophil-to-Lymphocyte ratio in patients with ischemic stroke or hemorrhagic stroke: An updated Meta-Analysis. *Front. Neurol.* **2019**, *10*, 1032. [CrossRef] [PubMed]
10. Neumann, J.; Henneberg, S.; von Kenne, S.; Nolte, N.; Müller, A.J.; Schraven, B.; Görtler, M.W.; Reymann, K.G.; Gunzer, M.; Riek-Burchardt, M. Beware the intruder: Real time observation of infiltrated neutrophils and neutrophil-Microglia interaction during stroke in vivo. *PLoS ONE* **2018**, *13*, e193970. [CrossRef] [PubMed]
11. Duan, Z.; Wang, H.; Wang, Z.; Hao, Y.; Zi, W.; Yang, D.; Zhou, Z.; Liu, W.; Lin, M.; Shi, Z.; et al. Neutrophil-Lymphocyte ratio predicts functional and safety outcomes after endovascular treatment for acute ischemic stroke. *Cerebrovasc. Dis.* **2018**, *45*, 221–227. [CrossRef] [PubMed]
12. Sharma, D.; Spring, K.J.; Bhaskar, S.M.M. Neutrophil-lymphocyte ratio in acute ischemic stroke: Immunopathology, management, and prognosis. *Acta Neurol. Scand.* **2021**, *144*, 486–499. [CrossRef]
13. Earls, R.H.; Lee, J.K. The role of natural killer cells in Parkinson's disease. *Exp. Mol. Med.* **2020**, *52*, 1517–1525. [CrossRef] [PubMed]
14. Huang, Z.; Fu, Z.; Huang, W.; Huang, K. Prognostic value of neutrophil-to-lymphocyte ratio in sepsis: A meta-analysis. *Am. J. Emerg. Medicine* **2020**, *38*, 641–647. [CrossRef] [PubMed]
15. Yin, Y.; Wang, J.; Wang, X.; Gu, L.; Pei, H.; Kuai, S.; Zhang, Y.; Shang, Z. Prognostic value of the neutrophil to lymphocyte ratio in lung cancer: A meta-analysis. *Clinics* **2015**, *70*, 524–530. [CrossRef] [PubMed]
16. Dong, C.-H.; Wang, Z.-M.; Chen, S.-Y. Neutrophil to lymphocyte ratio predict mortality and major adverse cardiac events in acute coronary syndrome: A systematic review and meta-analysis. *Clin. Biochem.* **2018**, *52*, 131–136. [CrossRef] [PubMed]
17. Lux, D.; Alakbarzade, V.; Bridge, L.; Clark, C.N.; Clarke, B.; Zhang, L.; Khan, U.; Pereira, A.C. The association of neutrophil-lymphocyte ratio and lymphocyte-monocyte ratio with 3-month clinical outcome after mechanical thrombectomy following stroke. *J. Neuroinflammation* **2020**, *17*, 60. [CrossRef] [PubMed]
18. Kakhki, R.D.; Dehghanei, M.; ArefNezhad, R.; Motedayyen, H. The predicting role of neutrophil- lymphocyte ratio in patients with acute ischemic and hemorrhagic stroke. *J. Stroke Cerebrovasc. Dis.* **2020**, *29*, 105233. [CrossRef]
19. Gong, P.; Liu, Y.; Gong, Y.; Chen, G.; Zhang, X.; Wang, S.; Zhou, F.; Duan, R.; Chen, W.; Huang, T.; et al. The association of neutrophil to lymphocyte ratio, platelet to lymphocyte ratio, and lymphocyte to monocyte ratio with post-thrombolysis early neurological outcomes in patients with acute ischemic stroke. *J. Neuroinflammation* **2021**, *18*, 51. [CrossRef] [PubMed]
20. Wang, L.; Song, Q.; Wang, C.; Wu, S.; Deng, L.; Li, Y.; Zheng, L.; Liu, M. Neutrophil to lymphocyte ratio predicts poor outcomes after acute ischemic stroke: A cohort study and systematic review. *J. Neurol. Sci.* **2019**, *406*, 116445. [CrossRef] [PubMed]
21. Lattanzi, S.; Brigo, F.; Trinka, E.; Cagnetti, C.; Di Napoli, M.; Silvestrini, M. Neutrophil- to-Lymphocyte ratio in acute cerebral hemorrhage: A system review. *Transl. Stroke Res.* **2019**, *10*, 137–145. [CrossRef] [PubMed]

22. Piri Cinar, B.; Kunt, R.; Yuksel, B.; Gulluoglu, H.; Sayilir, I.; Celiker Uslu, S.; Ozaydin Goksu, E.; Bülbül, N.G.; Yildiz, B.; Oz, D.; et al. Assessment of the relation between the neutrophil to lymphocyte ratio and severity of ischemic stroke in a large cohort. *Int. J. Clin. Pract.* **2021**, *75*, e13955. [CrossRef] [PubMed]
23. Aytac, E.; Akpinar, C.T.; Gürkas, E. Neutrophil to lymphocyte ratio: A simple and readily available independent marker of mortality in acute ischemic stroke. *Firat Med. J.* **2017**, *22*, 192–196.
24. Sarejloo, S.; Kheradjoo, H.; Haghi, S.E.; Hosseini, S.; Gargari, M.K.; Azarhomayoun, A.; Khanzadeh, S.; Sadeghvand, S. Neutrophil-to-Lymphocyte Ratio and Early Neurological Deterioration in Stroke Patients: A Systematic Review and Meta-Analysis. *Biomed. Res. Int.* **2022**, *2022*, 8656864. [CrossRef] [PubMed] [PubMed Central]
25. Li, S.J.; Cao, S.S.; Huang, P.S.; Nie, X.; Fu, Y.; Liu, J.R. Post-operative neutrophil-to-lymphocyte ratio and outcome after thrombectomy in acute ischemic stroke. *Front. Neurol.* **2022**, *13*, 990209. [CrossRef] [PubMed] [PubMed Central]
26. Xu, C.; Cai, L.; Yi, T.; Yi, X.; Hu, Y. Neutrophil-to-lymphocyte ratio is associated with stroke progression and functional outcome in patients with ischemic stroke. *Brain Behav.* **2023**, *13*, e3261. [CrossRef] [PubMed] [PubMed Central]
27. Aly, M.; Abdalla, R.N.; Batra, A.; Shaibani, A.; Hurley, M.C.; Jahromi, B.S.; Potts, M.B.; A Ansari, S. Follow-up neutrophil-lymphocyte ratio after stroke thrombectomy is an independent biomarker of clinical outcome. *J. NeuroInterventional Surg.* **2020**, *13*, 609–613. [CrossRef]
28. Cao, W.; Song, Y.; Bai, X.; Yang, B.; Li, L.; Wang, X.; Wang, Y.; Chang, W.; Chen, Y.; Wang, Y.; et al. Systemic-inflammatory indices and clinical outcomes in patients with anterior circulation acute ischemic stroke undergoing successful endovascular thrombectomy. *Heliyon* **2024**, *10*, e31122. [CrossRef] [PubMed] [PubMed Central]
29. Lattanzi, S.; Norata, D.; Broggi, S.; Meletti, S.; Świtońska, M.; Słomka, A.; Silvestrini, M. Neutrophil-to-Lymphocyte Ratio Predicts Early Neurological Deterioration after Endovascular Treatment in Patients with Ischemic Stroke. *Life* **2022**, *12*, 1415. [CrossRef] [PubMed] [PubMed Central]
30. Chen, S.; Cheng, J.; Ye, Q.; Ye, Z.; Zhang, Y.; Liu, Y.; Huang, G.; Chen, F.; Yang, M.; Wang, C.; et al. Day 1 neutrophil-to-lymphocyte ratio (NLR) predicts stroke outcome after intravenous thrombolysis and mechanical thrombectomy. *Front. Neurol.* **2022**, *13*, 941251. [CrossRef] [PubMed] [PubMed Central]
31. Sharma, D.; Bhaskar, S.M.M. Prognostic Role of the Platelet-Lymphocyte Ratio in Acute Ischemic Stroke Patients Undergoing Reperfusion Therapy: A Meta-Analysis. *J. Cent. Nerv. Syst. Dis.* **2022**, *14*, 11795735221110373. [CrossRef]
32. Sharma, D.; Spring, K.J.; Bhaskar, S.M.M. Role of Neutrophil-Lymphocyte Ratio in the Prognosis of Acute Ischaemic Stroke After Reperfusion Therapy: A Systematic Review and Meta-analysis. *J. Cent. Nerv. Syst. Dis.* **2022**, *14*, 11795735221092518. [CrossRef] [PubMed]
33. Li, S.; Hu, L.; Wang, J.; Zou, F.; Han, B.; Wang, Y.; Liu, K. Prolonged increased neutrophil-to-lymphocyte ratio is associated with mortality after successful revascularization for treatment of acute ischemic stroke. *BMC Neurol.* **2022**, *22*, 326. [CrossRef] [PubMed] [PubMed Central]
34. Zou, F.; Wang, J.; Han, B.; Bao, J.; Fu, Y.; Liu, K. Early Neutrophil-to-Lymphocyte Ratio Is a Prognostic Marker in Acute Ischemic Stroke After Successful Revascularization. *World Neurosurg.* **2022**, *157*, e401–e409. [CrossRef] [PubMed]
35. Lesmana, J.; Lefi, A. The Relationship between Neutrophil to Lymphocyte Ratio with Rehospitalisation and Mortality in Patients with Acute Coronary Syndrome. *Eur. Cardiol.* **2023**, *18*, e29. [CrossRef] [PubMed] [PubMed Central]
36. García-Escobar, A.; Vera-Vera, S.; Tébar-Márquez, D.; Rivero-Santana, B.; Jurado-Román, A.; Jiménez-Valero, S.; Galeote, G.; Cabrera, J.Á.; Moreno, R. Neutrophil-to-lymphocyte ratio an inflammatory biomarker, and prognostic marker in heart failure, cardiovascular disease and chronic inflammatory diseases: New insights for a potential predictor of anti-cytokine therapy responsiveness. *Microvasc. Res.* **2023**, *150*, 104598. [CrossRef] [PubMed]
37. Schupp, T.; Weidner, K.; Rusnak, J.; Jawhar, S.; Forner, J.; Dulatahu, F.; Brück, L.; Dudda, J.; Hoffmann, U.; Abba, M.; et al. The Neutrophil-to-Lymphocyte-Ratio as Diagnostic and Prognostic Tool in Sepsis and Septic Shock. *Clin Lab.* **2023**, *69*, 912–924. [CrossRef] [PubMed]
38. Sung, K.L.; Kuo, M.J.; Hsieh, C.Y.; Sung, S.F. High Neutrophil-to-Lymphocyte Ratio Predicts One-Year Risk of Pneumonia Post-Stroke Discharge. *Cerebrovasc. Dis.* **2023**, *52*, 567–574. [CrossRef] [PubMed]

diagnostics

MDPI

Article

Evaluating Machine Learning Models for Stroke Prognosis and Prediction in Atrial Fibrillation Patients: A Comprehensive Meta-Analysis

Bill Goh [1,2] and Sonu M. M. Bhaskar [1,2,3,4,5,6,*,†]

1 Global Health Neurology Lab, Sydney, NSW 2150, Australia
2 UNSW Medicine and Health, University of New South Wales (UNSW), South West Sydney Clinical Campuses, Sydney, NSW 2170, Australia
3 Clinical Sciences Stream, Ingham Institute for Applied Medical Research, Liverpool, NSW 2170, Australia
4 NSW Brain Clot Bank, NSW Health Pathology, Sydney, NSW 2170, Australia
5 Department of Neurology & Neurophysiology, Liverpool Hospital, South Western Sydney Local Health District, Liverpool, NSW 2170, Australia
6 Department of Neurology, Division of Cerebrovascular Medicine and Neurology, National Cerebral and Cardiovascular Center (NCVC), Suita 564-8565, Osaka, Japan
* Correspondence: sonu.bhaskar@globalhealthneurolab.org or bhaskar.sonu@ncvc.go.jp; Tel.: +81-6-6170-1070; Fax: +81-6-6170-1782
† Current address: National Cerebral and Cardiovascular Center (NCVC), Department of Neurology, 6-1 Kishibeshimmachi, Suita 564-8565, Osaka, Japan.

Abstract: Background/Objective: Atrial fibrillation (AF) complicates the management of acute ischemic stroke (AIS), necessitating precise predictive models to enhance clinical outcomes. This meta-analysis evaluates the efficacy of machine learning (ML) models in three key areas: stroke prognosis in AF patients, stroke prediction in AF patients, and AF prediction in stroke patients. The study aims to assess the accuracy and variability of ML models in forecasting AIS outcomes and detecting AF in stroke patients, while exploring the clinical benefits and limitations of integrating these models into practice. Methods: We conducted a systematic search of PubMed, Embase, and Cochrane databases up to June 2024, selecting studies that evaluated ML accuracy in stroke prognosis and prediction in AF patients and AF prediction in stroke patients. Data extraction and quality assessment were performed independently by two reviewers, with random-effects modeling applied to estimate pooled accuracy metrics. Results: The meta-analysis included twenty-four studies comprising 7,391,645 patients, categorized into groups for stroke prognosis in AF patients (eight studies), stroke prediction in AF patients (thirteen studies), and AF prediction in stroke patients (three studies). The pooled AUROC was 0.79 for stroke prognosis and 0.68 for stroke prediction in AF, with higher accuracy noted in short-term predictions. The mean AUROC across studies was 0.75, with models such as Extreme Gradient Boosting (XGB) and Random Forest (RF) showing superior performance. For stroke prognosis in AF, the mean AUROC was 0.78, whereas stroke prediction yielded a mean AUROC of 0.73. AF prediction post-stroke had an average AUROC of 0.75. These findings indicate moderate predictive capability of ML models, underscoring the need for further refinement and standardization. The absence of comprehensive sensitivity, specificity, positive predictive value (PPV), and negative predictive value (NPV) metrics limited the ability to conduct full meta-analytic modeling. Conclusions: While ML models demonstrate potential for enhancing stroke prognosis and AF prediction, they have yet to meet the clinical standards required for widespread adoption. Future efforts should focus on refining these models and validating them across diverse populations to improve their clinical utility.

Keywords: stroke; atrial fibrillation; machine learning; prognosis; acute ischemic stroke; prediction; artificial intelligence

1. Introduction

Stroke is the leading contributor to the global burden of neurological disease [1] and accounted for 7.25 million deaths in 2021, rendering it the second-leading cause of death worldwide [2,3]. For stroke survivors, its aftermath is debilitating; around half endure chronic disabilities, which impose significant financial, emotional, and societal costs [1]. Atrial fibrillation (AF) is a major contributor to cardioembolic strokes, responsible for about 15% of all strokes and increasing the risk by three to five times in affected patients [4,5]. The challenge of predicting and prognosing stroke in patients with AF persists due to the lack of robust, evidence-based approaches [6].

Despite advances in stroke treatment, such as intravenous thrombolysis and endovascular thrombectomy, AF patients continue to experience poor outcomes [7]. Machine learning (ML) models hold promise in surpassing traditional methods in stroke prediction and prognosis [8,9]. These models can identify high-risk patients and predict short-term outcomes [10], enabling timely and personalized interventions [10]. For instance, artificial intelligence (AI)-enhanced ECG interpretation can detect undiagnosed paroxysmal AF in embolic stroke of undetermined source (ESUS) patients, potentially improving patient prognosis through early intervention [5].

This meta-analysis evaluates the predictive and prognostic capabilities of ML models in pre- and post-stroke AF patients. By addressing critical clinical gaps, we aim to enhance patient outcomes through the integration of advanced AI technologies.

2. Objective

This meta-analysis systematically evaluates the predictive and prognostic capabilities of ML models in the context of AIS management among patients with AF. The study involves a comprehensive evaluation of ML models across three critical areas: stroke prognosis in AF patients, stroke prediction in AF patients, and AF prediction in stroke patients. Specifically, it aims to (1) assess the predictive accuracy of various ML models in forecasting AIS outcomes in AF patients; (2) evaluate the variability in model performance across different treatment regimens and patient demographics to identify areas requiring refinement for enhanced clinical application; and (3) explore the potential clinical benefits and limitations of integrating these ML models into practice to improve patient outcomes and guide treatment strategies effectively. By addressing these objectives, the study seeks to provide valuable insights into the strengths and gaps of current ML applications, informing future research directions and the development of more reliable and clinically relevant predictive models.

3. Materials and Methods

3.1. Literature Search: Study Identification and Selection

Studies were sourced from online databases, including PubMed, Cochrane Central Register of Controlled Trials (CENTRAL), and Embase, spanning the period from January 2019 to June 2024. An extensive search strategy was implemented, utilizing relevant terms such as "atrial fibrillation", "machine learning", "artificial intelligence", "acute ischemic stroke", "prognosis", "clinical outcomes", and "accuracy". All research was conducted with human participants and reported in the English language. A comprehensive search strategy is detailed in the Supplementary Materials. To identify further apposite studies, manual screening of reference lists from relevant articles, meta-analyses, and systematic reviews was conducted. The search process and inclusion of studies are illustrated in Figure 1, which follows the Preferred Reporting Items for Systematic Reviews and Meta-Analyses (PRISMA) flowchart, which adheres to reporting standards. This study was registered in Open Science (osf.io), registration number "e3cgf" (accessed on 19 August 2024).

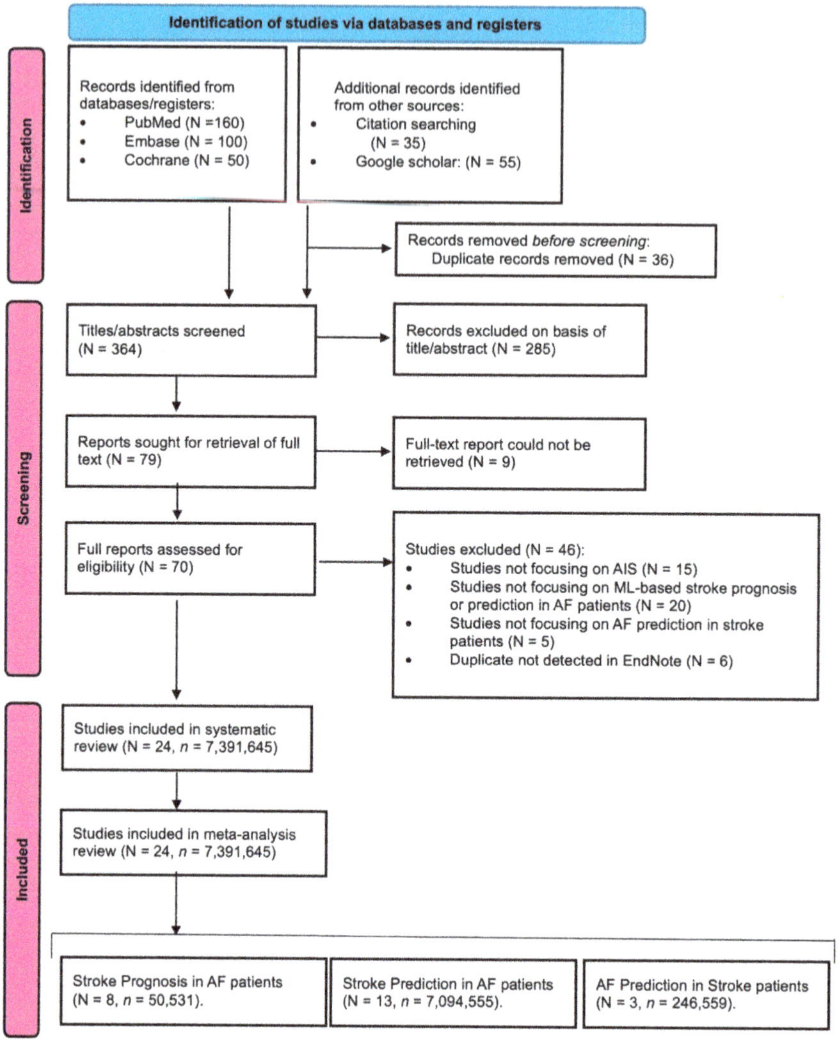

Figure 1. PRISMA flowchart depicting the process of study identification, screening, and inclusion. Abbreviations: PRISMA = Preferred Reporting Items for Systematic Reviews and Meta-Analyses; N = Number of Studies; n = Number of Patients.

3.2. Inclusion and Exclusion Criteria

The inclusion criteria for the study required that the research focused on the accuracy of ML models in detecting AF in patients with AIS and assessing the association between ML-prognosed AF or stroke and clinical outcomes. Eligible studies including adult patients aged 18 years and older were based on observational (cohort, cross-sectional) or randomized, controlled trial designs and were adaptive clinical trials, clinical studies, clinical trial protocols, or multicenter studies. Studies needed to involve human subjects, be published in English, and include both male and female participants. Conversely, the exclusion criteria eliminated studies that did not specifically focus on AIS or ML-based AF or stroke prognosis, including pediatric populations (age 18 years or younger) or that were case reports, reviews, or editorials.

3.3. Data Extraction

Data extraction was meticulously conducted by two independent reviewers using a standardized form to ensure consistency and accuracy. The extracted data included study characteristics such as the authors, publication year, country of origin, study design, and sample size. Key patient characteristics were also recorded, encompassing age, gender, stroke severity, and treatment types. Furthermore, outcome measures were carefully documented, focusing on the accuracy metrics of ML models in detecting AF and associated clinical outcomes, including functional status, mortality rates, and incidences of recurrent stroke. Any discrepancies between reviewers were resolved through discussion and consensus.

3.4. Data Modality Analysis

We retrieved and analyzed various data modalities utilized in the studies included in our meta-analysis to understand their impact on the performance of machine learning models. The studies incorporated diverse data types to enhance predictive accuracy, including clinical data (e.g., electronic health records detailing demographics, medical history, and treatment information), imaging data (e.g., MRI and CT scans for brain structure analysis), biomarker data from blood tests (e.g., levels of specific proteins or compounds indicative of disease states), and electrocardiogram (ECG) data (e.g., cardiac function and rhythm monitoring). Additionally, some studies employed genomic and proteomic data to explore genetic markers associated with stroke and atrial fibrillation.

3.5. Methodological Quality Assessment of Included Studies

The quality of the included studies was evaluated using multiple established checklists to ensure a comprehensive assessment. Observational studies were appraised with the Preferred Reporting Items for Systematic Reviews and Meta-Analyses (PRISMA) 2020 checklist (Supplementary Table S1) and the Meta-analysis of Observational Studies in Epidemiology (MOOSE) checklist (Supplementary Table S2). Additionally, the Standards for Reporting Diagnostic Accuracy Studies (STARD-2015) checklist (Supplementary Table S3) and the Modified Jadad Scale (MJS) (Supplementary Table S4) were employed to assess diagnostic study quality. Any disagreements or discrepancies identified during data extraction and quality assessment were resolved through mutual discussion and consensus among the reviewers, ensuring a robust evaluation process.

3.6. Statistical Analyses

Statistical analyses, including meta-analysis and advanced statistical procedures, were performed using STATA (Version 13.0, StataCorp., College Station, TX, USA). Heterogeneity across studies was assessed using the I^2 statistic to quantify variability. Potential sources of heterogeneity, such as differences in ML models, population demographics, and study designs, were explored. In cases of high heterogeneity, subgroup analyses or meta-regression were considered to account for variability. Prognostic accuracy analysis was undertaken to evaluate the role of ML in stroke outcome determination, focusing on sensitivity, specificity, and the area under the receiver operating characteristic (AUROC) curve as a prognostic indicator. AUROC scores and C-index values were considered interchangeable, with interpretations based on existing classification guidelines. We conducted a comprehensive meta-analysis to evaluate the performance of ML models, focusing on various performance metrics such as sensitivity, specificity, positive predictive value (PPV), negative predictive value (NPV), and AUROC. Packages metan and midas were installed. Standard errors were calculated from confidence intervals for sensitivity, specificity, PPV, NPV, and AUROC, where available. We conducted comprehensive statistical and error analyses to evaluate the predictive accuracy of the machine learning models. Specifically, we incorporated metrics, where available, such as mean squared error (MSE), root mean squared error (RMSE), and 95% confidence intervals for key performance indicators, including sensitivity, specificity, and AUROC. These analyses provide a clearer understanding of the uncertainty associated with these metrics. For instance, the AUROC confidence intervals for the stroke prognosis models ranged from 0.73 to 0.85, reflecting moderate variability. Additionally, we identified several sources of variability, such as differences in study populations, variations in data quality, and the complexity of the models. These factors contribute to variations in model performance and are critical in interpreting the predictive capabilities of each model.

Random-effects meta-analyses were performed, with forest plots generated for each metric. Heterogeneity was assessed using the I^2 statistic, followed by subgroup analyses based on model type. Meta-regression was used to explore study-level covariates. Publication bias was examined through funnel plots and Egger's test. The AUROC scores were interpreted based on the classification suggested by Çorbacıoğlu et al. (Excellent: 0.9 ≤ AUROC; Optimal: 0.8 ≤ AUROC < 0.9; Average: 0.7 ≤ AUROC < 0.8; Poor: 0.6 ≤ AUROC < 0.7; and Fail: 0.5 ≤ AUROC < 0.6) [11]. Sensitivity and specificity were also evaluated for clinical relevance, with a combined score benchmarked for clinical utility. Moreover, according to Power et al., scores should be at least 1.5 when added up to be considered clinically useful [12]. We also conducted a feature importance assessment using SHAP (SHapley Additive exPlanations) values to quantify the contribution of each feature in the machine learning models. This approach allowed us to interpret the model predictions by identifying the most influential variables, such as age, NIHSS score, and anticoagulation use. SHAP values were computed for all key features, providing insights into their relative impact on the model's accuracy.

4. Results

4.1. Description of Included Studies

This meta-analysis assessed the predictive capabilities of ML models in stroke and AF scenarios, incorporating data from 24 studies involving 7,391,645 patients.

These studies were categorized into three groups:

a. *Stroke Prognosis in AF Patients*: Eight studies examined the prognosis of stroke in patients with AF [10,13–19].
b. *Stroke Prediction in AF Patients*: 13 studies focused on predicting strokes in AF patients [20–32].
c. *AF Prediction in Stroke Patients*: Three studies investigated predicting AF in patients who had experienced a stroke [5,33,34].

The analysis spanned five single-center and 19 multicenter studies, reflecting diverse study designs and global reach across four continents. Excluded were studies not focusing on AIS or ML-based prediction in AF patients. Limited data and inconsistent accuracy reporting prevented a comprehensive meta-analysis for all metrics (sensitivity, specificity, PPV, NPV), but AUROC meta-analysis for stroke prognosis and stroke prediction in AF could be performed.

Our analysis, detailed in Tables 1–3, highlights the prevalence and characteristics of these modalities across the included studies, underscoring their crucial role in enhancing the robustness and interpretability of machine learning models in clinical settings. The tables in the meta-analysis provide a comprehensive overview of the studies and their findings related to ML applications in stroke and AF contexts. Table 1 lists clinical risk factors prevalent across studies, such as AF, hypertension, and diabetes, indicating a wide variability in prevalence rates. Table 2 details study characteristics, reflecting a mix of retrospective and prospective designs conducted across diverse geographic regions.

Table 1. Prevalence of clinical risk factors in included studies.

Author	Year	Clinical Risk Factors, *n* (%)							
		AF	HTN	HL	PS	CKD	CAD	Smoking	DM
Abedi et al. [13]	2021	2887 (26.5%)	8141 (75%)	6478 (60%)	NR	2387 (22%)	NR	NR	3595 (33%)
Abujaber et al. [14]	2023	NR	5486 (74%)	3295 (44%)	864 (12%)	NR	860 (12%)	1580 (21%)	3919 (53%)
Bernardini et al. [15]	2024	2340 (21%)	8889 (80%)	NR	1688 (15%)	NR	1869 (17%)	NR	2275 (21%)
Bisson et al. [33]	2023	NR	152,790 (64%)	NR	NR	NR	44,621 (19%)	30,255 (13%)	55,060 (23%)
Chen et al. [20]	2024	1738 (83%)	1287 (61%)	983 (47%)	NR	1060 (51%)	NR	531 (25%)	782 (37%)
Colangelo et al. [21]	2024	40%	83%	66%	89%	22%	18%	NR	40%
Gkantzios et al. [16]	2023	NR	NR	NR	NR	NR	NR	NR	NR
Goto et al. [17]	2020	4708 (100%)	NR	NR	NR	NR	878 (19%)	NR	NR
Han et al. [22]	2019	NR	NR	NR	NR	NR	NR	NR	NR
Handy et al. [23]	2021	NR	NR	NR	NR	NR	NR	NR	NR
Jeon et al. [10]	2023	1599 (52%)	2121 (69%)	NR	916 (30%)	NR	454 (15%)	NR	843 (27%)
Jung et al. [18]	2021	NR	NR	NR	NR	NR	NR	NR	NR
Jung et al. [24]	2022	100%	36,252 (58%)	4589 (7%)	100%	1210 (2%)	NR	7404 (12%)	13,802 (22%)
Choi et al. [35]	2021	NR	NR	NR	100%	NR	NR	NR	NR
Li et al. [25]	2016	100%	NR	NR	NR	NR	NR	NR	NR
Lip et al. [26]	2022	NR	475,894 (7.4%)	NR	38,050 (0.6%)	85,311 (1.3%)	131,405 (2%)	NR	222,013 (3.4%)
Lu et al. [27]	2022	NR	5423 (56%)	NR	167 (1.7%)	NR	NR	NR	2280 (24%)
Ming et al. [5]	2024	NR	108 (69%)	96 (61%)	14 (8.9%)	NR	NR	60 (38%)	60 (38%)
Nishi et al. [28]	2022	100%	1073 (61%)	752 (43%)	173 (9.8%)	542 (31%)	261 (15%)	NR	378 (22%)
Papadopoulou et al. [29]	2022	NR	22,019 (52%)	NR	12,317 (29%)	NR	NR	19,285 (45%)	6424 (15%)
Rebollo et al. [30]	2022	100%	NR	NR	NR	NR	NR	NR	NR
Sung et al. [34]	2022	NR	4824 (79%)	3522 (58%)	1417 (23%)	NR	663 (11%)	NR	2641 (43%)
Truong et al. [31]	2024	NR	13,259 (72%)	NR	1542 (8.4%)	NR	NR	NR	5621 (31%)
Zhang et al. [32]	2024	100%	18 (72%)	6 (24%)	100%	NR	NR	8 (32%)	9 (36%)

Abbreviations: AF: atrial fibrillation; HTN: hypertension; HL: hyperlipidemia; PS: prior stroke; CKD: chronic kidney disease; CAD: coronary artery disease; DM: diabetes mellitus; NR: not reported/available.

Table 2. Characteristics of studies evaluating machine learning models.

Author	Title	Year	Region	Center Type	Database	Study Design	Study Type	Centers
Abedi et al. [13]	Predicting short- and long-term mortality after acute ischemic stroke using EHR	2021	Multinationa	Multicenter	Geisinger Neuroscience Ischemic Stroke	Retrospective	Registry	NR
Abujaber et al. [14]	Predicting 90-day prognosis for patients with stroke: a machine learning approach	2023	Qatar	Single center	Stroke Registry of Hamad General Hospital	Retrospective	Registry	1
Bernardini et al. [15]	Machine learning approach for prediction of outcomes in anticoagulated patients with atrial fibrillation	2024	Italy	Multicenter	Survey on Anticoagulated Patients Register	Retrospective	Registry	NR

Table 2. *Cont.*

Author	Title	Year	Region	Center Type	Database	Study Design	Study Type	Centers
Bisson et al. [33]	Prediction of incident atrial fibrillation in post-stroke patients using machine learning: a French nationwide study	2023	France	Multicenter	Programme de Médicalisation des Systèmes d'Information	Retrospective	Registry	NR
Chen et al. [20]	Predicting stroke in Asian patients with atrial fibrillation using machine learning	2024	India	Multicenter	Training = KERALA-AF registry, Val = APHRS-AF registry	Retrospective	Registry	105 train/val
Colangelo et al. [21]	PRERISK: A Personalized, Artificial Intelligence-Based and Statistically-Based Stroke Recurrence Predictor for Recurrent Stroke	2024	Spain	Multicenter	Catalan Agency for Health Quality and Evaluation	Retrospective	Registry	88
Gkantzios et al. [16]	From Admission to Discharge: Predicting National Institutes of Health Stroke Scale Progression in Stroke Patients Using Biomarkers and Explainable Machine Learning	2023	Greece	Single center	Korgialeneio–Benakeio General Hospital, Athens	Retrospective	Registry	1
Goto et al. [17]	New artificial intelligence prediction model using serial prothrombin time international normalized ratio measurements in atrial fibrillation patients on vitamin K antagonists: GARFIELD-AF	2020	Multinational	Multicenter	GARFIELD-AF	Prospective	Registry	NR
Han et al. [22]	Atrial fibrillation burden signature and near-term prediction of stroke: A machine learning analysis	2019	USA	Single center	Veterans Health Administration	Retrospective	Registry	1
Handy et al. [23]	A nationwide deep learning pipeline to predict stroke and COVID-19 death in atrial fibrillation	2021	England	Multicenter	NHS Digital EHR	Prospective	Registry	NR
Jeon et al. [10]	Predicting short-term outcomes in atrial-fibrillation-related stroke using machine learning	2023	Korea	Multicenter	Train = K-ATTENTION, Val = KUSR	Retrospective	Registry	14
Jung et al. [18]	Outcome predictions using machine learning in atrial fibrillation-related stroke	2021	Korea	Multicenter	N/A	Prospective	Registry	11
Jung et al. [24]	Predicting Ischemic Stroke in Patients with Atrial Fibrillation Using Machine Learning	2022	Korea	Multicenter	KNHIS	Retrospective	Registry	NR

Table 2. *Cont.*

Author	Title	Year	Region	Center Type	Database	Study Design	Study Type	Centers
Choi et al. [35]	Interpretable machine learning for early neurological deterioration prediction in atrial fibrillation-related stroke	2021	Korea	Multicenter	K-ATTENTION	Prospective	Registry	11
Li et al. [25]	Integrated Machine Learning Approaches for Predicting Ischemic Stroke and Thromboembolism in Atrial Fibrillation	2016	China	Multicenter	Chinese Atrial Fibrillation Registry	Prospective	Registry	32
Lip et al. [26]	Improving dynamic stroke risk prediction in non-Anticoagulated patients with and without atrial fibrillation: Comparing common clinical risk scores and machine learning algorithms	2022	USA	Multicenter	Commercial plan + Medicare plan USA	Retrospective	Registry	2
Lu et al. [27]	Performance of multilabel machine learning models and risk stratification schemas for predicting stroke and bleeding risk in patients with non-valvular atrial fibrillation	2022	Australia	Multicenter	Sir Charles Gairdner + Osborne Park teaching hospitals	Retrospective	Registry	2
Ming et al. [5]	Machine Learning Modelling to Predict Atrial Fibrillation Detection in Embolic Stroke of Undetermined Source Patients	2024	Singapore	Single center	Stroke unit of undisclosed tertiary care hospital	Retrospective	Registry	1
Nishi et al. [28]	Predicting cerebral infarction in patients with atrial fibrillation using machine learning: The Fushimi AF registry	2022	Japan	Multicenter	Fushimi AF Registry	Retrospective	Registry	81
Papadopoulou et al. [29]	Prediction of atrial fibrillation and stroke using machine learning models in UK Biobank	2022	UK	Multicenter	UK Biobank	Prospective	Registry	NR
Rebollo et al. [30]	Development of a Machine Learning Predictive Model for Stroke Among Patients with Non-Valvular Atrial Fibrillation Receiving Oral Anticoagulant Treatment	2022	Spain	Multicenter	Spanish IQVIA's Longitudinal Patient Database	Retrospective	Registry	NR
Sung et al. [34]	Automated risk assessment of newly detected atrial fibrillation poststroke from electronic health record data using machine learning and natural language processing	2022	Taiwan	Multicenter	Ditmanson Medical Foundation Chia-Yi Christian Hospital + Ditmanson Research Database	Retrospective	Registry	NR

Table 2. *Cont.*

Author	Title	Year	Region	Center Type	Database	Study Design	Study Type	Centers
Truong et al. [31]	Development and Validation of Machine Learning Algorithms to Predict 1-Year Ischemic Stroke and Bleeding Events in Patients with Atrial Fibrillation and Cancer	2024	USA	Multicenter	Surveillance, Epidemiology, and End Results (SEER) registry	Retrospective	Registry	NR
Zhang et al. [32]	Using machine learning to identify proteomic and metabolomic signatures of stroke in atrial fibrillation	2024	China	Single center	First Affiliated Hospital of Harbin Medical University	Prospective	Registry	1

Abbreviations: APHRS: Asia Pacific Heart Rhythm Society; GARFIELD: Global Anticoagulant Registry in the Field; K-ATTENTION: Korean Atrial Fibrillation Evaluation Registry in Ischemic Stroke Patients; KUSR: Korea University Stroke Registry; KNHIS: Korean National Health Insurance Service; N/A: Not Available.

Table 3. Study hypotheses, machine learning models, and cohort details.

Author	Study	Hypothesis	Total Cohort	Training Cohort	External Validation Cohort	Highest Performing Model	Total Features Included
Abedi et al. [13]	Predicting short and long-term mortality after acute ischemic stroke using EHR	1 month mortality SP	7144	80%	20%	XGB	37
		3-month mortality SP	7144	80%	20%	XGB	
		6-month mortality SP	7144	80%	20%	XGB	
		12-month mortality SP	7144	80%	20%	XGB	
		18-month mortality SP	7144	80%	20%	XGB	
		24-month mortality SP	7144	80%	20%	XGB	
Abujaber et al. [14]	Predicting 90-day prognosis for patients with stroke: a machine learning approach	90-day SP	7452	100%	none	RF	19
Bernardini et al. [15]	Machine learning approach for prediction of outcomes in anticoagulated patients with atrial fibrillation	SP in AF patients	11,078	100%	none	Naive Multi-Task NN	68
		SP in AF patients on DOAC	5943	100%	none	GBDT NN	
		SP in AF patients on VKA	5135	100%	none	Naive Multi-Task NN	

Table 3. *Cont.*

Author	Study	Hypothesis	Total Cohort	Training Cohort	External Validation Cohort	Highest Perform-ing Model	Total Features Included
Bisson et al. [33]	Prediction of incident atrial fibrillation in post-stroke patients using machine learning: a French nationwide study	AF prediction in PS patients	240,306	70%	30%	DNN	16
Chen et al. [20]	Predicting stroke in Asian patients with atrial fibrillation using machine learning	Stroke prediction in AF patients	8065	3401	4664	LightGBM	12
Colangelo et al. [21]	PRERISK: A Personalized, Artificial Intelligence-Based and Statistically-Based Stroke Recurrence Predictor for Recurrent Stroke	Early recurrent stroke prediction	41,975	80%	20%	RF	13
		Late recurrent stroke prediction	41,975	80%	20%	RF	
		Long-term recurrent stroke prediction	41,975	80%	20%	RF	
Gkantzios et al. [16]	From Admission to Discharge: Predicting National Institutes of Health Stroke Scale Progression in Stroke Patients Using Biomarkers and Explainable Machine Learning	NIHSS score prediction in acute stroke	413	70%	30%	RF	32
Goto et al. [17]	New artificial intelligence prediction model using serial prothrombin time international normalized ratio measurements in atrial fibrillation patients on vitamin K antagonists: GARFIELD-AF	Major bleed prognosis in AF patients on VKA	4708	3185	1523	NN	10
		SP in AF patients on VKA	4708	3185	1523	NN	
		All cause death prognosis in AF patients on VKA	4708	3185	1523	NN	
Han et al. [22]	Atrial fibrillation burden signature and near-term prediction of stroke: A machine learning analysis	Stroke prediction in AF patients	9836	50% (train), 20% (val)	30% (test)	CNN	11

Table 3. *Cont.*

Author	Study	Hypothesis	Total Cohort	Training Cohort	External Validation Cohort	Highest Performing Model	Total Features Included
Handy et al. [23]	A nationwide deep learning pipeline to predict stroke and COVID-19 death in atrial fibrillation	Stroke prediction in AF patients	16,563	80%	20%	XGBoost	N/A
Jeon et al. [10]	Predicting short-term outcomes in atrial-fibrillation-related stroke using machine learning	UFO SP in AF patients	3205	2307	898	LightGBM	43
		Mortality SP in AF patients	3205	2307	898	LightGBM	
Jung et al. [18]	Outcome predictions using machine learning in atrial fibrillation-related stroke	All cause death SP in AF patients	3090	75%	25%	XGB	N/A
Jung et al. [24]	Predicting Ischemic Stroke in Patients with Atrial Fibrillation Using Machine Learning	Stroke prediction in AF patients	62,226	N/A	N/A	DNN	58
Choi et al. [35]	Interpretable machine learning for early neurological deterioration prediction in atrial fibrillation-related stroke	END SP in AF patients	2363	75%	25%	LightGBM	23
Li et al. [25]	Integrated Machine Learning Approaches for Predicting Ischemic Stroke and Thromboembolism in Atrial Fibrillation	2-year stroke prediction in AF patients	1864	60%	40%	LR + Wrapper	167
Lip et al. [26]	Improving dynamic stroke risk prediction in non-Anticoagulated patients with and without atrial fibrillation: Comparing common clinical risk scores and machine learning algorithms	Stroke prediction in AF patients	6,457,412	2/3 of data	1/3 of data	LR	19

Table 3. *Cont.*

Author	Study	Hypothesis	Total Cohort	Training Cohort	External Validation Cohort	Highest Performing Model	Total Features Included
Lu et al. [27]	Performance of multilabel machine learning models and risk stratification schemas for predicting stroke and bleeding risk in patients with non-valvular atrial fibrillation	Stroke prediction in AF patients	9670	75%	25%	Multilabel GBM	43
		Major bleed prediction in AF patients	9670	75%	25%	Multilabel GBM	
		Stroke prediction in AF patients without AT	9670	75%	25%	Multilabel GBM	
		Major bleed prediction in AF patients without AT	9670	75%	25%	Multilabel GBM	
Ming et al. [5]	Machine Learning Modelling to Predict Atrial Fibrillation Detection in Embolic Stroke of Undetermined Source Patients	Paroxysmal AF prediction in ESUS patients	157	70%	30%	SVM + Random oversampling	N/A
Nishi et al. [28]	Predicting cerebral infarction in patients with atrial fibrillation using machine learning: The Fushimi AF registry	Stroke prediction in AF patients without AT	1757	1005	752	CatBoost	14
Papadopoulou et al. [29]	Prediction of atrial fibrillation and stroke using machine learning models in UK Biobank	Stroke prediction in AF patients	454,118	80%	20%	XGB	99
Rebollo et al. [30]	Development of a Machine Learning Predictive Model for Stroke Among Patients With Non-Valvular Atrial Fibrillation Receiving Oral Anticoagulant Treatment	180-day stroke prediction in AF patients with AT	8028	NR	NR	XGB	N/A
		360-day stroke prediction in AF patients with AT	4628	NR	NR	XGB	
Sung et al. [34]	Automated risk assessment of newly detected atrial fibrillation poststroke from electronic health record data using machine learning and natural language processing	AF prediction in post-stroke patients	6096	4604	1492	XGB	20

Table 3. *Cont.*

Author	Study	Hypothesis	Total Cohort	Training Cohort	External Validation Cohort	Highest Performing Model	Total Features Included
Truong et al. [31]	Development and Validation of Machine Learning Algorithms to Predict 1-Year Ischemic Stroke and Bleeding Events in Patients with Atrial Fibrillation and Cancer	1-year stroke prediction in AF patients	18,388	70%	30%	RF	53
Zhang et al. [32]	Using machine learning to identify proteomic and metabolomic signatures of stroke in atrial fibrillation	Stroke prediction in AF patients	25	NR	NR	RF	N/A

Abbreviations: XGB: Extreme Gradient Boosting; RF: Random Forest; NN: neural network; GBDT: Gradient-Boosted Decision Trees; DNN: Deep Neural Network; LightGBM: Light Gradient-Boosting Machine; CNN: Convolutional Neural Network; LR: Logistic Regression; SVM: Support Vector Machine; NR: Not Reported; AF: atrial fibrillation; SP: stroke prognosis; DOAC: direct-acting oral anticoagulants; VKA: vitamin K antagonist; ESUS: embolic stroke of unknown source; UFO: Unfavorable Functional Outcome; END: early neurological deterioration; AT: anticoagulation therapy.

Table 3 summarizes study hypotheses, the ML models used, and cohort data, highlighting models like Extreme Gradient Boosting (XGB) and Random Forest (RF) for their robust performance.

Table 4 presents performance metrics like sensitivity, specificity, and AUROC, showcasing variability in predictive accuracy across studies. Table 5 focuses on stroke prognosis in AF patients, identifying models like XGB and RF for their predictive strengths.

Table 4. Performance metrics and outcomes by study hypotheses.

Author	Hypothesis	Sensitivity	Specificity	Cohort with Condition	PPV	NPV	AUROC (Internal Validation)	AUROC (External Validation)	C-Index	p-Value
Abedi et al. [13]	1-month mortality SP	NR	NR	NR	0.58	0.91	NR	0.81	NR	NR
	3-month mortality SP	NR	NR	NR	0.76	0.86	NR	0.79	NR	NR
	6-month mortality SP	NR	NR	NR	0.67	0.84	NR	0.79	NR	NR
	12-month mortality SP	NR	NR	NR	0.65	0.82	NR	0.77	NR	NR
	18-month mortality SP	NR	NR	NR	0.67	0.82	NR	0.80	NR	NR
	24-month mortality SP	NR	NR	NR	0.65	0.78	NR	0.76	NR	NR
Abujaber et al. [14]	90-day SP	NR	NR	NR	NR	NR	0.84	NR	NR	NR
Bernardini et al. [15]	SP in AF patients	NR	NR	NR	NR	NR	0.58 (0.54–0.63)	NR	NR	0.34
	SP in AF patients on DOAC	NR	NR	NR	NR	NR	0.61 (0.52–0.69)	NR	NR	0.53
	SP in AF patients on VKA	NR	NR	NR	NR	NR	0.57 (0.47–0.66)	NR	NR	0.41
Bisson et al. [33]	AF prediction in PS patients	NR	NR	NR	NR	NR	NR	NR	0.77	NR
Chen et al. [20]	Stroke prediction in AF patients	0.42	0.82	83	0.09	0.97	NR	0.67 (0.67–0.67)	NR	NR

Table 4. *Cont.*

Author	Hypothesis	Sensitivity	Specificity	Cohort with Condition	PPV	NPV	AUROC (Internal Validation)	AUROC (External Validation)	C-Index	*p*-Value
Colangelo et al. [21]	Early recurrent stroke prediction	NR	NR	41,975	NR	NR	NR	0.76 (0.74–0.77)	NR	<0.05
	Late recurrent stroke prediction	NR	NR	41,975	NR	NR	NR	0.60 (0.58–0.61)	NR	<0.05
	Long-term recurrent stroke prediction	NR	NR	41,975	NR	NR	NR	0.71 (0.69–0.72)	NR	<0.05
Gkantzios et al. [16]	NIHSS score prediction in acute stroke	NR	NR	413	NR	NR	NR	0.78	NR	
Goto et al. [17]	Major bleed prognosis in AF patients on VKA	0.79 (0.50–1.00)	0.78 (0.39–0.93)	4708	0.63	0.88	NR	0.75 (0.62–0.67)	NR	< 0.01
	SP in AF patients on VKA	0.85 (0.31–1.00)	0.53 (0.23–0.99)	4708	0.47	0.88	NR	0.70 (0.56–0.83)	NR	0.08
	All cause death prognosis in AF patients on VKA	0.63 (0.50–0.76)	0.65 (0.50–0.70)	4708	0.47	0.78	NR	0.61 (0.54–0.67)	NR	0.01
Han et al. [22]	Stroke prediction in AF patients	0.43	0.67	9836	NR	NR	NR	0.60	NR	0.64
Handy et al. [23]	Stroke prediction in AF patients	0.68 (0.62–0.74)	0.53 (0.51–0.55)	16,563	0.00	1.00	NR	0.61 (0.57–0.65)	NR	NR
Jeon et al. [10]	UFO SP in AF patients	NR	NR	3205	NR	NR	NR	0.87	NR	NR
	Mortality SP in AF patients	NR	NR	3205	NR	NR	NR	0.81	NR	NR
Jung et al. [18]	All cause death SP in AF patients	NR	NR	3090	NR	NR	NR	0.90 (0.89–0.91)	NR	NR
Jung et al. [24]	Stroke prediction in AF patients	NR	NR	62,226	NR	NR	NR	0.73 (0.72–0.73)	NR	NR
Choi et al. [35]	END SP in AF patients	NR	NR	2363	NR	NR	NR	0.77 (0.72–0.83)	NR	0.003
Li et al. [25]	2-year stroke prediction in AF patients	NR	NR	1864	NR	NR	NR	0.76	NR	NR
Lip et al. [26]	Stroke prediction in AF patients	NR	NR	6,457,412	NR	NR	NR	NR	0.89	NR
Lu et al. [27]	Stroke prediction in AF patients	NR	NR	9670	NR	NR	NR	0.69 (0.68–0.69)	NR	NR
	Major bleed prediction in AF patients	NR	NR	9670	NR	NR	NR	0.71 (0.70–0.72)	NR	NR
	Stroke prediction in AF patients without AT	NR	NR	9670	NR	NR	NR	0.67 (0.66–0.68)	NR	NR
	Major bleed prediction in AF patients without AT	NR	NR	9670	NR	NR	NR	0.63 (0.62–0.64)	NR	NR
Ming et al. [5]	Paroxysmal AF prediction in ESUS patients	0.6 (0.6–0.6)	0.82 (0.82–0.82)	157	NR	NR	NR	0.74 (0.74–0.74)	NR	NR
Nishi et al. [28]	Stroke prediction in AF patients without AT	0.74 (0.59–0.87)	0.54 (0.48–0.59)	1757	NR	NR	NR	0.72 (0.66–0.79)	NR	NR
Papadopoulou et al. [29]	Stroke prediction in AF patients	NR	NR	454,118	NR	NR	NR	0.63 (0.60–0.66)	NR	NR
Rebollo et al. [30]	180-day stroke prediction in AF patients with AT	NR	NR	8028	NR	NR	NR	0.87	NR	NR
	360-day stroke prediction in AF patients with AT	NR	NR	4628	NR	NR	NR	0.81	NR	NR
Sung et al. [34]	AF prediction in post-stroke patients	NR	NR	6096	NR	NR	NR	0.76	NR	NR
Truong et al. [31]	1-year stroke prediction in AF patients	0.52	0.67	523	0.04	0.98	NR	0.62 (0.55–0.69)	NR	0.0003
Zhang et al. [32]	Stroke prediction in AF patients	NR	NR	25	NR	NR	NR	0.98	NR	NR

Abbreviations: PPV: positive predictive value; NPV: negative predictive value; AUROC: Area Under Receiver Operating Characteristic; NR: Not Reported.

Table 5. Stroke prognosis in atrial fibrillation patients: performance metrics.

Author	Hypothesis	Sensitivity	Specificity	PPV	NPV	AUROC (Internal Validation)	AUROC (External Validation)	Highest Performing ML Model
Abedi et al. [13]	1-month mortality SP			0.58	0.91		0.81	XGB
	3-month mortality SP			0.76	0.86		0.79	XGB
	6-month mortality SP			0.67	0.84		0.79	XGB
	12-month mortality SP			0.65	0.82		0.77	XGB
	18-month mortality SP			0.67	0.82		0.80	XGB
	24-month mortality SP			0.65	0.78		0.76	XGB
Abujaber et al. [14]	90-day SP					0.84		RF
Bernardini et al. [15]	SP in AF patients					0.58		Naive Multi-Task NN
	SP in AF patients on DOAC					0.61		GBDT NN
	SP in AF patients on VKA					0.57		Naive Multi-Task NN
Gkantzios et al. [16]	NIHSS score prediction in acute stroke						0.78	RF
Goto et al. [17]	Major bleed prognosis in AF patients on VKA	0.79	0.78	0.63	0.88		0.75	NN
	SP in AF patients on VKA	0.85	0.53	0.47	0.88		0.70	NN
	All cause death prognosis in AF patients on VKA	0.63	0.65	0.47	0.78		0.61	NN
Jeon et al. [10]	UFO SP in AF patients						0.87	LightGBM
	Mortality SP in AF patients						0.81	LightGBM
Jung et al. [18]	All cause death SP in AF patients						0.90	XGB
Choi et al. [35]	END SP in AF patients						0.77	LightGBM

Abbreviations: AUROC: Area Under Receiver Operating Characteristic; ML: machine learning; SP: stroke prognosis; NN: neural network; XGB: Extreme Gradient Boosting; RF: Random Forest; GBDT: Gradient-Boosted Decision Trees; DNN: Deep Neural Network; VKA: vitamin K antagonist; DOAC: direct-acting oral anticoagulants.

Table 6 discusses stroke prediction in AF patients, with models like XGB achieving high AUROC scores. Lastly, Table 7 examines AF prediction in stroke patients, noting moderate success with models such as DNN and XGB.

Table 6. Stroke prediction in atrial fibrillation patients: performance metrics.

Author	Hypothesis	Sensitivity	Specificity	PPV	NPV	AUROC (External Validation)	C-Value	Highest Performing ML Model
Chen et al. [20]	Stroke prediction in AF patients	0.42	0.82	0.09	0.97	0.67		LightGBM
Colangelo et al. [21]	Early recurrent stroke prediction					0.76		RF
	Late recurrent stroke prediction					0.60		RF
	Long-term recurrent stroke prediction					0.71		RF
Han et al. [22]	Stroke prediction in AF patients	0.43	0.67			0.60		CNN
Handy et al. [23]	Stroke prediction in AF patients	0.68	0.53	0.00	1.00	0.61		XGBoost
Jung et al. [24]	Stroke prediction in AF patients					0.73		DNN
Li et al. [25]	2-year stroke_prediction in AF patients					0.76		LR + Wrapper
Lip et al. [26]	Stroke prediction in AF patients						0.89	LR
Lu et al. [27]	Stroke prediction in AF patients					0.69		Multilabel GBM
	Major bleed prediction in AF patients					0.71		Multilabel GBM
	Stroke prediction in AF patients without AT					0.67		Multilabel GBM
	Major bleed prediction in AF patients without AT					0.63		Multilabel GBM

Table 6. *Cont.*

Author	Hypothesis	Sensitivity	Specificity	PPV	NPV	AUROC (External Validation)	C-Value	Highest Performing ML Model
Nishi et al. [28]	Stroke prediction in AF patients without AT	0.74	0.54			0.72		CatBoost
Papadopoulou et al. [29]	Stroke prediction in AF patients					0.63		XGB
Rebollo et al. [30]	180-day stroke prediction in AF patients with AT					0.87		XGB
	360-day stroke prediction in AF patients with AT					0.81		XGB
Truong et al. [31]	1-year stroke prediction in AF patients	0.52	0.67	0.04	0.98	0.62		RF
Zhang et al. [32]	Stroke prediction in AF patients					0.98		RF

Abbreviations: AUROC: Area Under Receiver Operating Characteristic; C-value: Concordance value; ML: machine learning; LightGBM: Light Gradient-Boosting Machine; RF: Random Forest; SVM: Support Vector Machine.

Table 7. Atrial fibrillation prediction in stroke patients: performance metrics.

Author	Hypothesis	Sensitivity	Specificity	AUROC (External Validation)	C-Value	Highest Performing ML Model
Bisson et al. [33]	AF prediction in PS patients				0.77	DNN
Ming et al. [5]	Paroxysmal AF prediction in ESUS patients	0.6	0.82	0.74		SVM + Random oversampling
Sung et al. [34]	AF prediction in post-stroke patients			0.76		XGB

Abbreviations: AUROC: Area Under Receiver Operating Characteristic; ML: machine learning; DNN: Deep Neural Network; SVM: Support Vector Machine; XGB: Extreme Gradient Boosting.

These tables collectively demonstrate the potential and limitations of ML models in these clinical scenarios.

4.2. Overall Findings

A meta-analysis could not be performed for overall analyses. However, crude analyses involving seven ML models across 24 studies revealed an overall mean external AUROC of 0.75. Among these, XGB and RF models were frequently validated and excelled in performance. Neural networks (NN) and Gradient Boosting Machines (GBM) also showed notable performance but were less consistent. Logistic Regression (LR) and Support Vector Machines (SVM) demonstrated strong results in fewer studies, with LR achieving an AUROC of 0.83 over two hypotheses. Sensitivity and specificity averaged 0.63 and 0.67, while PPV and NPV were 0.47 and 0.88, highlighting room for improvement in clinical application. These results suggest that while ML models show moderate predictive capability, enhancements in model training and validation are necessary for achieving optimal clinical standards.

4.3. Stroke Prognosis in AF

Whilst the meta-analysis could not be performed, in eight studies involving 50,531 patients [10,13–19], the crude mean external AUROC for stroke prognosis in AF patients was 0.78, with a PPV of 0.62 and an NPV of 0.85. General AF populations showed a slightly higher AUROC of 0.80, while those on direct-acting oral anticoagulants (DOACs) or vitamin K antagonist (VKA) treatment had a lower external AUROC of 0.69. Among ML models, LightGBM and XGB models achieved high AUROCs of 0.82 and 0.80, respectively. In contrast, NN showed weaker performance, with external AUROCs of 0.69, possibly due to smaller datasets. These findings highlight the potential of specific ML models in accurately predicting stroke prognosis in AF patients, though variability remains, based on treatment and model choice.

4.4. Stroke Prediction in AF

The analysis of stroke prediction in AF encompassed 13 studies with a total of 7,094,555 participants [20–32], revealing a mean external AUROC of 0.73, indicative of average predictive performance. Notably, models like XGB and LR demonstrated superior AUROC scores of 0.77 and 0.83, respectively, while NN achieved a lower mean AUROC of 0.67. The variability in sensitivity (mean 0.56) and specificity (mean 0.64) highlights challenges in consistent predictive accuracy. Outliers such as Lip et al. and Zhang et al. achieved AUROCs of 0.89 and 0.98, suggesting the potential for exceptional predictive capability under specific conditions. These findings underscore the need for model refinement and standardization in clinical implementations.

4.5. AF Prediction in Stroke Patients

Across three studies involving 246,559 patients [5,33,34], ML models achieved an average AUROC of 0.75 for predicting AF in stroke patients. The models demonstrated moderate success, with a sensitivity of 0.6 and specificity of 0.82 reported in one study. Among the ML models evaluated, Deep Neural Networks (DNN) achieved a C-value of 0.77, XGB showed an AUROC of 0.76, and SVM achieved 0.74. These results suggest a promising yet varied potential for ML in predicting AF post-stroke, indicating the need for further research to enhance predictive accuracy and clinical relevance.

4.6. Meta-Analysis of Stroke Prognostic Accuracy in AF

The meta-analysis of stroke prognostic accuracy in patients with AF revealed a pooled AUROC of 0.79 (95% CI: 0.77–0.80) across all time periods, indicating good predictive accuracy. The forest plot (Figure 2) highlights trends in AUROC values over different durations, ranging from 0.71 to 0.82 for periods between 1 and 24 months. Despite the consistency in prognostic accuracy, there is notable variation in heterogeneity, with some time points (such as 3 and 12 months) exhibiting substantial heterogeneity ($I^2 = 94.7\%$ and 87.5%, respectively), while others show none ($I^2 = 0\%$). This suggests that while overall accuracy is stable, differences in study methodologies and populations contribute to variability in the results. The significant difference in heterogeneity between subgroups ($p < 0.0001$) highlights the importance of considering study context when interpreting pooled estimates.

Figure 2 presents a forest plot showing the prognostic accuracy of machine learning models for stroke prognosis in patients with atrial fibrillation (AF), measured using the Area Under the Receiver Operating Characteristic (AUROC) curve. The figure illustrates AUROC estimates across different follow-up durations, ranging from 1 to 24 months. Each horizontal line represents a study's AUROC estimate with confidence intervals, while the diamond at the bottom indicates the pooled estimate. The results suggest that machine learning models provide a good level of prognostic accuracy, with AUROC values ranging from 0.71 to 0.82. Notably, there is variability in heterogeneity among studies, with some time points showing substantial heterogeneity.

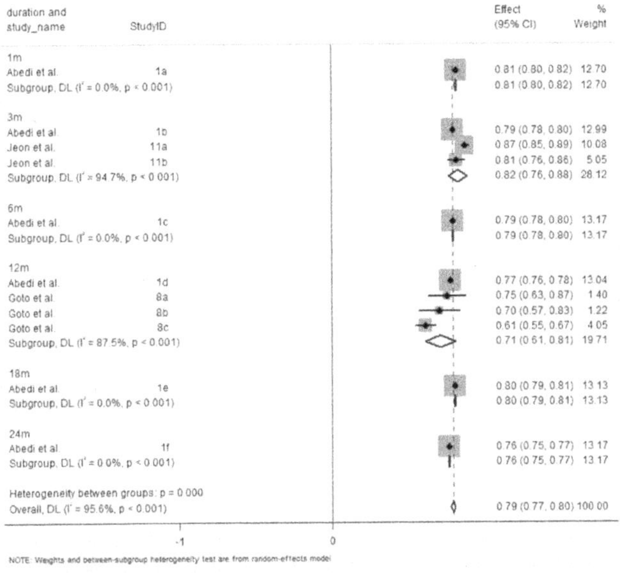

Figure 2. Forest plot of stroke prognosis accuracy in atrial fibrillation patients: AUROC estimates by follow-up duration. Abbreviations: AUROC = Area Under the Receiver Operating Characteristic; DL = DerSimonian and Laird Method; CI = confidence interval.

4.7. Meta-Analysis of Stroke Prediction Accuracy in AF

The meta-analysis of stroke prediction accuracy in AF revealed that the pooled AUROC across all time periods is 0.68 (95% CI: 0.63–0.72), indicating moderate predictive accuracy. The forest plot (Figure 3) illustrates the prediction accuracy based on AUROC values across various follow-up durations in patients with AF. Short-term predictions, such as at 3 months, generally demonstrate higher predictive accuracy, with AUROC values around 0.76 and minimal heterogeneity ($I^2 = 0\%$), indicating consistent results across studies. However, as the follow-up duration extends beyond 12 months, the AUROC tends to decline, with pooled estimates ranging from 0.66 to 0.71 and varying degrees of heterogeneity. While short-term predictions outperform those for longer durations, significant variation in heterogeneity between subgroups ($p < 0.0001$) underscores the impact of study design and population differences. This highlights the importance of considering study-specific factors when interpreting stroke prediction accuracy in AF patients over different time periods.

Figure 3 displays a forest plot of stroke prediction accuracy in patients with atrial fibrillation, evaluated using AUROC scores. The figure summarizes the predictive accuracy of machine learning models across various follow-up durations, with AUROC values spanning 0.66 to 0.76. Short-term predictions, particularly at 3 months, demonstrate higher predictive accuracy and minimal heterogeneity, indicating consistency across studies. Conversely, predictions for longer durations show decreased AUROC values and increased heterogeneity, highlighting the challenges in maintaining predictive accuracy over extended time frames. This figure emphasizes the superior performance of machine learning models in short-term stroke prediction and the influence of study design and duration on predictive accuracy.

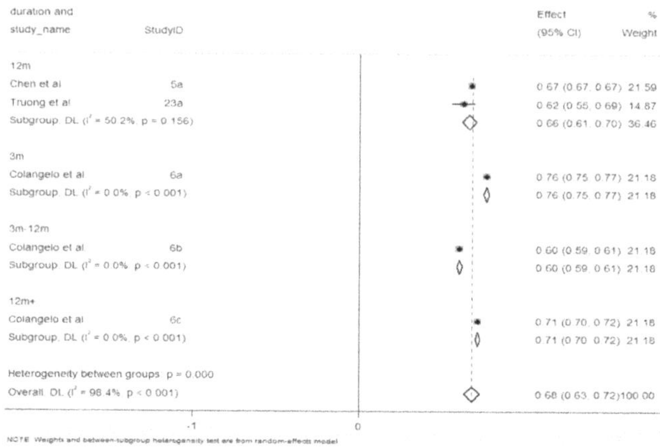

Figure 3. Forest Plot of Stroke Prediction Accuracy in Atrial Fibrillation Patients: AUROC Estimates by Follow-Up Duration. Abbreviations: AUROC = Area Under the Receiver Operating Characteristic; DL = DerSimonian and Laird Method; CI = confidence interval.

4.8. Feature Analysis and SHAP Values

To enhance the interpretability and explainability of the machine learning models, we applied SHAP values to assess the influence of individual features on model predictions. Key features such as age, NIHSS score, atrial fibrillation history, and anticoagulation use emerged as the most significant predictors. For example, higher NIHSS scores were strongly associated with poorer stroke outcomes [36], while anticoagulation therapy showed a positive influence on prognosis in patients with atrial fibrillation [37,38].

The SHAP-based feature analysis provided deeper insights into how specific clinical and demographic features impact the model's predictive performance, helping to clarify the underlying mechanics of the models [39]. This transparency enhances the potential clinical application of the models by providing explainable insights into the decision-making process. Detailed SHAP values for the key features are presented in Table 8.

Table 8. Key features and SHAP values in machine learning models for stroke prognosis and atrial fibrillation prediction.

Author	Total Number of Features	Features	Key Features and Their SHAP Values/Mean Absolute SHAP Values (If Included)
Abedi et al. [13]	37	Age, sex, BMI, DBP, SBP, creatinine, smoking, Hb, HbA1c, LDL, PLT, NIHSS score, AF, alcohol, TPA therapy, CHF, CKD, CLD, COPD, cirrhosis, diabetes, DLD, epilepsy, ESRD, thrombophilia, HTN, MI, migraine, mood disorder, cancer, PVD, PFO, rheumatological disease, syncope, thrombectomy, HD, stroke	1. Age 2. Hb 3. BMI 4. Thrombectomy 5. PLT 6. LDL 7. Creatinine 8. CHF
Abujaber et al. [14]	19	NIHSS score, age, pneumonia, pre-stroke mRS, LOS, UTI, stroke-type, sex, ethnicity, BMI, diabetes, HTN, DLD, stroke, AF, CAD, CHF, smoking, admission location	1. Pneumonia 2. LOS 3. Stroke type 4. NIHSS score

Table 8. *Cont.*

Author	Total Number of Features	Features	Key Features and Their SHAP Values/Mean Absolute SHAP Values (If Included)
Bernardini et al. [15]	68	Sex, age, paroxysmal AF, HTN, diabetes, CHF, CAD, PVD, COPD, cancer, dementia, anemia, falls, previous bleeds, stroke, eGFR, antiplatelet drugs, CHAD2DS2VASc score, HAS-BLED score, DOACs, VKAs, creatinine, RBC	1. Hb (−0.8–1.4) 2. Creatinine (−0.8–0.5) 3. RBC (−0.5–0.5) 4. eGFR (−0.6–0.6) 5. BMI (−0.5–0.6) 6. PLT (−1.2–1) 7. Age (−0.5–0.5) 8. Previous bleeds (0–1.9)
Bisson et al. [33]	16	Age, sex, HTN, diabetes, CHF, PVD, CHAD2DS2VASc score, SE, CAD, obesity, CKD, CLD, anemia, COPD, cancer, inflammatory disease, alcohol, thyroid disease, DLD, valvular disease, smoking	1. Age 2. HTN 3. CHF 4. CKD 5. Anemia 6. COPD 7. Thyroid disease 8. DLD
Chen et al. [20]	12	CKD, age, HTN, AF treatment, elevated AST, diuretic drugs, sex, diabetes, prior stroke/TIA/SE, enlarged LA, AF, mitral valve	1. CKD (0.049) 2. Age (0.046) 3. HTN (0.045) 4. AF treatment (0.04) 5. Elevated AST (0.035) 6. Diuretic drugs (0.032)
Colangelo et al. [21]	13	Previous stroke, Barthel index, AF, DLD, age, diabetes, sex, glycemia, BMI, HTN, cholesterol, smoking, alcohol	1. Previous stroke 2. Barthel index 3. AF 4. DLD 5. Age 6. Diabetes 7. Sex
Gkantzios et al. [16]	32	Age, sex, stroke type, SBP, glucose, CRP, ESR, HTN, smoking, diabetes, DLD, AF, previous stroke, previous MI, CHF, heart valve history, alcohol, antiplatelet drugs, anticoagulant drugs, 72 h data (SBP, intubation, cholesterol, LDL, T3 thyroid hormone, respiratory infection), stroke localization, NIHSS score	1. Respiratory infection (−0.3–0.21) 2. Intubation (−1.5–0.35) 3. Smoking (−0.22–0.1) 4. SBP (−0.15–0.05) 5. NIHSS score (−0.28–0.18) 6. Age (−0.15–0.08) 7. Stroke localization (−0.28–0.9)
Goto et al. [17]	10	Sex, age, DMI, LVEF, AF type, CHF, CAD, ACS, CHAD2DS2VASc score, HAS-BLED score	N/A
Han et al. [22]	11	Age, sex, MI, HTN, CHF, diabetes, PVD, CAD, Charlson comorbidity index, Selim comorbidity score, CHAD2DS2VASc score	N/A

Table 8. *Cont.*

Author	Total Number of Features	Features	Key Features and Their SHAP Values/Mean Absolute SHAP Values (If Included)
Handy et al. [23]	N/A	Age, age at AF diagnosis, follow-up time post AF diagnosis, sex, ethnicity	N/A
Jeon et al. [10]	43	Age, sex, BMI, initial clinical status (DBP, SBP, pulse rate, NIHSS score), pre-stroke mRS, CHAD2DS2VASc score, CHADS2 score, stroke onset time, stroke, AF, CHF, HTN, diabetes, CAD, PVD, imaging findings, laboratory findings (WBC, Hb, PLT, CRP, glucose, fasting glucose, HbA1c, cholesterol, TG, HDL, LDL, AST, ALT, ALP, bilirubin, uric acid, creatinine, CrCl, fibrinogen, homocysteine, CK-MB, FFA, thrombolytic treatment	1. NIHSS score (0.6) 2. Imaging findings—DWI lesion pattern (0.51) 3. Pre-stroke mRS (0.49) 4. CRP (0.49) 5. Age (0.42)
Jung et al. [18]	N/A	N/A	N/A
Jung et al. [24]	58	CHAD2DS2VASc score, age, sex, insurance type, income level, CHF, HTN, diabetes, PVD, surgical history, HLD, dorsopathies, vision loss, thyroid disease, arrhythmias, obesity, gout, prostatic hyperplasia CLD, COPD, gynecological problems, osteoporosis, CKD, PE, hearing loss, gallbladder/biliary tract/pancreatic disorders, hemorrhoids, diverticulitis, arthritis (OA, RA), heart valve disease, neuropathy, dizziness, incontinence, ureter stones, anemia, psoriasis, headache, Parkinson's disease, cancer, allergy, chronic gastritis, sexual dysfunction, insomnia, hypotension, depression, somatoform disorder, dementia, anxiety, afloqualone drugs, smoking, exercise, alcohol	1. Age 2. Sex 3. Thyroid disease 4. Arrhythmias 5. Hemolytic anemia 6. Cancer 7. Hemorrhoids 8. Diabetes 9. HTN 10. CKD 11. CHF 12. HLD 13. PVD 14. Gout 15. PE 16. Chronic gastritis
Choi et al. [35]	23	Fasting glucose, NIHSS score, mRS, initial glucose, lateralization of ischemic lesion, QRS axis, ALP, homocysteine, SVS, intracranial atherosclerosis, FDP = fibrin degradation product, LA diameter, DBP, D-dimer, hemorrhagic transformation, AST, hematocrit, uric acid, cholesterol, T-axis, bilirubin, LDL	1. Fasting glucose (0.14) 2. NIHSS score (0.125) 3. mRS (0.07) 4. Initial glucose (0.06) 5. Lateralization of ischemic lesion (0.045) 6. QRS axis (0.045) 7. ALP (0.044)
Li et al. [25]	167	N/A	N/A
Lip et al. [26]	19	Sex, age, CHF, HTN, diabetes, stroke, PVD, valvular disease, CAD, CKD, COPD, major bleeding, alcohol, alcohol disorders, inflammatory diseases, lipid disorders, CHADS2, CHAD2DS2VASc	N/A

Table 8. *Cont.*

Author	Total Number of Features	Features	Key Features and Their SHAP Values/Mean Absolute SHAP Values (If Included)
Lu et al. [27]	43	Previous stroke/TIA/embolism, age, cardiology admission, year of cohort entry, Hb, length of hospital stay, rate control drugs, bleeding history, eGFR, anti-arrhythmic drugs, gastric protective agents, diabetes, PE, AF, antiplatelet therapy, HTN, heart failure, DOACs, depression, falls	1. Previous stroke/TIA/embolism (0.20) 2. Age (0.135) 3. Cardiology admission (0.11) 4. Year of cohort entry (0.9) 5. Hb (0.8) 6. Length of hospital stay (0.06)
Ming et al. [5]	N/A	Peak mitral A-wave velocity, HDL, age, heart rate, LA volume, eGFR, SBP, mRS, cholesterol, sex, LDL, body surface area, triglycerides, BMI, DBP, height, NIHSS score	1. Peak mitral A-wave velocity (−0.5–1) 2. HDL (−0.4–1.5) 3. Age (−0.7–0.5) 4. Heart rate (−0.65–0.55) 5. LA volume (−0.5–0.8) 6. eGFR (−0.4–0.5) 7. SBP (−0.5–0.4)
Nishi et al. [28]	14	Age, height, weight, stroke, HTN, catheter ablation, antiplatelet drug, antiarrhythmic drug, creatinine, urea, mitral/aortic regurgitation, LV mass, relative wall thickness	N/A
Papadopoulou et al. [29]	99	Age, AF, height, weight, waist circumference, sex, no. of medications taken, albumin, aspirin, SBP, pulse rate, bilirubin, urate, seated height, FEV1, testosterone, cancer, hip circumference, triglycerides	1. Age (0.49) 2. AF (0.23) 3. Height (0.12) 4. Weight (0.1) 5. Waist circumference (0.08) 6. Sex (0.07) 7. No. of medications taken (0.06)
Rebollo et al. [30]	N/A	Previous stroke, age, antithrombotic drugs, HTN, BMI, lipid modifying agents, extrapyramidal disorders, gastritis/duodenitis, creatinine, corticosteroids, drugs for peptic ulcer and gastro-esophageal reflux disease, glucose	1. Previous stroke 2. Age 3. Antithrombotic drugs 4. HTN 5. BMI 6. Lipid modifying agents 7. Extrapyramidal disorders
Sung et al. [34]	20	Age, triglyceride, PLT, numbness, heart rate, pulse pressure, sex, creatinine, diabetes, urea, BMI	1. Age (0.5) 2. Triglyceride (0.23) 3. PLT (0.21) 4. Numbness (0.2) 5. Heart rate (0.18)

Table 8. *Cont.*

Author	Total Number of Features	Features	Key Features and Their SHAP Values/Mean Absolute SHAP Values (If Included)
Truong et al. [31]	53	Age, sex, ethnicity, year, geographical region, household median income, percentage of household with education level below high school, HTN, CHF, diabetes, previous stroke, PVD, previous bleeding, CKD, CLD, alcohol, COPD, hematological disorders, dementia, depression, thrombocytopenia, peptic ulcer disease, cancer, AFib, cancer type, cancer stage, tumor grade, active cancer status, antiplatelet drugs, anti-inflammatory drugs, ACE inhibitors, ARBs, calcium channel blockers, beta blockers, antiarrhythmic medications, diuretics, statins, pump proton inhibitors, serotonin reuptake inhibitors	1. No high school education 2. Household medium income 3. Time from cancer diagnosis 4. Previous stroke 5. ARBs 6. Anti-inflammatory drugs 7. Hematological disorder
Zhang et al. [32]	N/A	N/A	N/A

Abbreviations: BMI: body mass index; BP: blood pressure; DBP: diastolic blood pressure; SBP: systolic blood pressure; Hb: hemoglobin; LDL: low-density lipoprotein; HDL: high-density lipoprotein; PLT: platelet count; tPA: alteplase; CHF: congestive heart failure; CKD: chronic kidney disease; CLD: chronic liver disease; COPD: chronic obstructive pulmonary disease; DLD: dyslipidemia; ESRD: end stage renal disease; HTN: hypertension; MI: myocardial Infarction; PVD: peripheral vascular disease; PFO: patent foramen ovale; HD: heart disease; mRS: modified Rankin Scale; LOS: length of stay; UTI: urinary tract infection; CAD: coronary artery disease; eGFR: estimated glomerular filtration rate; CHAD2DS2VASc: congestive heart failure, hypertension, age ≥75 (doubled), diabetes, stroke (doubled), vascular disease, age 65 to 74, and sex category (female); HAS-BLED: hypertension, abnormal renal/liver function, stroke, bleeding history or predisposition, labile INR, elderly, drugs/alcohol concomitantly; FDP: fibrin degradation product.

5. Discussion

This meta-analysis provides insights into the current status of ML applications in managing stroke and AF. While ML models show potential in prognosis and prediction, they currently fall short of clinical standards required for widespread clinical use. This study synthesizes findings from recent studies to highlight the performance of ML models and the need for standardized methodologies to enhance their clinical utility. The average AUROC scores of 0.68 for stroke prediction and 0.79 for stroke prognosis suggest moderate accuracy, with models performing better in short-term predictions (3 months, AUROC 0.76) compared to long-term (12+ months, AUROC 0.71). A high heterogeneity is observed across studies, indicating significant variability in the results. The variability in model performance, particularly among neural networks, underscores the importance of consistent application and parameter settings. Neural networks can be highly sensitive to the choice of parameters and the quality of the training data, leading to variability in their performance [19,40,41]. Differences in data inclusion and the clinical risk factors used across studies demand a more standardized approach to enhance predictive accuracy [40]. The inconsistency in sensitivity, specificity, PPV, and NPV metrics across studies highlights the necessity for a more comprehensive analysis of these parameters in future research [41].

Our analysis indicates that models such as LightGBM and XGB demonstrate optimal performance in specific contexts; however, their efficacy is influenced by the treatment regimens administered, particularly DOACs and VKAs [10,13–19]. Specifically, we observed that the AUROC values were lower for patients receiving DOAC and VKA therapies, which may point to a reduced predictive capability of these models under such conditions. This variability in performance could be attributed to differences in dosing and patient response to anticoagulation therapies [42], which can significantly impact stroke risk and, consequently, model predictions. For instance, the pharmacokinetics of DOACs and VKAs can vary widely among individuals due to factors like age, renal function, and concomitant medications, potentially leading to discrepancies in stroke risk profiles [43]. To enhance predictive model accuracy, finer stratification of patient data is crucial, incorporating treatment types alongside individual characteristics and clinical contexts. This tailored approach will better address the complexities of anticoagulation therapy and improve stroke prevention outcomes in AF.

For broader clinical utility, ML models must consistently demonstrate high sensitivity, specificity, PPV, and NPV. The current inconsistency in these metrics across studies points to the need for a more comprehensive analysis of these parameters in future research. For stroke prediction, ML models achieved an average AUROC of 0.73, indicating average performance (Figure 4) [20–32]. While XGB and LR models showed higher AUROC scores (0.77 and 0.83, respectively), the variation in clinical risk factors across studies suggests a need for standardized approaches to data inclusion and analysis to enhance predictive accuracy.

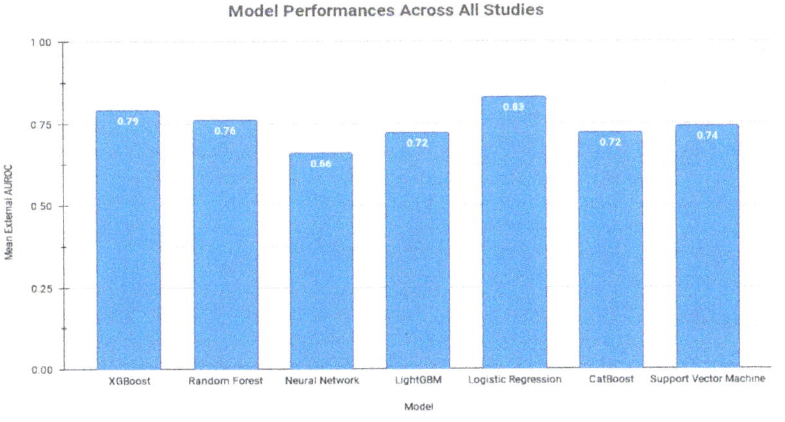

Figure 4. Distribution of mean external AUROC scores for machine learning models evaluated across multiple studies. Abbreviations: AUROC: Area Under the Receiver Operating Characteristic; XGBoost: Extreme Gradient Boosting; LightGBM: Light Gradient-Boosting Machine; CatBoost: Categorical Boosting.

ML models exhibited an average AUROC of 0.75 for AF prediction in stroke patients, with DNN achieving the highest C-value of 0.77 [5,33,34]. The performance of these models improved with larger datasets, as seen in studies with significantly larger training cohorts. Potential biases such as selection, publication, and reporting biases must be acknowledged. Addressing these biases through robust methodological frameworks and conducting publication bias assessments, where possible, will improve accuracy and generalizability.

In our analysis, we observed that the architecture of machine learning models significantly influences their predictive performance in stroke prognosis and prediction in AF patients. Particularly, models using XGB and RF demonstrated superior performance compared to other methods. XGB, a decision-tree-based ensemble model, leverages gradient boosting principles to optimize predictive accuracy through sequential model training [44]. Its ability to handle complex data structures and interactions between features contributes to its high AUROC scores across studies. Similarly, RF, which constructs multiple decision trees during training and outputs the mode of their predictions, excels in managing overfitting and improving prediction robustness [45]. The selection of features, such as age, NIHSS score, and anticoagulation use, was crucial in enhancing these models' performance, as indicated by the SHAP analysis. These architectural features and strategic feature selections underscore the potential of XGB and RF in clinical applications, providing a framework for future model development and refinement.

This study has several limitations. Firstly, the scarcity of studies in this emerging field weakens the robustness of the meta-analysis. Many available studies lack critical clinical indicators, such as sensitivity, specificity, PPV, and NPV, limiting the representation of mean clinical scores and predictive power. Furthermore, variability in the application of machine learning models, such as neural networks, across studies may affect their performance, complicating comparisons of their prognostic capabilities. Inherent challenges of meta-analysis also arise, including heterogeneity from pooling data across diverse study designs, populations, and methodologies. Although addressed with I^2 statistics and subgroup analyses, this variability still limits the generalizability of the findings. Additionally, potential biases, such as selection, publication, and reporting bias, could skew the overall conclusions. We aimed to assess publication bias through funnel plots and Egger's test, but limited studies and insufficient data hindered this effort. Finally, the limitations of the machine learning models themselves—such as risks of overfitting, underfitting, and the need for external validation—further restrict the applicability of our findings.

This meta-analysis underscores the promise of ML models in enhancing prognostic accuracy for stroke outcomes in patients with AF. Nonetheless, the current evidence is hindered by significant variability in model performance and methodological inconsistencies across studies. Future research must prioritize the standardization of ML applications, ensuring consistent parameter settings and data preprocessing protocols. Larger, multicenter studies encompassing diverse patient populations are crucial for improving the generalizability of findings [46]. Moreover, investigating hybrid models that combine traditional clinical risk factors with ML predictions could yield more accurate and clinically relevant tools [47]. Validation in real-world settings remains imperative to ascertain practical utility [48]. While ML models hold substantial potential to stratify AF patients based on risk, their integration into clinical practice should proceed with caution [49]. Clinicians must receive training to effectively interpret ML outputs, ensuring that these models complement, rather than replace, clinical judgment [50]. By harmonizing the predictive power of ML with clinician expertise, we can advance toward more precise and individualized stroke care [51]. Integrating comprehensive datasets that reflect diverse populations and treatment regimens will further facilitate robust model validation and refinement [52].

6. Conclusions

In conclusion, although ML models show considerable promise for improving stroke management and AF prediction, further research is needed to refine these models and validate their use in diverse clinical settings [8]. Despite this potential, current models do not yet meet optimal clinical standards for prediction and diagnosis. This necessitates further development and rigorous validation of ML models using diverse and representative populations. Short-term predictions show greater reliability than long-term ones, highlighting an area for improvement. Future studies should focus on standardizing methodologies, improving data quality, and exploring hybrid models that integrate traditional and ML-based risk assessments to achieve optimal clinical outcomes [52]. While ML offers valuable insights and tools for stroke management, its integration into clinical practice requires careful consideration, ongoing validation, and a commitment to enhancing model performance to meet the complex demands of stroke prognostication and prediction. By overcoming these challenges, ML models can enhance personalized stroke care for AF patients, improving outcomes through precision medicine.

Supplementary Materials: The following supporting information can be downloaded at: https://www.mdpi.com/article/10.3390/diagnostics14212391/s1, Search Strategy; Table S1: Preferred Reporting Items for Systematic Reviews and Meta-Analyses (PRISMA) 2020 checklist. Table S2: Meta-analysis of Observational Studies in Epidemiology (MOOSE) checklist. Table S3: STARD-2015 Checklist for Diagnostic Accuracy Studies. Table S4: Methodological quality assessment of included studies using the modified Jadad scale and assessment of funding bias.

Author Contributions: S.M.M.B. conceived the study, contributed to the planning, drafting, and revision of the manuscript, and supervised the student. S.M.M.B. encouraged B.G. to investigate and supervised the findings of this work. B.G. and S.M.M.B. contributed to the data collection and analyses. B.G. and S.M.M.B. wrote the first draft of this paper. All authors contributed to the revision of the manuscript. All authors have read and agreed to the published version of the manuscript.

Funding: We acknowledge the financial support received from the Grant-in-Aid for Scientific Research (KAKENHI) (PI: SMMB) by the Japan Society for the Promotion of Science (JSPS), Japanese Ministry of Education, Culture, Sports, Science and Technology (MEXT), Japan (Grant ID: P23712). We also extend our gratitude for the JSPS International Fellowship, supported by MEXT and the Australian Academy of Science, awarded to SMMB for the period 2023–2025.

Institutional Review Board Statement: As this study involves secondary analysis of published data, no additional ethical approval is required. However, all included studies will have been conducted in accordance with ethical guidelines and standards.

Informed Consent Statement: Not applicable.

Data Availability Statement: The original contributions presented in the study are included in the article and online supplemental information, and further inquiries can be directed to the corresponding author.

Conflicts of Interest: SMMB reports leadership or fiduciary roles in various organizations, including the *National Cerebral and Cardiovascular Center (Osaka, Japan)* as Visiting Director (2023–2025); *Rotary District 9675 (Sydney, Australia)* as District Chair for Diversity, Equity, and Inclusion; the *Global Health and Migration Hub Community, Global Health Hub Germany (Berlin, Germany)* as Chair, Founding Member, and Manager; and editorial board memberships at *PLOS One, BMC Neurology, Frontiers in Neurology, Frontiers in Stroke, Frontiers in Public Health, Journal of Aging Research, Neurology International, Diagnostics, and BMC Medical Research Methodology*. Additionally, SMMB serves as a Member of the College of Reviewers for the *Canadian Institutes of Health Research (CIHR), Government of Canada*; Director of Research for the *World Headache Society (Bengaluru, India)*; a member of the Scientific Review Committee at *Cardiff University Biobank (Cardiff, UK)*; and as an Expert Adviser/Reviewer for the Cariplo Foundation (Milan, Italy). These roles are unrelated to the submitted work. BG reports no conflicts of interest. The funding body had no influence on the study design, data collection, analysis, interpretation of findings, or manuscript preparation. The content is solely the responsibility of the authors and does not necessarily represent the official views of the affiliated or funding organizations.

References

1. Prust, M.L.; Forman, R.; Ovbiagele, B. Addressing disparities in the global epidemiology of stroke. *Nat. Rev. Neurol.* **2024**, *20*, 207–221. [CrossRef] [PubMed]
2. Steinmetz, J.D.; Seeher, K.M.; Schiess, N.; Nichols, E.; Cao, B.; Servili, C.; Cavallera, V.; Cousin, E.; Hagins, H.; Moberg, M.E.; et al. Global, regional, and national burden of disorders affecting the nervous system, 1990–2021: A systematic analysis for the Global Burden of Disease Study 2021. *Lancet Neurol.* **2024**, *23*, 344–381. [CrossRef] [PubMed]
3. Feigin, V.L.; Abate, M.D.; Abate, Y.H.; Abd ElHafeez, S.; Abd-Allah, F.; Abdelalim, A.; Abdelkader, A.; Abdelmasseh, M.; Abd-Elsalam, S.; Abdi, P.; et al. Global, regional, and national burden of stroke and its risk factors, 1990–2021: A systematic analysis for the Global Burden of Disease Study 2021. *Lancet Neurol.* **2024**, *23*, 973–1003. [CrossRef] [PubMed]
4. Escudero-Martínez, I.; Morales-Caba, L.; Segura, T. Atrial fibrillation and stroke: A review and new insights. *Trends Cardiovasc. Med.* **2023**, *33*, 23–29. [CrossRef] [PubMed]
5. Ming, C.; Lee, G.J.W.; Teo, Y.H.; Teo, Y.N.; Toh, E.M.S.; Li, T.Y.W.; Guo, C.Y.; Ding, J.; Zhou, X.; Teoh, H.L.; et al. Machine Learning Modeling to Predict Atrial Fibrillation Detection in Embolic Stroke of Undetermined Source Patients. *J. Pers. Med.* **2024**, *14*, 534. [CrossRef]
6. Patel, J.; Bhaskar, S.M.M. Diagnosis and Management of Atrial Fibrillation in Acute Ischemic Stroke in the Setting of Reperfusion Therapy: Insights and Strategies for Optimized Care. *J. Cardiovasc. Dev. Dis.* **2023**, *10*, 458. [CrossRef]
7. Patel, J.; Bhaskar, S.M.M. Atrial Fibrillation and Reperfusion Therapy in Acute Ischaemic Stroke Patients: Prevalence and Outcomes—A Comprehensive Systematic Review and Meta-Analysis. *Neurol. Int.* **2023**, *15*, 1014–1043. [CrossRef]
8. Goh, B.; Bhaskar, S.M.M. The role of artificial intelligence in optimizing management of atrial fibrillation in acute ischemic stroke. *Ann. N. Y. Acad. Sci.* **2024**, 1–12. [CrossRef]
9. Shen, H.; Huasen, B.B.; Killingsworth, M.C.; Bhaskar, S.M.M. Introducing the Futile Recanalization Prediction Score (FRPS): A Novel Approach to Predict and Mitigate Ineffective Recanalization after Endovascular Treatment of Acute Ischemic Stroke. *Neurol. Int.* **2024**, *16*, 605–619. [CrossRef]

10. Jeon, E.T.; Jung, S.J.; Yeo, T.Y.; Seo, W.K.; Jung, J.M. Predicting short-term outcomes in atrial-fibrillation-related stroke using machine learning. *Front. Neurol.* **2023**, *14*, 1243700. [CrossRef]
11. Çorbacıoğlu, Ş.K.; Aksel, G. Receiver operating characteristic curve analysis in diagnostic accuracy studies: A guide to interpreting the area under the curve value. *Turk. J. Emerg. Med.* **2023**, *23*, 195–198. [CrossRef] [PubMed]
12. Power, M.; Fell, G.; Wright, M. Principles for high-quality, high-value testing. *Evid. Based Med.* **2013**, *18*, 5–10. [CrossRef] [PubMed]
13. Abedi, V.; Avula, V.; Razavi, S.M.; Bavishi, S.; Chaudhary, D.; Shahjouei, S.; Wang, M.; Griessenauer, C.J.; Li, J.; Zand, R. Predicting short and long-term mortality after acute ischemic stroke using EHR. *J. Neurol. Sci.* **2021**, *427*, 117560. [CrossRef] [PubMed]
14. Abujaber, A.A.; Alkhawaldeh, I.M.; Imam, Y.; Nashwan, A.J.; Akhtar, N.; Own, A.; Tarawneh, A.S.; Hassanat, A.B. Predicting 90-day prognosis for patients with stroke: A machine learning approach. *Front. Neurol.* **2023**, *14*, 1270767. [CrossRef]
15. Bernardini, A.; Bindini, L.; Antonucci, E.; Berteotti, M.; Giusti, B.; Testa, S.; Palareti, G.; Poli, D.; Frasconi, P.; Marcucci, R. Machine learning approach for prediction of outcomes in anticoagulated patients with atrial fibrillation. *Int. J. Cardiol.* **2024**, *407*, 132088. [CrossRef]
16. Gkantzios, A.; Kokkotis, C.; Tsiptsios, D.; Moustakidis, S.; Gkartzonika, E.; Avramidis, T.; Tripsianis, G.; Iliopoulos, I.; Aggelousis, N.; Vadikolias, K. From Admission to Discharge: Predicting National Institutes of Health Stroke Scale Progression in Stroke Patients Using Biomarkers and Explainable Machine Learning. *J. Pers. Med.* **2023**, *13*, 1375. [CrossRef]
17. Goto, S.; Goto, S.; Pieper, K.S.; Bassand, J.P.; Camm, A.J.; Fitzmaurice, D.A.; Goldhaber, S.Z.; Haas, S.; Parkhomenko, A.; Oto, A.; et al. New artificial intelligence prediction model using serial prothrombin time international normalized ratio measurements in atrial fibrillation patients on vitamin K antagonists: GARFIELD-AF. *Eur. Heart J. Cardiovasc. Pharmacother.* **2020**, *6*, 301–309. [CrossRef]
18. Jung, J.M.; Jeon, E.T. Outcome predictions using machine learning in atrial fibrillation-related stroke. *Circ. Conf. Am. Hearts Assoc.* **2021**, *144*, A11932. [CrossRef]
19. Kim, S.H.; Jeon, E.T.; Yu, S.; Oh, K.; Kim, C.K.; Song, T.J.; Kim, Y.J.; Heo, S.H.; Park, K.Y.; Kim, J.M.; et al. Interpretable machine learning for early neurological deterioration prediction in atrial fibrillation-related stroke. *Sci. Rep.* **2021**, *11*, 20610. [CrossRef]
20. Chen, Y.; Gue, Y.; Calvert, P.; Gupta, D.; McDowell, G.; Azariah, J.L.; Namboodiri, N.; Bucci, T.; Jabir, A.; Tse, H.F.; et al. Predicting stroke in Asian patients with atrial fibrillation using machine learning: A report from the KERALA-AF registry, with external validation in the APHRS-AF registry. *Curr. Probl. Cardiol.* **2024**, *49*, 102456. [CrossRef]
21. Colangelo, G.; Ribo, M.; Montiel, E.; Dominguez, D.; Olive-Gadea, M.; Muchada, M.; Garcia-Tornel, A.; Requena, M.; Pagola, J.; Juega, J.; et al. PRERISK: A Personalized, Artificial Intelligence-Based and Statistically-Based Stroke Recurrence Predictor for Recurrent Stroke. *Stroke* **2024**, *55*, 1200–1209. [CrossRef] [PubMed]
22. Han, L.; Askari, M.; Altman, R.B.; Schmitt, S.K.; Fan, J.; Bentley, J.P.; Narayan, S.M.; Turakhia, M.P. Atrial fibrillation burden signature and near-term prediction of stroke: A machine learning analysis. *Circ. Cardiovasc. Qual. Outcomes* **2019**, *12*, e005595. [CrossRef] [PubMed]
23. Handy, A.; Wood, A.; Sudlow, C.; Tomlinson, C.; Kee, F.; Thygesen, J.H.; Mohammad, M.; Sofat, R.; Dobson, R.; Ip, S.; et al. A nationwide deep learning pipeline to predict stroke and COVID-19 death in atrial fibrillation. *Medrxiv* **2021**, *21*. [CrossRef]
24. Jung, S.; Song, M.K.; Lee, E.; Bae, S.; Kim, Y.Y.; Lee, D.; Lee, M.J.; Yoo, S. Predicting Ischemic Stroke in Patients with Atrial Fibrillation Using Machine Learning. *Front. Biosci. Landmark Ed.* **2022**, *27*, 80. [CrossRef]
25. Li, X.; Liu, H.; Du, X.; Zhang, P.; Hu, G.; Xie, G.; Guo, S.; Xu, M.; Xie, X. Integrated Machine Learning Approaches for Predicting Ischemic Stroke and Thromboembolism in Atrial Fibrillation. *AMIA Annu. Symp. Proc.* **2016**, *2016*, 799–807.
26. Lip, G.Y.H.; Tran, G.; Genaidy, A.; Marroquin, P.; Estes, C.; Landsheft, J. Improving dynamic stroke risk prediction in non-Anticoagulated patients with and without atrial fibrillation: Comparing common clinical risk scores and machine learning algorithms. *Eur. Heart J. Qual. Care Clin. Outcomes* **2022**, *8*, 548–556. [CrossRef]
27. Lu, J.; Hutchens, R.; Hung, J.; Bennamoun, M.; McQuillan, B.; Briffa, T.; Sohel, F.; Murray, K.; Stewart, J.; Chow, B.; et al. Performance of multilabel machine learning models and risk stratification schemas for predicting stroke and bleeding risk in patients with non-valvular atrial fibrillation. *Comput. Biol. Med.* **2022**, *150*, 106126. [CrossRef]
28. Nishi, H.; Oishi, N.; Ogawa, H.; Natsue, K.; Doi, K.; Kawakami, O.; Aoki, T.; Fukuda, S.; Akao, M.; Tsukahara, T. Predicting cerebral infarction in patients with atrial fibrillation using machine learning: The Fushimi AF registry. *J. Cereb. Blood Flow. Metab.* **2022**, *42*, 746–756. [CrossRef]
29. Papadopoulou, A.; Harding, D.; Slabaugh, G.; Marouli, E.; Deloukas, P. Prediction of atrial fibrillation and stroke using machine learning models in UK Biobank. *Heliyon* **2024**, *10*, e28034. [CrossRef]
30. Rebollo, P.; Wolk, A.; Luczko, M.; Tang, J.P. MSR124 Development of a Machine Learning Predictive Model for Stroke Among Patients With Non-Valvular Atrial Fibrillation Receiving Oral Anticoagulant Treatment. *Value Health* **2022**, *25*, S374. [CrossRef]
31. Truong, B.; Zheng, J.; Hornsby, L.; Fox, B.; Chou, C.; Qian, J. Development and Validation of Machine Learning Algorithms to Predict 1-Year Ischemic Stroke and Bleeding Events in Patients with Atrial Fibrillation and Cancer. *Cardiovasc. Toxicol.* **2024**, *24*, 365–374. [CrossRef] [PubMed]
32. Zhang, F.; Zhang, Y.; Zhou, Q.; Shi, Y.; Gao, X.; Zhai, S.; Zhang, H. Using machine learning to identify proteomic and metabolomic signatures of stroke in atrial fibrillation. *Comput. Biol. Med.* **2024**, *173*, 108375. [CrossRef] [PubMed]

33. Bisson, A.; Lemrini, Y.; El-Bouri, W.; Bodin, A.; Angoulvant, D.; Lip, G.Y.H.; Fauchier, L. Prediction of incident atrial fibrillation in post-stroke patients using machine learning: A French nationwide study. *Clin. Res. Cardiol.* **2023**, *112*, 815–823. [CrossRef] [PubMed]
34. Sung, S.F.; Sung, K.L.; Pan, R.C.; Lee, P.J.; Hu, Y.H. Automated risk assessment of newly detected atrial fibrillation poststroke from electronic health record data using machine learning and natural language processing. *Front. Cardiovasc. Med.* **2022**, *9*, 941237. [CrossRef] [PubMed]
35. Choi, K.-H.; Kim, J.-H.; Kang, K.-W.; Kim, J.-T.; Choi, S.-M.; Lee, S.-H.; Park, M.-S.; Kim, B.-C.; Kim, M.-K.; Cho, K.-H. Impact of Microbleeds on Outcome Following Recanalization in Patients With Acute Ischemic Stroke. *Stroke* **2019**, *50*, 127–134. [CrossRef]
36. Bhaskar, S.; Stanwell, P.; Bivard, A.; Spratt, N.; Walker, R.; Kitsos, G.H.; Parsons, M.W.; Evans, M.; Jordan, L.; Nilsson, M.; et al. The influence of initial stroke severity on mortality, overall functional outcome and in-hospital placement at 90 days following acute ischemic stroke: A tertiary hospital stroke register study. *Neurol. India* **2017**, *65*, 1252–1259. [CrossRef]
37. Marzec, L.N.; Wang, J.; Shah, N.D.; Chan, P.S.; Ting, H.H.; Gosch, K.L.; Hsu, J.C.; Maddox, T.M. Influence of Direct Oral Anticoagulants on Rates of Oral Anticoagulation for Atrial Fibrillation. *J. Am. Coll. Cardiol.* **2017**, *69*, 2475–2484. [CrossRef]
38. Kattoor, A.J.; Pothineni, N.V.; Goel, A.; Syed, M.; Syed, S.; Paydak, H.; Mehta, J.L. Prescription Patterns and Outcomes of Patients With Atrial Fibrillation Treated With Direct Oral Anticoagulants and Warfarin: A Real-World Analysis. *J. Cardiovasc. Pharmacol. Ther.* **2019**, *24*, 428–434. [CrossRef]
39. Nohara, Y.; Matsumoto, K.; Soejima, H.; Nakashima, N. Explanation of Machine Learning Models Using Improved Shapley Additive Explanation. In Proceedings of the 10th ACM International Conference on Bioinformatics, Computational Biology and Health Informatics, New York, NY, USA, 7–10 September 2019; p. 546.
40. Xiong, Z.; Liu, T.; Tse, G.; Gong, M.; Gladding, P.A.; Smaill, B.H.; Stiles, M.K.; Gillis, A.M.; Zhao, J. A Machine Learning Aided Systematic Review and Meta-Analysis of the Relative Risk of Atrial Fibrillation in Patients With Diabetes Mellitus. *Front. Physiol.* **2018**, *9*, 835. [CrossRef]
41. Wang, K.; Hong, T.; Liu, W.; Xu, C.; Yin, C.; Liu, H.; Wei, X.E.; Wu, S.-N.; Li, W.; Rong, L. Development and validation of a machine learning-based prognostic risk stratification model for acute ischemic stroke. *Sci. Rep.* **2023**, *13*, 13782. [CrossRef]
42. Seiffge, D.J.; De Marchis, G.M.; Koga, M.; Paciaroni, M.; Wilson, D.; Cappellari, M.; Macha Md, K.; Tsivgoulis, G.; Ambler, G.; Arihiro, S.; et al. Ischemic Stroke despite Oral Anticoagulant Therapy in Patients with Atrial Fibrillation. *Ann. Neurol.* **2020**, *87*, 677–687. [CrossRef] [PubMed]
43. Cross, B.; Turner, R.M.; Zhang, J.E.; Pirmohamed, M. Being precise with anticoagulation to reduce adverse drug reactions: Are we there yet? *Pharmacogenom. J.* **2024**, *24*, 7. [CrossRef] [PubMed]
44. Chen, T.; Guestrin, C. XGBoost: A Scalable Tree Boosting System. In Proceedings of the 22nd ACM SIGKDD International Conference on Knowledge Discovery and Data Mining, San Francisco, CA, USA, 13–17 August 2016; pp. 785–794.
45. Liaw, A.; Wiener, M. Classification and Regression by RandomForest. *Forest* **2001**, *23*, 18–22.
46. Subbiah, V. The next generation of evidence-based medicine. *Nat. Med.* **2023**, *29*, 49–58. [CrossRef] [PubMed]
47. Yagin, F.H.; Colak, C.; Algarni, A.; Gormez, Y.; Guldogan, E.; Ardigò, L.P. Hybrid Explainable Artificial Intelligence Models for Targeted Metabolomics Analysis of Diabetic Retinopathy. *Diagnostics* **2024**, *14*, 1364. [CrossRef]
48. Ramspek, C.L.; Jager, K.J.; Dekker, F.W.; Zoccali, C.; van Diepen, M. External validation of prognostic models: What, why, how, when and where? *Clin. Kidney J.* **2021**, *14*, 49–58. [CrossRef]
49. Vasey, B.; Ursprung, S.; Beddoe, B.; Taylor, E.H.; Marlow, N.; Bilbro, N.; Watkinson, P.; McCulloch, P. Association of Clinician Diagnostic Performance With Machine Learning–Based Decision Support Systems: A Systematic Review. *JAMA Netw. Open* **2021**, *4*, e211276. [CrossRef]
50. Anjara, S.G.; Janik, A.; Dunford-Stenger, A.; Mc Kenzie, K.; Collazo-Lorduy, A.; Torrente, M.; Costabello, L.; Provencio, M. Examining explainable clinical decision support systems with think aloud protocols. *PLoS ONE* **2023**, *18*, e0291443. [CrossRef]
51. Mainali, S.; Darsie, M.E.; Smetana, K.S. Machine Learning in Action: Stroke Diagnosis and Outcome Prediction. *Front. Neurol.* **2021**, *12*, 734345. [CrossRef]
52. Arora, A.; Alderman, J.E.; Palmer, J.; Ganapathi, S.; Laws, E.; McCradden, M.D.; Oakden-Rayner, L.; Pfohl, S.R.; Ghassemi, M.; McKay, F.; et al. The value of standards for health datasets in artificial intelligence-based applications. *Nat. Med.* **2023**, *29*, 2929–2938. [CrossRef]

diagnostics

MDPI

Article

Usefulness of Atherogenic Indices for Predicting High Values of Avoidable Lost Life Years Heart Age in 139,634 Spanish Workers

Tomás Sastre-Alzamora [1], Pedro J. Tárraga López [2,3], Ángel Arturo López-González [1,4,5,6,*], Daniela Vallejos [1], Hernán Paublini [1] and José Ignacio Ramírez Manent [1,5,6,7]

1 Research ADEMA SALUD Group, University Institute for Research in Health Sciences (IUNICS), 07010 Palma, Spain; tsastre04@sonrie.com (T.S.-A.); d.vallejos@eua.edu.es (D.V.); h.paublini@eua.edu.es (H.P.); joseignacio.ramirez@ibsalut.es (J.I.R.M.)
2 Faculty of Medicine, UCLM (University of Castilla La Mancha), 02008 Albacete, Spain; pjtarraga@sescam.jccm.es
3 SESCAM (Health Service of Castilla La Mancha), 02008 Albacete, Spain
4 Faculty of Dentistry, ADEMA University School, 07010 Palma, Spain
5 Institut d'Investigació Sanitària de les Illes Balears (IDISBA), Health Research Institute of the Balearic Islands, 07010 Palma, Spain
6 Health Service of the Balearic Islands, 07010 Palma, Spain
7 Faculty of Medicine, University of the Balearic Islands, 07010 Palma, Spain
* Correspondence: angarturo@gmail.com; Tel.: +34-687548105

Abstract: Background: Cardiovascular diseases (CVDs) are the leading cause of morbidity and mortality worldwide, accounting for one-third of all global deaths. The World Health Organization (WHO) asserts that prevention is the most effective strategy to combat CVD, emphasizing the need for non-invasive, low-cost tools to identify individuals at high risk of CVD. Atherogenic indices and heart age (HA) are valuable tools for assessing cardiovascular risk (CVR). The aim of our study was to evaluate the association between atherogenic indices and HA. Methods: A cross-sectional study was conducted involving 139,634 Spanish workers to determine the association between three atherogenic indices and HA. ROC curves were employed to identify the cut-off values for the various atherogenic indices used to estimate high HA. The cut-off points, along with their sensitivity, specificity, and Youden index, were determined, and the area under the curve (AUC) was calculated. Results: As the values of the atherogenic indices increased, so did the risk of having elevated avoidable lost life years (ALLY) HA. In the ROC curve analysis, the AUC with the best results corresponded to the total cholesterol/HDL-c atherogenic index, with an AUC of 0.803 in females and 0.790 in males. The LDL-c/HDL-c atherogenic index showed an AUC of 0.780 in women and 0.750 in men, with Youden indices around 0.4. When analyzing the AUC of the atherogenic index for triglycerides/HDL-c, the results were 0.760 in women and 0.746 in men. Conclusions: Atherogenic indices and HA show a close relationship, with an increase in these indices leading to a rise in HA values. Raising patient awareness that as their CVR levels increase, so does their HA may be useful in achieving some benefit in reducing CVR.

Keywords: cardiovascular diseases; atherogenic indices; heart age; preventive activities; lifestyle habits

1. Introduction

Cardiovascular diseases (CVDs) are the leading cause of morbidity and mortality worldwide [1], accounting for approximately one-third of all global deaths [2,3], with about three-quarters of these deaths occurring in low- and middle-income countries. In this context, the identification and management of cardiovascular risk factors (CVRFs) are crucial for the prevention and treatment of these diseases [4]. Traditional risk factors include hypertension [5], dyslipidaemia [6], diabetes mellitus [7], smoking [8], physical inactivity [9], and obesity [10]. However, in recent decades, new markers and tools have been developed and validated to more accurately assess individual cardiovascular risk (CVR). Among these, atherogenic indices [11,12] and the concept of "heart age" [13,14] have gained prominence as complementary indicators in cardiovascular risk assessment.

Atherogenic indices are biochemical parameters that reflect the relationship between different lipid fractions in the blood and are used to assess the risk of atherosclerosis [15,16] and cardiovascular events [17,18]. The most common include the total cholesterol/HDL-c index [19], the atherogenic index of plasma (AIP) [20,21], and the non-HDL/HDL cholesterol ratio [22]. These indices provide a more integrated measure of lipid profiles compared to the traditional measurements of total cholesterol, LDL, and HDL separately [17].

The total cholesterol/HDL-c index is one of the oldest and most widely used atherogenic indices. An elevated TC/HDL ratio has been shown to be associated with a greater risk of cardiovascular events [23]. A high value indicates a predominance of atherogenic particles in the blood, suggesting greater potential for the formation of atheromatous plaques in the arteries [24].

The AIP is calculated as the ratio between triglycerides and HDL cholesterol (TG/HDL-C). This index has been proposed as a marker for the size and number of LDL particles, with higher values being associated with smaller and denser LDL particles, which are more atherogenic [25]. Several studies have demonstrated that AIP is an independent predictor of cardiovascular disease and can be particularly useful in assessing risk in patients with metabolic syndrome [26,27] and type 2 diabetes mellitus [28,29].

The concept of "heart age" (HA) refers to a CVR assessment tool that translates an individual's absolute risk into an equivalent age of their cardiovascular system [30]. This approach aims to communicate risk in a more understandable and motivating way for patients [31]. HA is calculated using algorithms based on traditional risk factors such as age, sex, blood pressure, cholesterol levels, the presence of diabetes, and smoking habits [32].

Several online calculators and mobile applications are available for healthcare professionals and patients to calculate HA [33]. However, interpreting the results can be challenging, especially for patients, as entering the same data can yield different outcomes [34]. One of the most commonly used algorithms is the one developed by the Framingham Heart Study [35], which estimates HA based on the Framingham Risk Score. This tool not only helps estimate CVR, but can also be used to monitor the impact of therapeutic interventions and lifestyle changes.

HA is particularly useful in clinical practice because it translates abstract risk data into a more tangible message. For instance, informing a 40-year-old patient that they have a HA of 60 can serve as a strong motivator for adopting lifestyle changes or adhering to medical treatments. Some studies have shown that communicating CVR in terms of HA can improve adherence to lifestyle and therapeutic recommendations, ultimately reducing the risk of cardiovascular events [36,37]. In recent decades, the World Health Organization (WHO) has emphasized that the best strategy to prevent CVDs is prevention [38], making it crucial to have non-invasive, low-cost tools to identify individuals at high risk of developing CVDs.

The integration of atherogenic indices and HA into clinical practice can provide a more comprehensive and accurate assessment of CVR. While atherogenic indices offer detailed insights into lipid profile and atherosclerosis risk, HA translates overall CVR into a format that is easily understandable and motivating.

To effectively implement these approaches in clinical practice, a multifaceted strategy is recommended, encompassing patient education, the use of digital tools, and the integration of these parameters into electronic medical record systems. Patient education is crucial for understanding the importance of these indices and the relevance of heart age. Employing mobile applications and online calculators can facilitate access to and the use of these tools for both healthcare professionals and patients.

The benefits of using atherogenic indices and HA include more accurate risk assessment, improved risk communication to patients, and increased motivation to adopt lifestyle changes and adhere to therapies. However, challenges in implementation include the need for education and training for healthcare professionals, integration into existing health systems, and variability in available calculators and algorithms.

The aim of our study was to assess whether there is an association between atherogenic indices and HA in order to influence patients' engagement in preventive activities, as recommended by the WHO, and to encourage patients to modify harmful lifestyle habits.

If we find an association between atherogenic indices and heart age, we believe that one of the most effective ways to motivate patients to change harmful lifestyle habits is by making them aware of the potential years of life lost. The clinical utility of this study lies in the fact that by calculating the patient's atherogenic index during consultation, we can make them aware of the potential years of life lost, thereby emphasizing the urgency of adopting healthier lifestyle habits.

2. Materials and Methods

2.1. Participants

A total of 139,634 Spanish workers from various areas and labor sectors (mainly public administration, health, hospitality, construction, and commerce) participated in a descriptive, cross-sectional survey. Employees who underwent occupational medicine exams between January 2019 and June 2020 were chosen to be part of the study. Examine the flowchart in Figure 1.

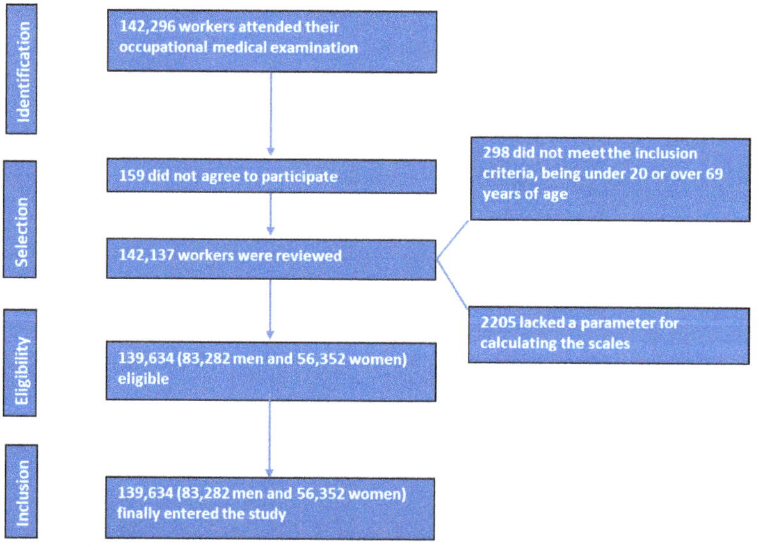

Figure 1. PRISMA flowchart of participants in the study.

The inclusion criteria to select the sample are detailed below:

- Individuals aged 20 to 69 years old.
- Consent to take part in the research.
- Permission given for the data to be used for epidemiological research.
- Being employed by one of the companies participating in the research and not being temporarily disabled at the time of the study.

2.2. Determination of Variables

Following process standardization to prevent interobserver bias, occupational health professionals from the participating firms conducted all measurements, whether anthropometric (height, weight, and waist circumference), analytical, or clinical.

2.2.1. Anthropometric Determinations

Measurements of height (in cm) and weight (in kg) were taken using a SECA 700 scale with a stadia meter. Measurements were taken with the person wearing underwear according to the international standards for anthropometric evaluation of the ISAK [39].

When the subject was standing straight, feet together and abdomen relaxed, waist circumference was measured using a tape measure parallel to the floor at the level of the last floating rib.

2.2.2. Clinical Determinations

Blood pressure was measured after 10 min of rest, with the subject seated. Three measurements were made at one-minute intervals, and the average of the three was calculated.

2.2.3. Analytical Determinations

The blood sample was taken after a minimum of 12 h of fasting, and processed in 48 to 72 h. Measurements of triglycerides, total cholesterol, and blood sugar were performed automatically by enzymatic procedures. The Cl2Mg dextran sulphate precipitation process was employed for HDL-cholesterol.

By using the Friedewald formula, which is only reliable when triglycerides do not exceed 400, LDL-cholesterol can be calculated indirectly. Each and every analytical parameter is stated in mg/dL [40].

2.2.4. Risk Scales

The following formulae were used to calculate the different atherogenic indices:

- Total cholesterol/HDL-c [19].
- LDL-c/HDL-c [41].
- Triglycerides/HDL-c [42].

Various cut-off thresholds were determined for each index based on data already available in the literature.

Total cholesterol/HDL-c was categorized as high risk if it was above 7% or 9%; moderate if it was between 4.5 and 7% or 5 and 9%; and low risk below 4.5% or 5% for women and men, respectively. The LDL-c/HDL-c index risk was low if below 3% and high above 3%. The triglyceride/HDL ratio was considered to be elevated at 3% and above [43].

HA values were found using the "Heart Age Calculator" feature on the website http://www.heartage.me, accessed on 7 July 2020. We have verified that the website is no longer available. The formula used to calculate cardiac age included total cholesterol, HDL cholesterol, fasting glucose, smoking, BMI, and systolic and diastolic blood pressure, with scores similar to the original formula. Based on these values, a series of points were added or subtracted from the chronological age, which were also adjusted by sex. The final result corresponded to the individual's heart age. In Table 1, we provide the values assigned to each of the variables used to calculate HA. The scale can be used with people who are between 18 and 80 years old. The highest number of years that can be gained or maintained compared to chronological age (CA) is 20, while the maximum age cut-off for HA is 19 years [44]. Avoidable lost life years (ALLY) [45] are defined as the difference between CA and HA. CA and HA are allies. This concept also addresses heart aging, with 17 as the cut-off age for high ALLY heart age [42].

Anyone who had smoked at least one cigarette in the previous thirty days (or its equivalent in other forms of consumption) or who had stopped smoking less than a year before was considered a smoker.

The 2011 national classification of occupations (CNAE) and the Spanish Society of Epidemiology's criteria were used to divide social class into three groups: class I, which included directors/managers, university professionals, athletes, and artists; class II, which included intermediate occupations and self-employed workers without employees; and class III, which included unskilled workers. This was carried out in compliance with the suggestion made by the Spanish Society of Epidemiology's social determinants committee [46].

Table 1. Variables and scores used to calculate cardiac age.

Total Cholesterol (mg/dL)	Points	HDL-c (md/dL)	Points	Glucose (mg/dL)	Points	Smokers	Points	BMI (kg/m^2)	Points	SBP/DBP mm Hg	Points
Men											
<130	−5	<35	2	<70	−5	Yes	5	<20	−2	<120/<80	−3
130–145	−4	35–39	1	70–80	−3	Non	0	20–22.5	−1	120–139/80–89	0
146–160	−3	40–45	0	81–90	−2			22.6–24.9	0	140–159/90–99	2
161–175	−2	46–50	−1	91–99	0			25–29.9	1	≥160/≥100 *	3
176–190	−1	51–55	−2	100–109	2			30–34.9	2		
191–199	0	56–60	−3	110–125	3			≥35	3		
200–239	1	61–65	−4	>125 *	5						
>240	2	>65	−5								
Women											
Total Cholesterol (mg/dL)	**Points**	**HDL-c (md/dL)**	**Points**	**Glucose (mg/dL)**	**Points**	**Smokers**	**Points**	**BMI (kg/m^2)**	**Points**	**SBP/DBP mm Hg**	**Points**
<130	−5	<40	2	<70	−5	Yes	5	<20	−2	<120/<80	−3
130–145	−4	40–49	1	70–80	−3	Non	0	20–22.5	−1	120–139/80–89	0
146–160	−3	50–55	0	81–90	−2			22.6–24.9	0	140–159/90–99	2
161–175	−2	56–60	−1	91–99	0			25–29.9	1	≥160/≥100 *	3
176–190	−1	61–65	−2	100–109	2			30–34.9	2		
191–199	0	66–70	−3	110–125	3			≥35	3		
200–239	1	71–75	−4	>125 *	5						
>240	2	<75	−5								

(*) Or in treatment.

There were three designated categories for educational attainment: elementary, high school, and university.

Strong adherence was defined as results equal to or higher than 9 in the 14-question Mediterranean diet adherence questionnaire [47], which has a score range of 0 to 1 for each question. Amount of physical activity was measured using the International Physical Activity Questionnaire, or IPAQ [48]. Alcohol consumption was measured in alcohol units (AUs)—in Spain, one AU is equivalent to 10 g of pure ethanol. It is considered high when a woman consumes 14 AUs and a man 21 AUs each week [49].

2.3. Statistical Analysis

For each of the categorical variables, a descriptive analysis was carried out to determine the frequency and distribution of the responses. The mean and standard deviation for quantitative data were computed using a normal distribution. Student's t test for independent samples (for mean comparison) and the chi^2 test (with Fisher's exact statistic correction when necessary) were used for bivariate association analysis. The variables linked to the most important risk factors were identified using multivariate approaches. Multivariate analysis was conducted using logistic regression, and Hosmer–Lemeshow goodness-of-fit and odds ratios were computed. ROC curves were used to determine the cut-off values for the various atherogenic indices used to estimate high cardiac age. The cut-off points with their sensitivity, specificity, and Youden index were determined, along with the area under the curve (AUC). The Statistical Package for the Social Sciences (SPSS) version 28.0 for Windows was used to conduct the statistical analysis, with an accepted statistical significance level of 0.05.

3. Results

The anthropometric and clinical details of the study participants are displayed in Table 2. The analyses comprised 139,634 individuals, of whom 83,282 were men (59.6%) and 56,352 were women (40.4%). The average age was just over 40, with the majority of the group being in the 30-to-49-year-old age range. Men had more negative anthropometric, clinical, and analytical findings. With only an elementary education, the majority of the labor force belonged to social class III. Most men did not follow a balanced diet or engage in regular physical activity (the situation was better for women). Approximately one-third smoked.

Table 2. Characteristics of the population.

	Men n = 83,282	Women n = 56,352	All n = 139,634	
	Mean (SD)	Mean (SD)		p-Value
Age (years)	41.4 (10.7)	40.1 (10.4)	40.9 (10.6)	<0.0001
Height (cm)	173.8 (7.1)	161.2 (6.5)	168.7 (9.2)	<0.0001
Weight (kg)	83.2 (14.6)	66.3 (13.9)	76.3 (16.5)	<0.0001
Body mass index (kg/m^2)	27.5 (4.5)	25.5 (5.3)	26.7 (4.9)	<0.0001
Waist circumference (cm)	90.2 (10.3)	76.3 (10.5)	84.6 (12.4)	<0.0001
Waist-to-height ratio	0.52 (0.06)	0.47 (0.07)	0.50 (0.07)	<0.0001
Systolic blood pressure (mmHg)	126.2 (15.9)	115.6 (15.7)	121.9 (16.7)	<0.0001
Diastolic blood pressure (mmHg)	76.6 (10.9)	71.1 (10.7)	74.4 (11.1)	<0.0001
Total cholesterol (mg/dL)	199.6 (38.6)	194.6 (36.9)	197.6 (38.0)	<0.0001
HDL/cholesterol (mg/dL)	50.0 (7.7)	54.7 (9.2)	51.9 (8.6)	<0.0001
LDL/cholesterol (mg/dL)	122.6 (37.4)	121.5 (37.1)	122.2 (37.3)	<0.0001
Triglycerides (mg/dL)	133.8 (95.6)	90.8 (49.7)	116.4 (83.0)	<0.0001
Glycaemia (mg/dL)	93.0 (25.4)	86.8 (18.1)	90.5 (22.9)	<0.0001
	n (%)	n (%)		p-Value
20–29 years	12,558 (15.1)	10,110 (18.0)	22,668 (16.2)	<0.0001
30–39 years	24,648 (29.6)	17,460 (31.0)	42,108 (30.2)	
40–49 years	25,178 (30.2)	17,094 (30.3)	42,272 (30.3)	
50–59 years	17,370 (20.9)	9984 (17.7)	27,354 (19.6)	
60–70 years	3528 (4.2)	1704 (3.0)	5232 (3.7)	
Social class I	6234 (7.5)	7632 (13.6)	13,866 (9.9)	<0.0001
Social class II	19,856 (23.8)	18,112 (32.1)	37,968 (27.2)	
Social class III	57,192 (68.7)	30,608 (54.3)	87,800 (62.9)	
Elementary school	55,306 (66.4)	27,086 (48.1)	82,392 (59.0)	
High school	22,408 (26.9)	22,574 (40.0)	44,982 (32.2)	
University	5568 (6.7)	6692 (11.9)	12,260 (8.8)	
Non-smokers	55,618 (66.8)	38,252 (67.9)	93,870 (67.2)	<0.0001
Smokers	27,664 (33.2)	18,100 (32.1)	45,764 (32.8)	
No physical activity	51,984 (62.4)	28,962 (51.4)	80,946 (58.0)	<0.0001
Yes physical activity	31,298 (37.6)	27,390 (48.6)	58,688 (42.0)	
No Mediterranean diet	54,792 (65.8)	29,764 (52.8)	84,556 (60.6)	<0.0001
Yes Mediterranean diet	28,490 (34.2)	26,588 (47.2)	55,078 (39.4)	
No alcohol consumption	56,022 (67.3)	47,536 (84.4)	103,558 (74.2)	<0.0001
Yes alcohol consumption	27,260 (32.7)	8816 (15.6)	36,076 (25.8)	

HDL—high-density lipoprotein. LDL—low-density lipoprotein.

The mean values, in both sexes, of ALLY heart age increased in parallel with the increase in values of the different atherogenic indices. In all cases, ALLY HA values were higher in men. The differences observed in all cases showed statistical significance (see Table 3).

Similarly, when stratifying by age, the mean values of ALLY heart age increased in both sexes as age increased. The same trend is observed in the values of the different atherogenic indices (Table 4).

Something similar was observed when assessing the prevalence of high ALLY HA values, i.e., the prevalence increased as the values of all atherogenic indices increased. Prevalence was always higher in men. In all cases, as can be seen in Table 5, the differences observed were statistically significant.

Table 3. Mean values of ALLY heart age according to levels of different atherogenic indices by sex.

	Men			Women		
ALLY HA	*n*	**Mean (SD)**	***p*-Value**	*n*	**Mean (SD)**	***p*-Value**
Low TC/HDL-c	6,8020	5.2 (7.6)	<0.001	45,174	−0.4 (8.8)	<0.001
Moderate TC/HDL-c	14,988	13.7 (6.3)		10,730	9.7 (8.6)	
High TC/HDL-c	274	18.8 (3.0)		448	15.2 (7.3)	
Normal LDL-c/HDL-c	70,700	5.5 (7.7)	<0.001	42,996	−0.6 (8.8)	<0.001
High LDL-c/HDL-c	12,582	13.7 (6.3)		13,356	9.0 (8.8)	
Normal Triglycerides/HDL-c	59,156	4.9 (7.6)	<0.001	51,302	0.8 (9.3)	<0.001
High Triglycerides/HDL-c	24,126	11.3 (7.5)		5050	10.6 (8.8)	

HDL—high-density lipoprotein. LDL—low-density lipoprotein. TC—total cholesterol. ALLY HA—avoidable lost life years heart age. SD—standard deviation.

Table 4. The influence of age on the relationship between atherogenic indices and heart age.

		Total Cholesterol/HDL-c *	LDL-c/HDL-c *	TG/HDL-c *	ALLY HA *
Men	*n*	**Mean (SD)**	**Mean (SD)**	**Mean (SD)**	**Mean (SD)**
20–29 years	12,558	3.3 (0.9)	1.9 (0.8)	1.9 (1.4)	1.3 (4.9)
30–39 years	24,648	3.9 (1.1)	2.4 (0.9)	2.5 (2.3)	4.2 (6.7)
40–49 years	25,178	4.3 (1.2)	2.7 (1.0)	3.1 (2.8)	7.9 (8.1)
50–59 years	17,370	4.6 (1.2)	2.9 (1.0)	3.4 (2.8)	10.7 (7.9)
60–70 years	3528	4.7 (1.2)	3.0 (1.1)	3.5 (2.2)	11.8 (7.4)
		Total Cholesterol/HDL-c *	**LDL-c/HDL-c ***	**TG/HDL-c ***	**ALLY HA ***
Women	*n*	**Mean (SD)**	**Mean (SD)**	**Mean (SD)**	**Mean (SD)**
20–29 years	10,110	3.1 (0.9)	1.8 (0.8)	1.4 (0.8)	−2.1 (5.0)
30–39 years	17,640	3.4 (1.0)	2.1 (0.9)	1.6 (0.9)	−1.8 (7.7)
40–49 years	17,094	3.8 (1.0)	2.5 (0.9)	1.8 (1.2)	2.8 (10.2)
50–59 years	9984	4.3 (1.1)	2.8 (1.0)	2.1 (1.4)	8.4 (10.6)
60–70 years	1704	4.4 (1.1)	2.9 (1.0)	2.3 (1.2)	8.6 (9.9)

(*) Statistical significance ($p < 0.001$) HDL-c—high-density lipoprotein–cholesterol. LDL-c—low-density lipoprotein–cholesterol. TG—triglycerides. ALLY HA—avoidable lost life years heart age. SD—standard deviation.

Table 5. Prevalence of high values of heart age according to levels of different atherogenic indices by sex.

	Men			Women		
High ALLY HA	*n*	%	***p*-Value**	*n*	%	***p*-Value**
Low TC/HDL-c	68,020	11.6	<0.001	45,174	7.0	<0.001
Moderate TC/HDL-c	14,988	44.9		10,730	32.0	
High TC/HDL-c	274	86.1		448	63.8	
Normal LDL-c/HDL-c	70,700	12.9	<0.001	42,996	7.6	<0.001
High LDL-c/HDL-c	12,582	45.4		13,356	29.7	
Normal Triglycerides/HDL-c	59,156	10.9	<0.001	51,302	9.7	<0.001
High Triglycerides/HDL-c	24,126	70.1		5050	37.7	

HDL—high-density lipoprotein. LDL—low-density lipoprotein. TC—total cholesterol. ALLY—avoidable lost life years heart age.

Table 6 shows the results of the multinomial logistic regression analysis stratified by age groups. Sex, social class, educational level, tobacco and alcohol consumption, physical activity, adherence to the Mediterranean diet, and levels of various atherogenic indices have been established as independent variables.

Table 6. Multinomial logistic regression.

	20–29 Years	30–39 Years	40–49 Years	50–59 Years	60–70 Years
High ALLY Heart Age	**OR (95% CI)**	**OR (95% CI)**	**OR (95% CI)**	**OR (95% CI)**	**OR (95% CI)**
Women	1	1	1	1	1
Men	1.25 (1.20–1.31)	1.67 (1.46–1.88)	1.54 (1.43–1.65)	1.12 (1.05–0.20)	1.32 (1.24–1.40)
Social class I	1	1	1	1	1
Social class II	2.01 (1.86–2.16)	1.96 (1.63–2.30)	1.88 (1.71–2.05)	1.87 (1.71–2.04)	1.97 (1.69–2.25)
Social class III	2.41 (2.20–2.62)	3.30 (2.67–3.94)	2.37 (2.00–2.75)	4.20 (3.72–4.69)	5.21 (4.70–5.72)
University	1	1	1	1	1
High school	1.85 (1.69–2.02)	1.49 (1.29–1.70)	1.38 (1.29–1.48)	1.52 (1.38–1.66)	1.48 (1.30–1.67)
Elementary school	2.58 (2.30–2.67)	2.62 (2.30–2.95)	2.68 (2.40–2.97)	2.46 (2.09–2.82)	2.40 (2.11–2.70)
Non-smokers	1	1	1	1	1
Smokers	29.15 (25.45–32.85)	22.33 (21.0–23.67)	21.50 (19.93–23.08)	13.23 (12.28–14.19)	4.97 (4.23–5.70)
Physical activity	1	1	1	1	1
No physical activity	1.63 (1.50–1.76)	3.32 (2.68–3.96)	4.15 (3.70–4.60)	4.45 (3.65–5.26)	5.98 (5.01–6.95)
Mediterranean diet	1	1	1	1	1
No Mediterranean diet	1.34 (1.25–1.44)	1.89 (1.50–2.29)	2.27 (2.01–2.53)	2.39 (2.09–2.70)	3.11 (2.60–3.62)
No alcohol consumption	1	1	1	1	1
Alcohol consumption	3.43 (2.98–3.88)	4.04 (3.51–4.58)	4.33 (3.90–4.77)	7.34 (6.04–8.65)	7.45 (6.02–7.88)
TC/HDL-c low	1	1	1	1	1
TC/HDL-c moderate	7.13 (6.60–7.67)	6.45 (5.01–7.90)	5.44 (5.07–5.82)	4.02 (3.60–4.44)	3.00 (2.60–3.41)
TC/HDL-c high	29.98 (27.02–32.94)	20.24 (18.01–22.47)	16.52 (14.99–18.05)	7.34 (6.00–8.68)	6.90 (6.02–7.79)
LDL-c/HDL-c low	1	1	1	1	1
LDL-c/HDL-c high	2.99 (2.51–3.47)	2.61 (2.37–2.85)	2.46 (2.11–2.82)	2.29 (2.02–2.57)	2.22 (1.80–2.64)
Triglycerides/HDL-c low	1	1	1	1	1
Triglycerides/HDL-c high	2.45 (2.20–2.30)	1.90 (1.70–2.11)	1.72 (1.49–1.95)	1.54 (1.41–1.67)	1.41 (1.31–1.52)

HDL—high-density lipoprotein. LDL—low-density lipoprotein. TC—total cholesterol. ALLY—avoidable lost life years. OR—odds ratio. In all cases, the differences are statistically significant.

It is observed that the effect of the different independent variables on the appearance of elevated heart age values is not always the same across the different age groups.

Sex, social class, and educational level behave similarly in all groups. Tobacco consumption and the various atherogenic indices show higher odds ratios in younger people, with their influence decreasing as a person's age increases. On the other hand, alcohol consumption, physical inactivity, and low adherence to the Mediterranean diet increase their influence (odds ratio values) as a person ages.

In Figure 2 and Table 7, we see how the areas under the curve of all atherogenic indices for predicting the presence of high ALLY HA are moderate–high, especially in women.

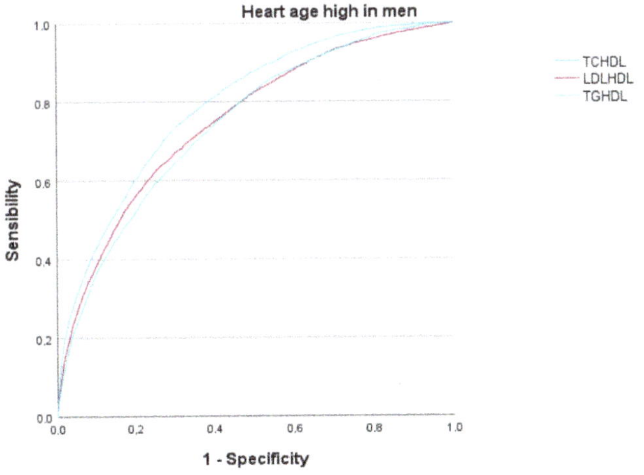

Figure 2. *Cont.*

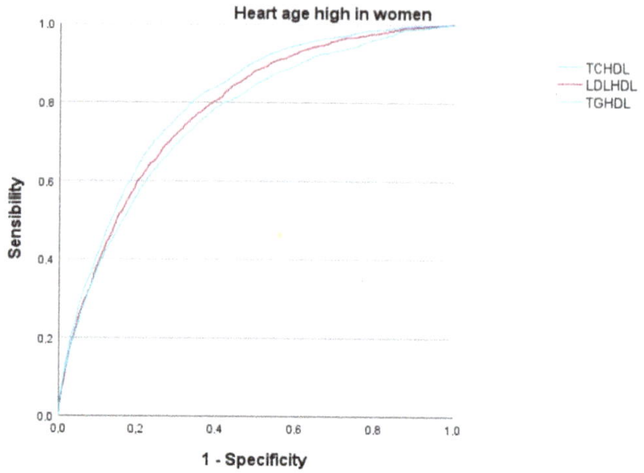

Figure 2. ROC curve.

Table 7. Area under the curve by sex.

	Men	Women
	AUC (95% CI)	**AUC (95% CI)**
Total cholesterol/HDL-c	0.790 (0.786–0.794)	0.803 (0.798–0.808)
LDL-c/HDL-c	0.750 (0.746–0.754)	0.780 (0.774–0.785)
Triglycerides/HDL-c	0.742 (0.737–0.746)	0.760 (0.754–0.766)

HDL—high-density lipoprotein. LDL—low-density lipoprotein. AUC—area under the curve.

Table 8 displays the cut-off points, along with their sensitivity, specificity, and Youden index, for the various atherogenic indices in predicting high ALLY heart age (HA). In all cases, Youden indices are higher in women.

Table 8. Cut-off for atherogenic indices to predict high values of heart age.

	Men	Women
	Cut-Off–Sensitivity–Specificity–Youden	**Cut-Off–Sensitivity–Specificity–Youden**
Total cholesterol/HDL-c	4.30–71.6–71.5–0.431	4.00–73.8–72.4–0.464
LDL-c/HDL-c	2.71–68.2–68.2–0.364	2.60–71.2–71.0–0.422
Triglycerides/HDL-c	2.53–67.0–67.0–0.340	1.73–70.0–69.5–0.395

HDL—high-density lipoprotein. LDL—low-density lipoprotein.

4. Discussion

Cardiovascular diseases remain the leading cause of death worldwide and have been increasing in recent years. Atherosclerosis is the primary pathophysiological factor underlying these diseases. Another CVRF is so-called HA, which estimates cardiac aging. The rise in CVDs is significantly influenced by population aging, harmful lifestyle habits, urban living conditions, and obesity [50]. In recent decades, the WHO has emphasized that preventive measures are the most effective strategy to reduce cardiovascular morbidity and mortality. Therefore, the early identification of CVRF and the implementation of motivating interventions are essential to encourage patients to adopt healthy lifestyle habits [51].

Changes in diet within modern society—particularly in large cities—such as the rapid consumption of food due to time constraints and excessive reliance on prepared meals, processed products and snacks (including popcorn, potato chips, salted crackers, sunflower seeds, desserts, sweets, sugary or carbonated drinks, etc.), coupled with a lack of physical exercise and increased screen time (computer, mobile phones, television, video games, etc.), along with the rising rates of obesity [52] driven by these factors, contribute to lipid disorders that create the conditions necessary for the formation of atherosclerotic plaques in arteries [53]. In our sample, 65.8% of men and 52.8% of women did not follow a Mediterranean diet. A similar pattern was observed with regular physical activity, where 48.6% of women adhered to it, compared to only 37.6% of men. In the multinomial logistic regression analysis stratified by age groups in our study, the influence of these two variables increases as the population ages. Physical activity exerts a significantly greater influence (OR) than the Mediterranean diet on high ALLY HA.

Among harmful habits, smoking is one of the primary risk factors for developing atherosclerosis, as it exacerbates dyslipidaemias [54]. Smokers have a three to six times higher risk of developing atherosclerosis compared to non-smokers, and they typically develop it ten years earlier [55,56]. In our study, 33.2% of men and 32.1% of women were smokers despite anti-smoking campaigns in our country and public health education programs. In the multinomial logistic regression analysis, we observe that as age increases, smoking shows lower ORs for high ALLY HA, indicating a reduced influence in older populations. This highlights the critical importance of early intervention to mitigate the impact of smoking on HA.

Excessive alcohol consumption has also been shown to be detrimental to cardiovascular health. One challenge is defining what constitutes excessive alcohol consumption, as there is no international consensus on this definition. Various studies have described the effects of alcohol on the cardiovascular system as a J-shaped curve, where moderate drinkers might have a lower CVR compared to abstainers, while heavy drinkers have a higher CVR [57,58]. However, there is insufficient evidence to recommend reducing alcohol consumption as a preventive measure for CVDs [59,60]. In this variable, our population showed that the percentage of men who consumed alcohol regularly was twice that of women, with 32.7% of men compared to 15.6% of women. In our study, we found that alcohol consumption was detrimental across all age groups, with a 343% increased risk in the youngest cohort (20–29 years), progressively worsening across older age groups. As the individual's age increases, alcohol exerts a greater influence on high ALLY HA, with the OR rising from 3.43 in the youngest group to more than double at 7.45 in the oldest age group (60–70 years).

These lifestyle habits, already proven as CVRFs, show worse outcomes in men than in women, as reflected in all of our results. Both the mean values of ALLY heart age according to levels of different atherogenic indices by sex and the prevalence of high heart age values according to levels of different atherogenic indices by sex consistently demonstrate worse outcomes for men compared to women. Among the variables influencing healthy lifestyle habits, social class and education level have been described as key factors [61]. Our results highlight the impact of education level on high ALLY heart age, where we observe a consistent influence across all age groups studied. Individuals with only elementary school education show an OR ranging between 2.4 and 2.68 for high ALLY HA, underscoring the role of education in cardiovascular health.

Low socioeconomic status—which comprises factors such as educational attainment, income level, stable employment, and access to housing [62]—has also been associated with increased cardiovascular morbidity and mortality. Increased CVR in individuals from lower socioeconomic backgrounds has been linked to chronic stress, as well as a higher prevalence of other CVRFs (hypertension, obesity, smoking, dyslipidemia, diabetes), which are often associated with lower educational levels and limited purchasing power [63,64]. These findings are consistent with the data obtained in our study, where social class similarly influences all age groups analyzed. We observe that individuals in social class III have an OR more than double that of those in social class I. Additionally, this variable worsens with age; for instance, the OR in social class III increases from 2.41 in the 20–29 age group to 5.21 in the oldest group (60–70 years). A similar pattern is seen between social classes I and

II, where social class II consistently shows higher ORs across all age groups. Interestingly, in social class II, the highest ORs are observed in the youngest population group, with differences ranging from 4% to 14% compared to the other age groups.

All these CVRFs lead to changes in an individual's metabolism that facilitate the development of prediabetes, insulin resistance, diabetes, dyslipidemia, obesity, and metabolic syndrome [65–67]. These conditions themselves are also risk factors for CVDs and contribute to the progression of atherosclerosis and increased cardiovascular mortality [68–71].

The formation and progression of atherosclerotic plaques are strongly related to the presence of CVRFs [72]. Failure to intervene early and in a multidisciplinary manner on these risk factors allows for the progression of atherosclerotic lesions over time and leads to complications that increase mortality [73].

Among the latest tools developed to assess CVR are atherogenic indices, which evaluate the risk of atherosclerosis at a low cost and are non-invasive, unlike molecular biomarkers, which, although highly specific, are expensive and therefore not efficient for the general population [74].

In our study, we evaluated a sample of 139,634 individuals from different autonomous communities in Spain. We evaluated 83,282 men and 56,352 women, making it a representative sample of the Spanish population. Most participants had attained only an elementary education and belonged to social class III. The mean age of the sample was slightly over 40 years, with around 60% within the age range of 30 to 49 years. Anthropometric, clinical, and analytical parameters revealed worse results in men. Among unhealthy lifestyle habits, it is noteworthy that one-third of the sample were smokers and two-thirds of the men did not engage in regular physical activity or follow a Mediterranean diet, compared to approximately 50% of the women.

When studying the association between atherogenic indices and ALLY HA, we found that the latter increased as the mean values of atherogenic indices rose. A similar pattern was observed when evaluating the prevalence of elevated ALLY HA values; that is, prevalence increased as the values of all atherogenic indices increased. Prevalence was consistently higher in men. In all cases, the observed differences were statistically significant.

Multinomial logistic regression analysis revealed that both sex and higher values of the different atherogenic indices increase the risk of having elevated ALLY HA values. The significant elevation in the odds ratio indicates a strong association between exposure and the event. These results reinforce the findings of other studies that have linked atherogenic indices with cardiovascular mortality [75,76].

In the ROC curve analysis for the three atherogenic risk formulas evaluated, the areas under the curve (AUC) with the best results correspond to the total cholesterol/HDL-c index, with a value of 0.803 in females and 0.790 in males, and the LDL-c/HDL-c atherogenic index, with a value of 0.780 in females and 0.750 in males. Both are above 0.75, indicating good performance. However, the Youden indices are not excessively high, around 0.4.

Upon analyzing the AUC of the triglycerides/HDL-C atherogenic index, the result was found to be 0.760 in females (considered good) and 0.746 in males, which is close to 0.75 but does not quite reach this value. This suggests that, strictly speaking, the test result should be regarded as moderate. The Youden index was lower in both cases, indicating lower sensitivity and specificity.

We found no studies in the literature that evaluate this specific association, so we cannot compare our results with those obtained by other authors. To address this gap, we attempt to compare our findings with articles assessing the relationship between atherogenic indices and cardiovascular risk.

Clearly, another risk factor for CVDs is an aging population [77], which constitutes a non-modifiable risk factor. Therefore, our efforts should focus on promoting healthy lifestyle habits that enhance quality of life and facilitate healthy aging.

The study by Hernáez et al. (2019) highlights that while aging has been associated with lower LDL cholesterol levels in various studies, CVR increases with age as LDL cholesterol is highly atherogenic [78]. This could correlate with our findings, where an increase in atherogenic risk corresponds with a rise in HA. Additionally, a study involving 178,282 working women examined various CVRFs and atherogenic indices, observing in a multivariate analysis that age was the only factor affecting all the CVR scales analyzed [79].

Other studies evaluating cardiovascular risk using various scales such as REGICOR, SCORE, and atherogenic indices also found an increase in CVR with age, despite this not being the primary objective of the studies [80–82].

Our findings show that as age increases, so do the three atherogenic indices evaluated. The same pattern is observed with ALLY HA, with the mean standard deviation increasing nearly tenfold between the first decade (20–29 years) and the last decade (60–70 years) analyzed. Notably, across all age groups, the ALLY HA values are lower in women than in men, which could be influenced by the less atherogenic lipid profile observed in women in our study. This aligns with the publication by Fappi et al. (2020), which details the relationship between cardiometabolic profile and chronological age [83]·

Although chronological heart age is undeniably associated with an individual's CA, it is clear that various factors influence myocardial cells at the molecular and cellular level. Therefore, the heart grows and ages under the influence of these factors. This means that the biological age of the heart may not align with CA. Consequently, the biological age of the heart could be greater than the individual's CA, potentially increasing cardiovascular morbidity and mortality [84].

As previously discussed, the best measure to reduce cardiovascular mortality is prevention. To achieve this, it is essential to be able to assess CVR easily and have sufficiently motivating and incentivizing tools to encourage patients to change their unhealthy lifestyle habits. This way, patients can understand their risk in a straightforward manner and learn how their actions can reduce it. Several studies have already demonstrated the effectiveness of different tools in motivating patients, such as the concept of "pulmonary age" for smoking cessation [85]; "vascular age" to reduce cardiovascular risk factors [86]; and "heart age" via electrocardiogram to lower cardiovascular risk factors [87].

One meta-analysis published in January 2024, examining the effect of communicating CVR to patients, its influence on their risk perception, and changes in their behaviors and risk factors, concluded that informing patients about their CVR has a positive impact on risk awareness and the intention to change risk factors. Patients experienced a reduction in overall risk scores between 6 and 12 months of follow-up, along with improvements in blood pressure and cholesterol levels. These results were more pronounced in patients with no established CVDs. There was a significant shift towards a healthier diet and increased motivation for initiating and adhering to necessary medication. The authors concluded that effectively informing patients about their cardiovascular risk levels can lead to meaningful reductions in overall vascular risk [88].

According to the results of our study, there is a close relationship between atherogenic indices, which assess the risk of atherosclerosis, and HA, which evaluates cardiac aging. As the values of these indices increase, there is also an increase in ALLY HA.

The ability to communicate a patient's cardiovascular risk is of paramount importance, particularly in public health and preventive activities. We believe that explaining to patients that as their atherogenic index increases, so does their chances of experiencing a cardiovascular event, along with the aging of their heart, may serve as a more graphic and comprehensible method. This approach could be highly beneficial in clinical practice, acting as a motivating tool to encourage lifestyle changes and reduce the CVR of our patients.

5. Strengths and Limitations

The strengths of this study include the large sample size—which exceeds 139,000 individuals—and the lack of existing studies addressing a similar topic, making this study a potential reference for future research.

The main limitation is that HA is assessed through validated risk calculators rather than objective methods, due to the impracticality of performing objective measurements in such a large sample.

Another significant limitation of our study is the lack of records for patients undergoing treatment with lipid-lowering medications, which means that some results could potentially be underestimated.

Further, our data were derived from routine occupational health screenings among the working population in Spain aged 20 to 69 years. Consequently, the findings of this study may not be generalizable to other populations with different demographic characteristics.

6. Conclusions

Atherogenic indices and HA show a close relationship, with increases in these indices being reflected by a rise in HA values.

The effect of different dependent variables on the occurrence of elevated HA values is not consistent across all age groups.

Sex, social class, and educational level behave similarly across all age groups.

Smoking and the studied atherogenic indices show higher odds ratios in younger individuals, with their influence decreasing as the person's age increases.

Alcohol intake, physical inactivity, and low adherence to the Mediterranean diet increase their influence as the individual ages.

Raising patients' awareness that their HA increases as their cardiovascular risk levels rise could be useful in achieving benefits in cardiovascular risk reduction.

Author Contributions: Conceptualization: T.S.-A., P.J.T.L., J.I.R.M. and Á.A.L.-G.; Data collection and analysis: D.V., H.P. and T.S.-A.; Data curation: H.P.; Methodology: T.S.-A., P.J.T.L. and Á.A.L.-G.; Validation: D.V., T.S.-A. and J.I.R.M.; Formal analysis: T.S.-A. and Á.A.L.-G.; Investigation: T.S.-A. and D.V.; Draft: T.S.-A. and P.J.T.L.; Revision: J.I.R.M. and Á.A.L.-G. All authors have read and agreed to the published version of the manuscript.

Funding: This research received no external funding.

Institutional Review Board Statement: The 2013 Declaration of Helsinki and the institutional research committee's ethical guidelines were upheld in this investigation. The confidentiality and anonymity of the information gathered will always be ensured. The Research Ethics Committee of the Balearic Islands (CEI-IB) approved the study (IB 4383/20) on 26 November 2020. Every worker's data that was part of the study were coded, so only the person in charge of the study could determine who was who. Study participants will be able to exercise their rights of access, rectification, cancellation, and opposition of the collected data, as guaranteed by the research team's strict adherence to Organic Law 3/2018 of 5 December, on the protection of personal data and guarantee of digital rights.

Informed Consent Statement: Informed consent was obtained from all subjects involved in the study.

Data Availability Statement: This study's data are stored in a database that complies with all security measures at ADEMA-Escuela Universitaria. The Data Protection Delegate is Ángel Arturo López González.

Conflicts of Interest: The authors declare that they have no conflicts of interest.

References

1. Roth, G.A.; Mensah, G.A.; Johnson, C.O.; Addolorato, G.; Ammirati, E.; Baddour, L.M.; Barengo, N.C.; Beaton, A.Z.; Benjamin, E.J.; Benziger, C.P.; et al. GBD-NHLBI-JACC Global Burden of Cardiovascular Diseases Writing Group. Global Burden of Cardiovascular Diseases and Risk Factors, 1990–2019: Update From the GBD 2019 Study. *J. Am. Coll. Cardiol.* **2020**, *76*, 2982–3021, Erratum in: *J. Am. Coll. Cardiol.* **2021**, *77*, 1958–1959. [CrossRef] [PubMed] [PubMed Central]
2. Global Cardiovascular Risk Consortium; Magnussen, C.; Ojeda, F.M.; Leong, D.P.; Alegre-Diaz, J.; Amouyel, P.; Aviles-Santa, L.; De Bacquer, D.; Ballantyne, C.M.; Bernabé-Ortiz, A.; et al. Global Effect of Modifiable Risk Factors on Cardiovascular Disease and Mortality. *N. Engl. J. Med.* **2023**, *389*, 1273–1285. [CrossRef] [PubMed] [PubMed Central]
3. van der Bijl, K.; Elgendi, M.; Menon, C. Automatic ECG Quality Assessment Techniques: A Systematic Review. *Diagnostics* **2022**, *12*, 2578. [CrossRef] [PubMed] [PubMed Central]

4. Teo, K.K.; Rafiq, T. Cardiovascular Risk Factors and Prevention: A Perspective From Developing Countries. *Can. J. Cardiol.* **2021**, *37*, 733–743. [CrossRef] [PubMed]
5. van Oort, S.; Beulens, J.W.J.; van Ballegooijen, A.J.; Grobbee, D.E.; Larsson, S.C. Association of Cardiovascular Risk Factors and Lifestyle Behaviors With Hypertension: A Mendelian Randomization Study. *Hypertension* **2020**, *76*, 1971–1979. [CrossRef] [PubMed]
6. Pirillo, A.; Casula, M.; Catapano, A.L. European guidelines for the treatment of dyslipidaemias: New concepts and future challenges. *Pharmacol. Res.* **2023**, *196*, 106936. [CrossRef] [PubMed]
7. Wong, N.D.; Sattar, N. Cardiovascular risk in diabetes mellitus: Epidemiology, assessment and prevention. *Nat. Rev. Cardiol.* **2023**, *20*, 685–695. [CrossRef] [PubMed]
8. Minué-Lorenzo, C.; Olano-Espinosa, E. Tobacco consumption, the forgotten factor in the calculation and approach of cardiovascular risk. *Med. Clin.* **2019**, *152*, 154–158. [CrossRef] [PubMed]
9. Turcu, A.M.; Ilie, A.C.; Ștefăniu, R.; Țăranu, S.M.; Sandu, I.A.; Alexa-Stratulat, T.; Pîslaru, A.I.; Alexa, I.D. The Impact of Heart Rate Variability Monitoring on Preventing Severe Cardiovascular Events. *Diagnostics* **2023**, *13*, 2382. [CrossRef]
10. Preda, A.; Carbone, F.; Tirandi, A.; Montecucco, F.; Liberale, L. Obesity phenotypes and cardiovascular risk: From pathophysiology to clinical management. *Rev. Endocr. Metab. Disord.* **2023**, *24*, 901–919. [CrossRef] [PubMed] [PubMed Central]
11. Kurklu, H.A.; Tan, T.S.; Ozyuncu, N.; Baskovski, E.; Ozdol, C. Atherogenic Index of Plasma Predicts Obstructive Coronary Artery Disease in Patients with Stable Angina Pectoris. *Diagnostics* **2023**, *13*, 3249. [CrossRef] [PubMed] [PubMed Central]
12. Shin, H.R.; Song, S.; Cho, J.A.; Ly, S.Y. Atherogenic Index of Plasma and Its Association with Risk Factors of Coronary Artery Disease and Nutrient Intake in Korean Adult Men: The 2013-2014 KNHANES. *Nutrients* **2022**, *14*, 1071. [CrossRef] [PubMed] [PubMed Central]
13. Lindow, T.; Maanja, M.; Schelbert, E.B.; Ribeiro, A.H.; Ribeiro, A.L.P.; Schlegel, T.T.; Ugander, M. Heart age gap estimated by explainable advanced electrocardiography is associated with cardiovascular risk factors and survival. *Eur. Heart J. Digit. Health* **2023**, *4*, 384–392. [CrossRef] [PubMed] [PubMed Central]
14. Gray, B.J.; Craddock, C.; Couzens, Z.; Dunseath, G.J.; Shankar, A.G.; Luzio, S.D.; Perrett, S.E. Quantifying cardiovascular disease risk and heart age predictions for men in the prison environment. *Public Health* **2023**, *225*, 285–290. [CrossRef] [PubMed]
15. Ramírez-Manent, J.I.; Tomás-Gil, P.; Coll Villalonga, J.L.; Marti-Lliteras, P.; López-González, A.A.; Paublini, H. Relationship between atherogenic dyslipidemia and lipid triad with scales that assess non alcoholic liver disease in 418,343 spanish workers. *Acad. J. Health Sci.* **2023**, *38*, 66–73. [CrossRef]
16. Araújo, Y.B.; Almeida, A.B.R.; Viana, M.F.M.; Meneguz-Moreno, R.A. Use of Atherogenic Indices as Assessment Methods of Clinical Atherosclerotic Diseases. *Arq. Bras. Cardiol.* **2023**, *120*, e20230418. [CrossRef]
17. Fernández-Macías, J.C.; Ochoa-Martínez, A.C.; Varela-Silva, J.A.; Pérez-Maldonado, I.N. Atherogenic Index of Plasma: Novel Predictive Biomarker for Cardiovascular Illnesses. *Arch. Med. Res.* **2019**, *50*, 285–294. [CrossRef] [PubMed]
18. García, J.J.T. Atherogenic indices: Usefulness as predictors of cardiovascular disease. *Clin. Investig. Arterioscler.* **2022**, *34*, 269–270. [CrossRef]
19. Drwiła, D.; Rostoff, P.; Nessler, J.; Konduracka, E. Prognostic value of non-traditional lipid parameters: Castelli Risk Index I, Castelli Risk Index II, and triglycerides to high-density lipoprotein cholesterol ratio among patients with non-ST-segment elevation myocardial infarction during 1 year follow-up. *Kardiologiia* **2022**, *62*, 60–66. [CrossRef]
20. Min, Q.; Wu, Z.; Yao, J.; Wang, S.; Duan, L.; Liu, S.; Zhang, M.; Luo, Y.; Ye, D.; Huang, Y.; et al. Association between atherogenic index of plasma control level and incident cardiovascular disease in middle-aged and elderly Chinese individuals with abnormal glucose metabolism. *Cardiovasc. Diabetol.* **2024**, *23*, 54. [CrossRef]
21. Okan, T.; Doruk, M.; Ozturk, A.; Topaloglu, C.; Dogdus, M.; Yilmaz, M.B. Evaluation of Plasma Atherogenic Index, Triglyceride-Glucose Index and Other Lipid Ratios as Predictive Biomarkers of Coronary Artery Disease in Different Age Groups. *Diagnostics* **2024**, *14*, 1495. [CrossRef] [PubMed]
22. Eliasson, B.; Gudbjörnsdottir, S.; Zethelius, B.; Eeg-Olofsson, K.; Cederholm, J.; National Diabetes Register (NDR). LDL-cholesterol versus non-HDL-to-HDL-cholesterol ratio and risk for coronary heart disease in type 2 diabetes. *Eur. J. Prev. Cardiol.* **2014**, *21*, 1420–1428. [CrossRef] [PubMed]
23. Nawrocka-Rutkowska, J.; Szydłowska, I.; Jakubowska, K.; Olszewska, M.; Chlubek, D.; Szczuko, M.; Starczewski, A. The Role of Oxidative Stress in the Risk of Cardiovascular Disease and Identification of Risk Factors Using AIP and Castelli Atherogenicity Indicators in Patients with PCOS. *Biomedicines* **2022**, *10*, 1700. [CrossRef]
24. Uzunget, S.B.; Sahin, K.E. Atherogenic index of plasma is an independent predictor of mitral annular calcification. *BMC Cardiovasc. Disord.* **2022**, *22*, 511. [CrossRef]
25. Khosravi, A.; Sadeghi, M.; Farsani, E.S.; Danesh, M.; Heshmat-Ghahdarijani, K.; Roohafza, H.; Safaei, A. Atherogenic index of plasma: A valuable novel index to distinguish patients with unstable atherogenic plaques. *J. Res. Med. Sci.* **2022**, *27*, 45. [CrossRef] [PubMed] [PubMed Central]
26. Essiarab, F.; Taki, H.; Lebrazi, H.; Sabri, M.; Saïle, R. Usefulness of lipid ratios and atherogenic index of plasma in obese Moroccan women with or without metabolic syndrome. *Ethn. Dis.* **2014**, *24*, 207–212.
27. Sabarinathan, M.; Rajan, D.S.D.; Ananthi, N.; Krishnan, M. Atherogenic index of plasma, lipid accumulation and visceral adiposity in metabolic syndrome patients. *Bioinformation* **2022**, *18*, 1109–1113. [CrossRef] [PubMed] [PubMed Central]

28. Yin, B.; Wu, Z.; Xia, Y.; Xiao, S.; Chen, L.; Li, Y. Non-linear association of atherogenic index of plasma with insulin resistance and type 2 diabetes: A cross-sectional study. *Cardiovasc. Diabetol.* **2023**, *22*, 157. [CrossRef]
29. Shi, Y.; Wen, M. Sex-specific differences in the effect of the atherogenic index of plasma on prediabetes and diabetes in the NHANES 2011-2018 population. *Cardiovasc. Diabetol.* **2023**, *22*, 19. [CrossRef]
30. Sastre Alzamora, T.; Tomás-Gil, P.; Paublini, H.; Pallarés Ferreres, L.; Ramírez-Manent, L.; López-González, A.A. Relationship between different scales of overweight and obesity and heart age values in 139634 spanish workers. *Acad. J. Health Sci.* **2023**, *38*, 137–144. [CrossRef]
31. Bonner, C.; Batcup, C.; Cornell, S.; Fajardo, M.A.; Hawkes, A.L.; Trevena, L.; Doust, J. Interventions Using Heart Age for Cardiovascular Disease Risk Communication: Systematic Review of Psychological, Behavioral, and Clinical Effects. *JMIR Cardio.* **2021**, *5*, e31056. [CrossRef] [PubMed]
32. Bonner, C.; Raffoul, N.; Battaglia, T.; Mitchell, J.A.; Batcup, C.; Stavreski, B. Experiences of a National Web-Based Heart Age Calculator for Cardiovascular Disease Prevention: User Characteristics, Heart Age Results, and Behavior Change Survey. *J. Med. Internet Res.* **2020**, *22*, e19028. [CrossRef] [PubMed]
33. Starnecker, F.; Reimer, L.M.; Nissen, L.; Jovanović, M.; Kapsecker, M.; Rospleszcz, S.; von Scheidt, M.; Krefting, J.; Krüger, N.; Perl, B.; et al. Guideline-Based Cardiovascular Risk Assessment Delivered by an mHealth App: Development Study. *JMIR Cardio.* **2023**, *7*, e50813. [CrossRef] [PubMed] [PubMed Central]
34. Bonner, C.; Fajardo, M.A.; Hui, S.; Stubbs, R.; Trevena, L. Clinical Validity, Understandability, and Actionability of Online Cardiovascular Disease Risk Calculators: Systematic Review. *J. Med. Internet Res.* **2018**, *20*, e29. [CrossRef] [PubMed] [PubMed Central]
35. Lopez-Gonzalez, A.A.; Aguilo, A.; Frontera, M.; Bennasar-Veny, M.; Campos, I.; Vicente-Herrero, T.; Tomas-Salva, M.; De Pedro-Gomez, J.; Tauler, P. Effectiveness of the Heart Age tool for improving modifiable cardiovascular risk factors in a Southern European population: A randomized trial. *Eur. J. Prev. Cardiol.* **2015**, *22*, 389–396. [CrossRef]
36. Bonner, C.; Bell, K.; Jansen, J.; Glasziou, P.; Irwig, L.; Doust, J.; McCaffery, K. Should heart age calculators be used alongside absolute cardiovascular disease risk assessment? *BMC Cardiovasc. Disord.* **2018**, *18*, 19. [CrossRef]
37. Bonner, C.; Jansen, J.; Newell, B.R.; Irwig, L.; Glasziou, P.; Doust, J.; Dhillon, H.; McCaffery, K. I don't believe it, but i'd better do something about it: Patient experiences of online heart age risk calculators. *J. Med. Internet Res.* **2014**, *16*, e120. [CrossRef]
38. WHO. World Heart Federation. World Stroke Organization: Global Atlas on Cardiovascular Disease Prevention and Control. Published by the World Health Organization in Collaboration with the World Heart Federation and the World Stroke Organization. 2011. Available online: https://iris.who.int/handle/10665/44701 (accessed on 23 July 2024).
39. Stewart, A.; Marfell-Jones, M.; Olds, T.; Ridder, H. *International Standards for Anthropometric Assessment*; International Society for the Advancement of Kinanthropometry–ISAK: Lower Hutt, New Zealand, 2011.
40. Rosales-Rimache, J.; Apaza-Condori, J.; Rabanal-Sanchez, J.; Jari, L.; Soncco-Llulluy, F. Comparison of the Friedewald and Vujovic methods with the calculated LDL concentration in a biochemical auto-analyzer. *Medwave* **2024**, *24*, e2775. [CrossRef] [PubMed]
41. Li, Z.; Cheng, Q.; Liu, Y.; Cheng, X.; Wang, S.; He, Y.; Wang, X.; Huang, M.; Li, Y.; Xue, X.; et al. Low-/high-density lipoprotein cholesterol ratio and carotid plaques in patients with coronary heart disease: A Chinese cohort study. *Lipids Health Dis.* **2021**, *20*, 144. [CrossRef]
42. Morales, M.T.; Hijano-Villegas, S.; Martínez-Llamas, M.S.; López-Barba, J.; Díaz-Portillo, J. *Guía del paciente con trastornos lipídicos*; Ministerio de Sanidad y Consumo. Instituto Nacional de Gestión Sanitaria: Madrid, Spain, 2007.
43. López González, Á.A.; Rivero Ledo, Y.I.; Vicente Herrero, M.T.; Gil Llinás, M.; Tomás Salvá, M.; Riutord Fe, B. Índices aterogénicos en trabajadores de diferentes sectores laborales del área Mediterránea Española. *Clin. Investig. Arterioscler.* **2015**, *27*, 118–128. [CrossRef]
44. Sastre Alzamora, T.; Tomás-Gil, P.; Marti- Lliteras, P.; Pallarés Ferreres, L.; Ramírez-Manent, L.; López-González, A.A. Estimation of heart age in 139.634 spanish workers: Influence of sociodemographic variables and healthy habits and determination of cut-off points. *Acad. J. Health Sci.* **2023**, *38*, 24–30. [CrossRef]
45. Cuende, J.I. Vascular Age, RR, ALLY, RALLY and Vascular Speed, Based on SCORE: Relations Between New Concepts of Cardiovascular Prevention. *Rev. Esp. Cardiol.* **2018**, *71*, 399–400. [CrossRef] [PubMed]
46. Domingo-Salvany, A.; Bacigalupe, A.; Carrasco, J.M.; Espelt, A.; Ferrando, J.; Borrell, C.; del Grupo de Determinantes Sociales de la Sociedad Española de Epidemiología. Propuestas de clase social neoweberiana y neomarxista a partir de la Clasificación Nacional de Ocupaciones 2011. *Gac. Sanit.* **2013**, *27*, 263–272. [CrossRef]
47. Schröder, H.; Fitó, M.; Estruch, R.; Martínez-González, M.A.; Corella, D.; Salas-Salvadó, J.; Lamuela-Raventós, R.; Ros, E.; Salaverría, I.; Fiol, M.; et al. A short screener is valid for assessing Mediterranean diet adherence among older Spanish men and women. *J. Nutr.* **2011**, *141*, 1140–1145. [CrossRef]
48. Craig, C.L.; Marshall, A.L.; Sjöström, M.; Bauman, A.E.; Booth, M.L.; Ainsworth, B.E.; Pratt, M.; Ekelund, U.L.; Yngve, A.; Sallis, J.F.; et al. International physical activity questionnaire: 12-country reliability and validity. *Med. Sci. Sports Exerc.* **2003**, *35*, 1381–1395. [CrossRef] [PubMed]
49. Ruiz-Flores Bistuer, M.; Vicente Herrero, M.T.; Lladosa Marco, S.; Capdevila García, L.M.; López González, Á.A. Prevalencia de consumo de alcohol en trabajadores de la función pública [Alcohol consumption prevalence in civil servants]. *Arch. Prev. Riesgos. Labor.* **2017**, *20*, 111–114. [CrossRef]

50. Hsiao, C.L.; Chen, P.Y.; Hsu, P.J.; Lin, S.K. Nomogram and Carotid Risk Score for Predicting Moderate or High Carotid Atherosclerosis among Asymptomatic Elderly Recycling Volunteers. *Diagnostics* **2022**, *12*, 1407. [CrossRef] [PubMed] [PubMed Central]
51. Berra, K.; Franklin, B.; Jennings, C. Community-Based Healthy Living Interventions. *Prog. Cardiovasc. Dis.* **2017**, *59*, 430–439. [CrossRef] [PubMed]
52. Verde, L.; Docimo, A.; Chirico, G.; Savastano, S.; Colao, A.; Barrea, L.; Muscogiuri, G. How Fast Do "Owls" and "Larks" Eat? *Nutrients* **2023**, *15*, 1437. [CrossRef] [PubMed] [PubMed Central]
53. Lovren, F.; Teoh, H.; Verma, S. Obesity and atherosclerosis: Mechanistic insights. *Can. J. Cardiol.* **2015**, *31*, 177–183. [CrossRef] [PubMed]
54. Ramírez-Manent, J.I.; Tomás-Gil, P.; Coll Villalonga, J.L.; Marti-Lliteras, P.; López-González, A.A.; Paublini, H. Influence of sociodemographic variables and tobacco consumption on the prevalence of atherogenic dyslipidemia and lipid triad in 418.343 spanish workers. *Acad. J. Health Sci.* **2023**, *38*, 84–89. [CrossRef]
55. Group, T. working Management of Peripheral Arterial Disease: Transatlantic Inter-Society Consensus. *J. Vasc. Surg.* **2000**, *31*, S192–S273.
56. Mirończuk, A.; Kapica-Topczewska, K.; Socha, K.; Soroczyńska, J.; Jamiołkowski, J.; Chorąży, M.; Czarnowska, A.; Mitrosz, A.; Kułakowska, A.; Kochanowicz, J. Disturbed Ratios between Essential and Toxic Trace Elements as Potential Biomarkers of Acute Ischemic Stroke. *Nutrients* **2023**, *15*, 1434. [CrossRef] [PubMed] [PubMed Central]
57. Roerecke, M. Alcohol's Impact on the Cardiovascular System. *Nutrients* **2021**, *13*, 3419. [CrossRef] [PubMed] [PubMed Central]
58. Minzer, S.; Losno, R.A.; Casas, R. The Effect of Alcohol on Cardiovascular Risk Factors: Is There New Information? *Nutrients* **2020**, *12*, 912. [CrossRef] [PubMed] [PubMed Central]
59. Barbería-Latasa, M.; Gea, A.; Martínez-González, M.A. Alcohol, Drinking Pattern, and Chronic Disease. *Nutrients* **2022**, *14*, 1954. [CrossRef] [PubMed] [PubMed Central]
60. Hrelia, S.; Di Renzo, L.; Bavaresco, L.; Bernardi, E.; Malaguti, M.; Giacosa, A. Moderate Wine Consumption and Health: A Narrative Review. *Nutrients* **2022**, *15*, 175. [CrossRef] [PubMed] [PubMed Central]
61. Liao, W.; Liu, X.; Kang, N.; Song, Y.; Yuchi, Y.; Hu, Z.; Hou, J.; Wang, C.; Li, Y. Associations between healthy lifestyle score and health-related quality of life among Chinese rural adults: Variations in age, sex, education level, and income. *Qual. Life Res.* **2023**, *32*, 81–92. [CrossRef] [PubMed]
62. Teshale, A.B.; Htun, H.L.; Owen, A.; Gasevic, D.; Phyo, A.Z.Z.; Fancourt, D.; Ryan, J.; Steptoe, A.; Freak-Poli, R. The Role of Social Determinants of Health in Cardiovascular Diseases: An Umbrella Review. *J. Am. Heart Assoc.* **2023**, *12*, e029765. [CrossRef] [PubMed] [PubMed Central]
63. Gullón, P.; Díez, J.; Cainzos-Achirica, M.; Franco, M.; Bilal, U. Social inequities in cardiovascular risk factors in women and men by autonomous regions in Spain. *Gac. Sanit.* **2021**, *35*, 326–332. [CrossRef] [PubMed] [PubMed Central]
64. Basora, J.; Villalobos, F.; Pallejà-Millán, M.; Babio, N.; Goday, A.; Zomeño, M.D.; Pintó, X.; Sacanella, E.; Salas-Salvadó, J. Deprivation Index and Lifestyle: Baseline Cross-Sectional Analysis of the PREDIMED-Plus Catalonia Study. *Nutrients* **2021**, *13*, 3408. [CrossRef] [PubMed] [PubMed Central]
65. Peña-Jorquera, H.; Cid-Jofré, V.; Landaeta-Díaz, L.; Petermann-Rocha, F.; Martorell, M.; Zbinden-Foncea, H.; Ferrari, G.; Jorquera-Aguilera, C.; Cristi-Montero, C. Plant-Based Nutrition: Exploring Health Benefits for Atherosclerosis, Chronic Diseases, and Metabolic Syndrome—A Comprehensive Review. *Nutrients* **2023**, *15*, 3244. [CrossRef] [PubMed]
66. Riutord Sbert, P.; Riutord Fe, B.; Riutord Fe, N.; Arroyo Bote, S.; López González, A.A.; Ramírez Manent, J.I. Relationship between physical activity and adherence to the mediterranean diet with metabolic syndrome, hypertriglyceridemic waist phenotype and hypertensive waist. *Acad. J. Health Sci.* **2022**, *37*, 33–38. [CrossRef]
67. Puche-Juarez, M.; Toledano, J.M.; Ochoa, J.J.; Diaz-Castro, J.; Moreno-Fernandez, J. Influencia del tejido adiposo en la programación metabólica temprana: Factores condicionantes y detección temprana. *Diagnostics* **2023**, *13*, 1510. [CrossRef]
68. Huang, Y.; Cai, X.; Mai, W.; Li, M.; Hu, Y. Association between prediabetes and risk of cardiovascular disease and all cause mortality: Systematic review and meta-analysis. *BMJ* **2016**, *355*, i5953. [CrossRef] [PubMed] [PubMed Central]
69. Di Pino, A.; DeFronzo, R.A. Insulin Resistance and Atherosclerosis: Implications for Insulin-Sensitizing Agents. *Endocr. Rev.* **2019**, *40*, 1447–1467. [CrossRef] [PubMed] [PubMed Central]
70. Poznyak, A.; Grechko, A.V.; Poggio, P.; Myasoedova, V.A.; Alfieri, V.; Orekhov, A.N. The Diabetes Mellitus-Atherosclerosis Connection: The Role of Lipid and Glucose Metabolism and Chronic Inflammation. *Int. J. Mol. Sci.* **2020**, *21*, 1835. [CrossRef] [PubMed] [PubMed Central]
71. Aboonabi, A.; Meyer, R.R.; Singh, I. The association between metabolic syndrome components and the development of atherosclerosis. *J. Hum. Hypertens.* **2019**, *33*, 844–855. [CrossRef] [PubMed]
72. Buliga-Finis, O.N.; Ouatu, A.; Badescu, M.C.; Dima, N.; Tanase, D.M.; Richter, P.; Rezus, C. Beyond the Cardiorenal Syndrome: Pathophysiological Approaches and Biomarkers for Renal and Cardiac Crosstalk. *Diagnostics* **2022**, *12*, 773. [CrossRef] [PubMed] [PubMed Central]
73. Aursulesei Onofrei, V.; Ceasovschih, A.; Marcu, D.T.M.; Adam, C.A.; Mitu, O.; Mitu, F. Mortality Risk Assessment in Peripheral Arterial Disease-The Burden of Cardiovascular Risk Factors over the Years: A Single Center's Experience. *Diagnostics* **2022**, *12*, 2499. [CrossRef] [PubMed] [PubMed Central]

74. Bordeianu, G.; Mitu, I.; Stanescu, R.S.; Ciobanu, C.P.; Petrescu-Danila, E.; Marculescu, A.D.; Dimitriu, D.C. Circulating Biomarkers for Laboratory Diagnostics of Atherosclerosis-Literature Review. *Diagnostics* **2022**, *12*, 3141. [CrossRef] [PubMed] [PubMed Central]
75. Rodríguez-Martínez, B.S.; Hernández-González, M.A.; Borrayo-Sánchez, G.; Solorio-Meza, S.E.; Medina-Torres, Ó.M.; Gutiérrez-Galván, M.X.; De La Roca-Chiapas, J.M.; Cruz-Aceves, I.; Fernández-Hernández, J.P. Índice aterogénico y su relación con la mortalidad del síndrome isquémico coronario agudo [Atherogenic index and its association with acute coronary ischemic syndrome mortality]. *Rev. Med. Inst. Mex. Seguro. Soc.* **2023**, *61*, S226–S232. [PubMed] [PubMed Central]
76. Tamosiunas, A.; Luksiene, D.; Kranciukaite-Butylkiniene, D.; Radisauskas, R.; Sopagiene, D.; Bobak, M. Predictive importance of the visceral adiposity index and atherogenic index of plasma of all-cause and cardiovascular disease mortality in middle-aged and elderly Lithuanian population. *Front. Public Health* **2023**, *11*, 1150563. [CrossRef] [PubMed] [PubMed Central]
77. United Nations, Department of Economic and Social Affairs, Population Division World Population Ageing 2019: Highlights. ST/ESA/SER.A/430. Available online: https://www.un.org/en/development/desa/population/publications/pdf/ageing/WorldPopulationAgeing2019-Highlights.pdf (accessed on 3 August 2024).
78. Hernáez, Á.; Soria-Florido, M.T.; Schröder, H.; Ros, E.; Pintó, X.; Estruch, R.; Salas-Salvadó, J.; Corella, D.; Arós, F.; Serra-Majem, L.; et al. Role of HDL function and LDL atherogenicity on cardiovascular risk: A comprehensive examination. *PLoS ONE* **2019**, *14*, e0218533. [CrossRef] [PubMed] [PubMed Central]
79. López-González, Á.A.; Albaladejo Blanco, M.; Vidal Ribas, C.; Tomás-Gil, P.; Riutord Sbert, P.; Ramírez-Manent, J.I. Determination of the Level of Cardiovascular Risk in 172,282 Spanish Working Women. *Diagnostics* **2023**, *13*, 2734. [CrossRef] [PubMed] [PubMed Central]
80. López-González, Á.A.; Manzanero, Z.; Vicente-Herrero, M.T.; García-Agudo, S.; Gil-Llinás, M.; Moreno-Morcillo, F. Relationship between blood glucose levels and cardiovascular risk in the Spanish Mediterranean population. *Turk. J. Med. Sci.* **2017**, *47*, 754–763. [CrossRef] [PubMed]
81. Elmugadam, A.; Elfadil, G.A.; Hamad, A.I.; El Shikieri, A.B.; Aledrissy, M.; Altayb, H.N. Atherogenic Index of Plasma and Anthropometric Measurements among Osteoporotic Postmenopausal Sudanese Women: Possible Risk for Cardiovascular Disease. *J. Aging Res.* **2022**, *2022*, 1545127. [CrossRef] [PubMed] [PubMed Central]
82. Arias Rebolledo, K.L.; Leguen, F.J.; González Casquero, R.; Roberto Martín, J.A.; Riutord Sbert, P.; Gil Llinás, M. Correlation between heart age and other scales and parameters related to cardiovascular risk. *Acad. J. Health Sci.* **2022**, *37*, 65–70. [CrossRef]
83. Fappi, A.; Mittendorfer, B. Different physiological mechanisms underlie an adverse cardiovascular disease risk profile in men and women. *Proc. Nutr. Soc.* **2020**, *79*, 210–218. [CrossRef] [PubMed] [PubMed Central]
84. Keller, K.M.; Howlett, S.E. Sex Differences in the Biology and Pathology of the Aging Heart. *Can. J. Cardiol.* **2016**, *32*, 1065–1073. [CrossRef] [PubMed]
85. Parkes, G.; Greenhalgh, T.; Griffin, M.; Dent, R. Effect on smoking quit rate of telling patients their lung age: The Step2quit randomised controlled trial. *BMJ* **2008**, *336*, 598–600. [CrossRef] [PubMed] [PubMed Central]
86. Groenewegen, K.A.; den Ruijter, H.M.; Pasterkamp, G.; Polak, J.F.; Bots, M.L.; Peters, S.A. Vascular age to determine cardiovascular disease risk: A systematic review of its concepts, definitions, and clinical applications. *Eur. J. Prev. Cardiol.* **2016**, *23*, 264–274. [CrossRef] [PubMed]
87. Lindow, T.; Palencia-Lamela, I.; Schlegel, T.T.; Ugander, M. Heart age estimated using explainable advanced electrocardiography. *Sci. Rep.* **2022**, *12*, 9840. [CrossRef] [PubMed] [PubMed Central]
88. Bakhit, M.; Fien, S.; Abukmail, E.; Jones, M.; Clark, J.; Scott, A.M.; Glasziou, P.; Cardona, M. Cardiovascular disease risk communication and prevention: A meta-analysis. *Eur. Heart J.* **2024**, *45*, 998–1013. [CrossRef] [PubMed] [PubMed Central]

diagnostics

Article

Identification and Functional Investigation of *SOX4* as a Novel Gene Underpinning Familial Atrial Fibrillation

Wei-Feng Jiang [1], Yu-Min Sun [2], Xing-Biao Qiu [3], Shao-Hui Wu [3], Yuan-Yuan Ding [4], Ning Li [5], Chen-Xi Yang [6,7,8], Ying-Jia Xu [6,7,8], Ting-Bo Jiang [1,*] and Yi-Qing Yang [6,7,8,9,*]

1 Department of Cardiology, The First Affiliated Hospital of Soochow University, Suzhou 215006, China; wf.jiang@hotmail.com
2 Department of Cardiology, Shanghai Jing'an District Central Hospital, Fudan University, Shanghai 200040, China; sunyumin@fudan.edu.cn
3 Department of Cardiology, Shanghai Chest Hospital, Shanghai Jiao Tong University School of Medicine, Shanghai 200030, China; qiuxingbiao@hotmail.com (X.-B.Q.); wushaohui18@163.com (S.-H.W.)
4 Shanghai Health Development Research Center, and Shanghai Medical Information Center, Shanghai 200031, China; dingyuanyuan@shdrc.org
5 Department of Cardiology, Putuo Hospital, Shanghai University of Traditional Chinese Medicine, Shanghai 200062, China; frankleelx@163.com
6 Department of Cardiology, Shanghai Fifth People's Hospital, Fudan University, Shanghai 200240, China; 13472626672@163.com (C.-X.Y.); xuyingjia@5thhospital.com (Y.-J.X.)
7 Center for Complex Cardiac Arrhythmias of Minhang District, Shanghai Fifth People's Hospital, Fudan University, Shanghai 200240, China
8 Department of Cardiovascular Research Laboratory, Shanghai Fifth People's Hospital, Fudan University, Shanghai 200240, China
9 Department of Central Laboratory, Shanghai Fifth People's Hospital, Fudan University, Shanghai 200240, China
* Correspondence: jtbsdfyy@163.com (T.-B.J.); yangyiqing@fudan.edu.cn (Y.-Q.Y.)

Abstract: Background: Atrial fibrillation (AF) signifies the most prevalent supraventricular arrhythmia in humans and may lead to cerebral stroke, cardiac failure, and even premature demise. Aggregating strong evidence points to genetic components as a cornerstone in the etiopathogenesis of familial AF. However, the genetic determinants for AF in most patients remain elusive. **Methods:** A 4-generation pedigree with idiopathic AF and another cohort of 196 unrelated patients with idiopathic AF as well as 278 unrelated healthy volunteers were recruited from the Chinese population of Han ethnicity. A family-based whole-exome sequencing examination followed by a Sanger sequencing assay in all research subjects was implemented. The functional impacts of the identified *SOX4* mutations were explored via a dual-reporter assay. **Results:** Two new heterozygous *SOX4* mutations, NM_003107.3: c.211C>T; p.(Gln71*) and NM_003107.3: c.290G>A; p.(Trp97*), were observed in the family and 1 of 196 patients with idiopathic AF, respectively. The two mutations were absent in the 278 control individuals. The biochemical measurements revealed that both Gln71*- and Trp97*-mutant SOX4 failed to transactivate *GJA1 (Cx43)*. Moreover, the two mutations nullified the synergistic activation of *SCN5A* by SOX4 and TBX5. **Conclusions:** The findings first indicate *SOX4* as a gene predisposing to AF, providing a novel target for antenatal genetic screening, individualized prophylaxis, and precision treatment of AF.

Keywords: atrial fibrillation; human genetics; sequencing examination; SOX4; biochemical assay

1. Introduction

Atrial fibrillation (AF) represents the most prevalent form of cardiac dysrhythmia in the settings of clinical practices, occurring in up to 1% of the general population globally [1,2]. The incidence of this complex heart rhythm disorder increases drastically with increasing age, rising from roughly 1% in cases under 60 years of age to about 9% in those aged over 65 years, and up to 12% in those 75 years old and above [3]. According to the estimated global burden of disease, AF affects at least 43 million people around the globe [4], and merely in the United States, AF was gauged to affect 5.2 million in the year 2010, with an anticipation of up to 12.1 million in the year 2030 [1]. The overall lifetime risk for developing AF is ~15% in Chinese individuals, ~20% in African American persons, and ~30–40% in White folks [1]. Considering that up to two-thirds of subjects with silent/subclinical AF are symptomless and probably undiagnosed, the true incidence of AF is evidently underreckoned [5–7]. It is currently estimated that the actual prevalence of AF is at least 3–4% when asymptomatic AF is included [8]. AF is a major contributor to an extensive variety of adverse clinical consequences, including degraded health-associated quality of life with impaired physical ability [9–13], thromboembolic/ischemic cerebral stroke [14–18], early-onset cognitive decline/dementia [19–23], chronic kidney disease/acute renal injury [24–26], myocardial infarction [27–30], congestive cardiac failure [31–33], life-threatening ventricular dysrhythmias [34–37], and even cardiovascular demise [38–41]. Rather, AF confers an increased risk of multiple adverse outcomes, encompassing a 6-fold risk of chronic renal disease, 1.5-fold risk of dementia/intellectual impairment, 2.4-fold risk of stroke, 1.5-fold risk of myocardial infarction, 1.3-fold risk of peripheral artery disease, 5-fold risk of cardiac failure, and 2-fold risk of cardiac lethality [1]. In retrospective clinical research of Medicare beneficiaries over 65 years of age, after the diagnosis of incident AF, the most common outcome during the first 5 years was mortality (48.8%); the next most frequent was heart failure (13.7%), followed by new-attack cerebral stroke (7.1%), hemorrhage of the gastrointestinal tract (5.7%), and myocardial infarction (3.9%) [1,42]. Consequently, AF has led to substantive morbidity and considerable mortality along with enormous economic incumbrance [1,43–47]. Despite the pronounced significance, the pathogenic mechanisms underlying the initiation and maintenance of AF remain incompletely understood.

It is well known that the etiological mechanisms initiating and sustaining AF are exceedingly complex and extremely distinct, and both non-inherited/environmental hazard factors and inheritable defective ingredients may result in the occurrence and sustainment of AF [1–3,48–51]. There is an increasing amount of non-genetic factors predicted to enhance the susceptibility to AF, encompassing advancing age, alcohol consumption, smoking, inhalation of airborne particulate substances, sedentary lifestyles, obesity, primary arterial hypertension, obstructive sleep apnea, rheumatic heart malady, acute myocardial infarction, dilated cardiomyopathy, peri-atrial inflammation/pericarditis/myocarditis, pneumonia/pulmonary embolism, cardiothoracic surgery, diabetes mellitus, hyperthyroidism, metabolic disorder, β-Thalassemia, and cardiac autonomic nerve imbalance [1,3,48–50,52–60]. However, accumulating epidemiological evidence has convincingly indicated that heritable causative components exert pivotal roles in the development and maintenance of AF, especially for familial AF [2,49–51]. To date, >60 genes causative for AF have been discovered, of which the vast majority code for ion channel subunits of potassium (such as KCNQ1), natrum (such as SCN5A) and calcium (such as CACNA1A), gap junction proteins/connexins (such as GJA1/Cx43), cardiac transcription factors (such as TBX5), signaling molecules (such as ANP), and myocardial structural proteins (such as TTN) [2,49–51,61–72]. Additionally, genome-wide genetic association investigations have unveiled that genetic variants at over 135 chromosomal loci are associated with an increased vulnerability to AF. However, the functionally pathogenic effects of most of these AF-associated variants remain to be validated experimentally [2,73]. Nevertheless, due to conspicuous genetic heterogeneity, the heritable determinants responsible for AF in most patients remain to be ascertained. In addition, there exist potential confounders in genetic studies, such as population stratification and environmental factors, which could affect the interpretation of the results. To address these gaps, the scope of the current research was to identify a novel gene responsible for AF and unravel its functional characteristics.

2. Materials and Methods

2.1. Enrollment and Clinical Study of Individuals

The present retrospective clinical investigation was fulfilled strictly in harmony with the ethical tenets summarized in the Declaration of Helsinki. The institutional ethical panel from Shanghai Chest Hospital approved the protocols applied to the human investigation (with an approved protocol code of KS1101 and an ethical approval date of 12 April 2011 for the study protocol). The research participants or their legal guardians provided signed informed consent forms at initial recruitment, before the commencement of clinical studies.

For the current human investigation, a 4-generation pedigree with a high incidence of idiopathic AF (arbitrarily termed Family AF-01) and another cohort of 196 unrelated patients with idiopathic AF as well as 278 unrelated healthy volunteers without a familial history of AF were recruited from the Chinese population of Han ethnicity between March 2012 and September 2024. The inclusion criteria for the family members and another cohort of AF patients were informed consent and no known causes for AF. The exclusion criteria for the family members and another cohort of AF patients encompassed no informed consent and known causes for AF. Individual-level phenotypic data, which comprised surveys of personal, familial, and medical histories, medical records, transthoracic echocardiographic images, 12-lead electrocardiograms, physical examination findings, and routine bioassay reports, were collected from each research participant in the clinical study. A 24-h or 48-h Holter/ambulatory electrocardiographic monitor was performed whenever indicated. Clinical diagnosis and categorization of AF were implemented as elaborated elsewhere [1]. A whole venous blood sample (3.0 mL) was collected from each study participant and utilized to routinely isolate genomic DNA from the blood leukocytes.

2.2. Genetic Examination

Whole-exome sequencing (WES) assays in three AF-affected and two healthy members from Family AF-01 were conducted as previously described [62,68,74–77]. In brief, a sample of 4 μg of genomic DNA from each family member chosen for the WES assay was used to construct an exome DNA library. Sequencing of exome DNA libraries was completed with the HiSeq Sequencing Kit (Illumina, San Diego, CA, USA) on the HiSeq4000 Genome Analyzer (Illumina, San Diego, CA, USA) as per the manufacturer's manuals. Raw exome-sequencing data were subject to a procedure using Pipeline version 1.9 (Illumina, San Diego, CA, USA) to read bases and align to the human genome (referential GRCh37/hg19) using BWA (version 0.7.12). Calling of sequence variations was implemented by employing SAMtools (version 1.9) and GATK (version 4.1.8), followed by annotation and filtering of genetic variants with ANNOVAR (an online version at http://www.openbioinformatics.org/annovar/) as previously described [62,68,74–77]. Briefly, based on the possible inheritance patterns of AF in the studied pedigree (Family AF-01), the genetic variations that did not match any reasonable inheritance fashion of AF were filtered out. Then the population genetics databases of gnomAD and dbSNP were applied to filter out the synonymous polymorphic variations and any genetic variation with a frequency of >0.001 in either of the two databases. The non-synonymous variations (encompassing nonsense, missense, readthrough, and small insertion/deletion variations as well as splicing donor/acceptor variations) with minor allele frequencies of <0.001 that passed the pedigree assay with reasonable inheritance modes of AF were annotated with ANNOVAR. The deleterious variants annotated by ANNOVAR were selected as candidate pathogenic variants to be subject to validation by Sanger sequencing analysis in the entire family (Family AF-01). Once a rare deleterious variation was confirmed to co-segregate with AF in the whole pedigree, hence a potentially AF-causative mutation, a Sanger sequencing assay of the coding exons along with flanking introns of the genes containing the identified AF-causing mutation was conducted in Family AF-01, and another cohort of 196 unrelated patients with idiopathic AF as well as the 278 healthy subjects served as controls. Additionally, such publicly accessible population genetics databases as dbSNP and gnomAD were retrieved as previously described [74] to check the novelty of the discovered AF-causing mutations.

2.3. Construction of Gene-Expressing Plasmids

Given that *SOX4* is an intronless gene, a 1496-bp fragment including the entire coding exon of wild-type human *SOX4* (Nucleotide accession version: NM_003107.3/GenBank accession version: NC_000006.12) was amplified from human genomic DNA via polymerase chain reaction (PCR) utilizing the DreamTaq™ DNA Polymerase Kit (Thermo Scientific, Carlsbad, CA, USA) with a *SOX4*-specific pair of oligonucleotide primers of 5′-GTAGAATTCCGCGAGGGTGTGAGCGCGC-3′ (forward) and 5′-GTAGTCGACGGCCCT TCTCCCTGCCTGC-3′ (backward). The amplicons containing a whole coding exon of wild-type human *SOX4* and the pCI mammalian expression vector (Promega, Madison, WI, USA) were doubly cut by *EcoRI* and *SalI*, isolated by gel electrophoresis, purified with the GeneJET™ Gel Extraction Kit (Thermo Scientific, Carlsbad, CA, USA), and ligated using T_4 DNA ligase to generate the wild-type human SOX4-pCI recombinant vector. With the wild-type human SOX4-pCI vector utilized as a template, the Gln71*-mutant human SOX4-pCI vector was procured by site-targeted mutagenesis with the Phusion™ Site-Directed Mutagenesis Kit (Thermo Scientific, Carlsbad, CA, USA) in addition to a complimentary pair of oligonucleotide primers (forward: 5′-TTCATGGTGTGGTCGTAGATCGAGCGGCGCA-3′; backward: 5′-TGCGCCGCTCGATCTACGACCACACCATGAA-3′). Similarly, the Trp97*-mutant human SOX4-pCI vector was yielded with a complimentary pair of primers of 5′-CTGGGCAAACGCTAGAAGCTGCTCAAAGA-3′ (forward) and 5′-TCTTTGAGCAGCTTC TAGCGTTTGCCCAG-3′ (backward). Additionally, a 1904-bp promoter fragment (from −97 to −2000 upstream of the transcriptional initial site, with the first transcribed nucleotide being numbered as +1) of the human *GJA1/Cx43* gene (GenBank accession version: NC_000006.12) was amplified from a template of human genomic DNA by PCR utilizing the DreamTaq™ DNA Polymerase Kit (Thermo Scientific, USA) and a *GJA1*-specific pair of oligonucleotide primers of 5′-GCAGAGCTCGACCTAACAAAACATAACAG-3′ (forward) and 5′-GCACTCGAGGAGGTAGGGAGGAGGCTGG-3′ (backward). The amplified *GJA1* promoter and the vector of pGL3-Basic (Promega, Madison, WI, USA) were doubly digested by *Sac*I and *Xho*I, isolated by gel electrophoresis, extracted with a gel extraction kit, and ligated with T_4 DNA ligase to construct the GJA1-luciferase (GJA1-luc) reporter vector expressing firefly luciferase. Similarly, a 2220-bp promoter fragment (from −751 to −2970 upstream of the transcriptional initiation site) of the human *SCN5A* gene (GenBank accession version: NC_000003.12) was amplified utilizing a *SCN5A*-specific pair of oligonucleotide primers of 5′-CTGGGTACCTGAACACTTAGGATTTGTGC-3′ (forward) and 5′-CTGCTCGAGTACACATCTACCTGAACAC-3′ (backward), doubly cut with *Kpn*I and *Xho*I, and inserted into the pGL3-Basic vector (Promega, USA) at the *Kpn*I-*Xho*I sites to construct the SCN5A-luc reporter vector expressing firefly luciferase. The TBX5-pcDNA3.1 plasmid expressing wild-type human TBX5 was produced as previously described [78]. The integrity of all the recombinant expression vectors was validated by DNA sequencing examination.

2.4. Cellular Transfection and Dual-Reporter Gene Measurement

COS-7 and HEK293 cells were routinely cultivated as described elsewhere [79,80]. Cells were counted with an automated hemocytometer (Thermo Scientific, Waltham, MA, USA) and seeded in a 24-well plate (Thermo Scientific, Rochester, NY, USA), maintained for 24 h, and then transiently transfected with various expressing plasmids utilizing the TurboFect Transfection Reagent (Thermo Scientific, Carlsbad, CA, USA). Specifically, COS-7 cells were transfected with 3 ng of pGL4.75 (Promega, Madison, WI, USA), 1.6 μg of GJA1-luc, and 0.4 μg of each recombinant expression plasmid (empty pCI, wild-type human SOX4-pCI, Gln71*-mutant human SOX4-pCI, or Trp97*-mutant human SOX4-pCI, singly or together). For the synergistic activation analysis, HEK293 cells were transfected with 1 ng of pGL4.75 (Promega, Madison, WI, USA), 0.8 μg of SCN5A-luc, and 0.4 μg of each recombinant expression plasmid (empty pCI, wild-type human TBX5-pcDNA3.1, wild-type human SOX4-pCI, Gln71*-mutant human SOX4-pCI, or Trp97*-mutant human SOX4-pCI, alone or in combination). Here, as an internal control, the plasmid of pGL4.75 (Promega, Madison, WI, USA) expressing renilla luciferase was applied to balance transfection efficiency. The plasmid of empty pCI (Promega, Madison, WI, USA) was employed as an external negative control. Cells were harvested 36 h post cellular transfection, lysed, and the firefly/renilla luciferase activities in the cellular lysates were measured on a luminometer (the GloMax® Discover System; Promega, Madison, WI, USA) with the Dual-Glo® Luciferase Assay System (Promega, Madison, WI, USA) as per the manuals and analyzed as described previously [81–83]. Briefly, the promoter activities were expressed as the ratios of firefly to renilla luminescence intensities. For every recombinant expression plasmid, at least three independent cellular transfection tests in vitro were implemented in triplicates, and the mean of three independent experimental results was used for comparison between or among groups.

2.5. Statistical Analysis

The statistical assay was conducted with the aid of SPSS version 17.0 (SPSS, Chicago, IL, USA) as described previously [84–86]. Briefly, such quantitative variables as ages and promoter activities were expressed as means ± standard deviations ($\overline{x} \pm$ SD). Such qualitative variables as sex and family history of AF were expressed as a frequency numeral (%). Unpaired Student's t-test was used to compare continuous parameters between the two groups. A one-way analysis of variance (ANOVA) along with the Tukey–Kramer post-hoc test was applied to the comparison of quantitative variables among three or more groups. A Chi-square or Fisher's exact test was utilized to compare categorical parameters between the two groups. A two-tailed $p < 0.05$ indicated a statistical difference.

3. Results

3.1. Baseline Demographic and Clinical Profiles of the Pedigree with AF and Other Study Participants

As shown clearly in Figure 1, a 27-member family spanning 4 generations with a high incidence of AF (arbitrarily termed Family AF-01) was enrolled, encompassing 26 living family members.

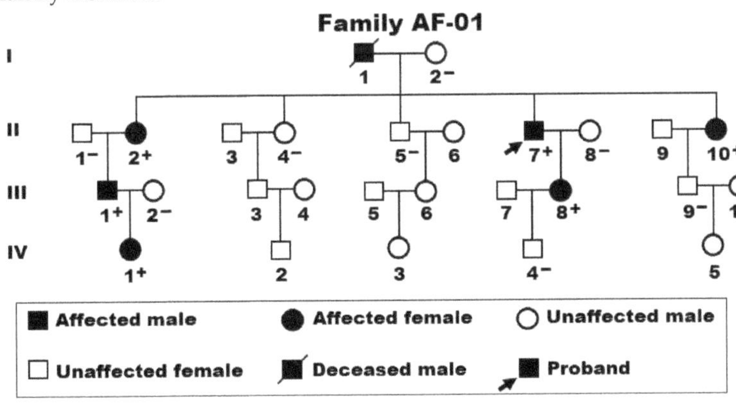

Figure 1. Pedigree suffering from atrial fibrillation. The pedigree was arbitrarily termed as Family AF-01. "+" signifies a member harboring the discovered heterogeneous *SOX4* mutation; "–" represents a non-carrier of the discovered *SOX4* mutation.

In Family AF-01, seven members, including three male members and four female members, were diagnosed with AF according to the electrocardiographic findings. No well-established nonheritable factors predisposing to AF were found in the family members, such as advancing age, obstructive sleep apnea, coronary heart disease, cardiac surgery, dilated cardiomyopathy, pulmonary heart disease, myocarditis, heart failure, primary hypertension, diabetes mellitus, hyperthyroidism, and chronic kidney disease. The proband (member II-7 in Family AF-01), a fifty-seven-year-old male member with twelve years of AF history, was admitted to the hospital because of recurrent syncope. He had an electrocardiogram-documented AF as shown in Figure 2 and then underwent a successful termination of AF by catheter-based radiofrequency ablation.

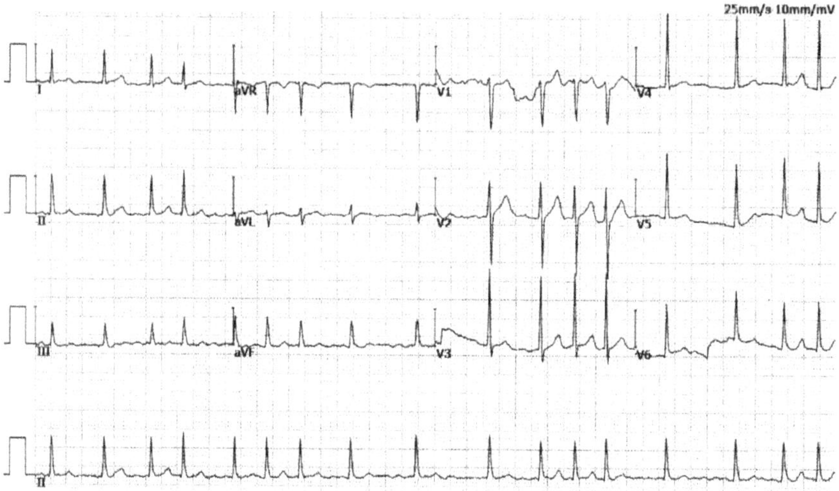

Figure 2. A representative electrocardiogram recorded from the proband (II-7) of Family AF-01. The standard 12-lead electrocardiogram illustrates atrial fibrillation.

The proband's elder sister (member II-2 in Family AF-01), a sixty-five-year-old female individual with seventeen years of AF history, experienced a successful termination of AF by radiofrequency ablation when she was 52 years old. The proband's father (member I-1 in Family AF-01) had thirty-six years of AF history and died of a thromboembolic cerebral stroke at 71 years of age. The index patient's other relatives with AF possessed a definite history of taking antiarrhythmic drugs, but none of them experienced surgical or interventional therapy for AF at the time of recruitment. The proband's unaffected relatives (ten male and ten female subjects) denied a history of AF occurrence, with their electrocardiograms being normal. The basic demographic and clinical characteristic data of the family members with different types of AF are summarized in Table 1.

Table 1. Baseline demographical and clinical characteristic profiles of the pedigree members with atrial fibrillation.

	Subject Information			Cardiac Phenotype	Electrocardiogram			Echocardiogram	
Identity (Family AF-01)	**Sex**	**Age at Initial Diagnosis of AF (years)**	**Age at Recruitment (years)**	**AF (Clinical Classification)**	**Heart Rate (Beats/min)**	**QRS Interval (ms)**	**QTc (ms)**	**LAD (mm)**	**LVEF (%)**
I-1	M	25	71 *	Permanent	NA	NA	NA	NA	NA
II-2	F	35	65	LSP	82	102	438	42	58
II-7	M	45	57	LSP	106	90	416	39	62
II-10	F	41	55	LSP	95	93	330	38	66
III-1	M	37	42	LSP	81	98	420	36	63
III-8	F	34	34	Persistent	84	88	432	33	67
IV-1	F	20	20	Paroxysmal	102	80	405	31	65

M, male; F, female; AF, atrial fibrillation; LAD, left atrial diameter; QTc, corrected QT interval; LVEF, left ventricular ejection fraction; LSP, long-standing persistent; NA, not available. * Age at death.

In addition, all the available family members affected by AF also had dental developmental abnormalities and hypoplastic fifth distal phalanges/nails, as well as mild facial dysmorphisms characteristic of micrognathia with a bulbous nasal tip.

In addition, another cohort of 196 unrelated patients with idiopathic AF and a total of 278 unrelated healthy volunteers without a familial history of AF were enrolled from the Chinese population of Han ethnicity and clinically evaluated. The baseline demographic and clinical characteristic profiles of the cohort of AF patients and the healthy volunteers employed as controls are provided in Table 2.

Table 2. Baseline demographic and clinical characteristic profiles of the cohort of 196 patients with idiopathic atrial fibrillation and the 278 control individuals.

Variable	Patient Group (n = 196)	Control Group (n = 238)	p-Value
Gender (female/male)	88/108	107/131	0.99000
Age (years)	56.93 ± 8.48	57.11 ± 9.27	0.8344
Family history of atrial fibrillation (%)	35 (18)	0 (0)	<0.0001 *
History of implanted pacemaker (%)	8 (4)	0 (0)	0.0032 *
History of ischemic stroke (%)	11 (6)	0 (0)	0.0003 *
Body mass index (kg/m^2)	23.97 ± 2.52	24.12 ± 2.67	0.5506
Total cholesterol (mmol/L)	3.81 ± 0.73	3.78 ± 0.65	0.6511
Fasting blood glucose (mmol/L)	4.58 ± 0.64	4.60 ± 0.71	0.7603
Triglyceride (mmol/L)	1.49 ± 0.47	1.50 ± 0.51	0.8333
Diastolic blood pressure (mmHg)	82.97 ± 6.58	83.05 ± 7.13	0.9042
Systolic blood pressure (mmHg)	125.61 ± 9.73	126.23 ± 10.28	0.5222
Left atrial diameter (mm)	38.75 ± 7.61	36.04 ± 6.83	0.0001 *
Resting heart rate (beats/min)	76.02 ± 14.51	75.84 ± 10.33	0.8804
Left ventricular ejection fraction (%)	62.78 ± 7.09	63.04 ± 6.82	0.6980
History of smoking (%)	9 (5)	11 (5)	0.9882
History of alcohol consumption (%)	18 (9)	23 (10)	0.8648

* $p < 0.05$.

3.2. Discovery of Two New AF-Causative Mutations in SOX4

The WES assay was fulfilled in three AF-affected members (II-7, II-10, and III-8) and two healthy members (III-9 and IV-4) from Family AF-01 (Figure 1), which yielded about 23 Gb of DNA sequences for each member, with approximately 96% mapped to the human reference genome (referential GRCh37/hg19) and about 74% aligned with the target sequences. An average of 17,094 non-synonymous genetic variations per member (ranging from 16,102 to 18,150) passed pedigree filtering based on the presumable transmission modes, among which 13 heterozygous missense/nonsense variations with a minor allele frequency of <0.1% went through filtering by ANNOVAR and were shared by all the three AF-inflicted members (II-7, II-10, and III-8 from Family AF-01) who were subject to WES, as provided in Table 3.

Sanger sequencing examination in all the pedigree members as indicated in Family AF-01 revealed that only the nonsense variation of chr6: 21,594,976C>T (GRCh37.p13/hg19: NC_000006.11), equivalent to chr6: 21,594,745C>T (GRCh38.p14/hg38: NC_000006.12) or NM_003107.3: c.211C>T; p.(Gln71*) in *SOX4*, was verified to be in co-segregation with AF in the whole pedigree. Furthermore, a Sanger sequencing assay of the complete coding region of the *SOX4* gene was executed in all the study subjects using the oligonucleotide primers given in Table 4, which uncovered that the mutation of NM_003107.3: c.211C>T; p.(Gln71*) in *SOX4* existed in all the AF-affected family members but in none of the healthy family members as well as the 278 unrelated control subjects. Genetic assay of Family AF-01 suggested that AF was inherited in an autosomal-dominant fashion.

The sequencing chromatograms showing the heterozygous c.211C>T mutation in *SOX4* together with its wild-type control are presented in Figure 3A. The schemas delineating the main structural domains of the wild-type and Gln71*-mutant human SOX4 proteins are provided in Figure 3B.

Table 3. Non-synonymous exon variants shared by all three members with atrial fibrillation (from Family AF-01) undergoing whole-exome sequencing assay.

Chr	Position (GRCh37)	Ref	Alt	Gene	Variation
1	67,185,057	A	G	*SGIP1*	NM_032291.4: c.1711A>G; p.(Asn571Asp)
1	210,267,717	C	T	*SYT14*	NM_001146261.4: c.628C>T; p.(Pro210Ser)
2	109,289,367	A	G	*LIMS1*	NM_001193485.3: c.562A>G; p.(Lys188Glu)
2	191,375,262	A	T	*NEMP2*	NM_001142645.2: c.955A>T; p.(Lys319*)
3	71,064,787	A	C	*FOXP1*	NM_032682.6: c.887A>C; p.(His296Pro)
4	139,980,574	G	A	*ELF2*	NM_201999.3: c.1311G>A; p.(Gly437Arg)
5	127,686,668	A	C	*FBN2*	NM_001999.4: c.2704A>C; p.(Ile902Leu)
6	21,594,976	C	T	*SOX4*	NM_003107.3: c.211C>T; p.(Gln71*)
7	64,292,159	C	A	*ZNF138*	NM_006524.4: c.461C>A; p.(Ser154*)
10	31,803,533	A	T	*ZEB1*	NM_001128128.3: c.639A>T; p.(Arg217Ser)
11	40,137,306	A	T	*LRRC4C*	NM_020929.3: c.537A>T; p.(Leu179Phe)
15	30,010,872	A	C	*TJP1*	NM_003257.5: c.3474A>C; p.(Glu1158Asp)
18	52,946,829	C	G	*TCF4*	NM_001083962.2: c.608C>G; p.(Ser203Cys)

Chr, chromosome; Alt, alteration; Ref, reference.

Table 4. Primers to amplify the complete coding exon of the human *SOX4* gene.

Coding Region	Forward Primer (5′→3′)	Reverse Primer (5′→3′)	Amplicon (bp)
Part 1	CTCTCTTTACCCACCTCCGC	GACCTTGTCTCCCTTCTCCC	643
Part 2	GCCCAGGAAGAAGGTGAAGT	GCGCCCTCCTCCTCGTACAG	603
Part 3	TGGCGGAGAAGAAGGTGAAG	TCGTCTGTCCTTTTCGTTTCT	647

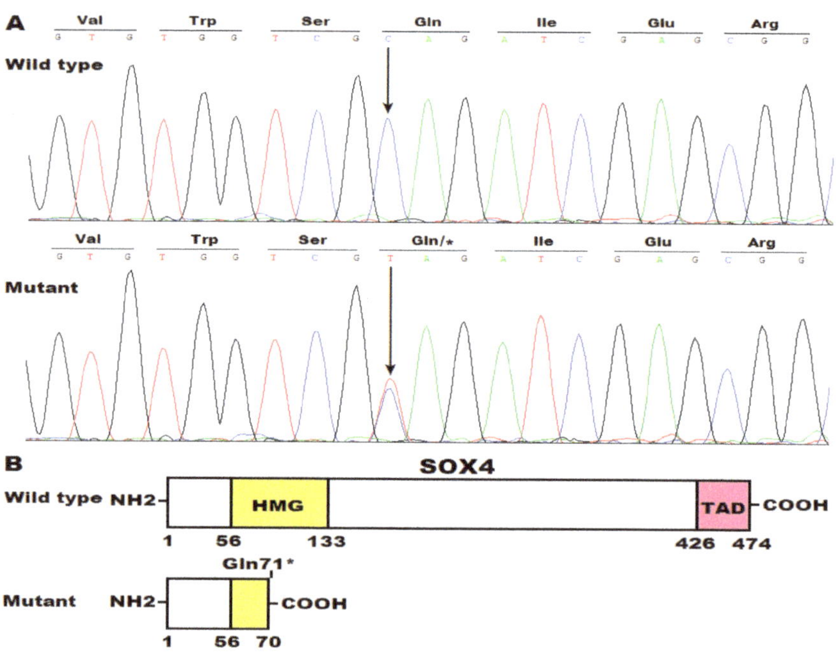

Figure 3. A novel *SOX4* mutation predisposing to atrial fibrillation. (**A**) Sequencing electropherograms exhibiting the heterogeneous SOX4 mutation identified in the index patient with atrial fibrillation (mutant) along with its wild-type control detected in an unaffected subject (wild type). An arrow pinpoints where the mutation occurs. (**B**) Schematic drawings describing the main structural domains of SOX4. TAD: transcriptional activation domain; HMG: high mobility group.

Furthermore, a Sanger sequencing assay of the whole coding region of the SOX4 gene was also conducted in another cohort of 196 unrelated patients with idiopathic AF, which unveiled a heterozygous SOX4 mutation of NM_003107.3: c.290G>A; p.(Trp97*) in one female patient aged 47 years with no family history of AF, who also had delayed intellectual development, bilateral 5th finger clinodactyly, and dysplastic 5th toenails, as well as mild facial dysmorphisms characteristic of upturned nares and a wide mouth with cupid bow. Echocardiographic images demonstrated her normal ventricular function and an enlarged left atrium (48 mm for the left atrial diameter). This mutation was neither found in her parents (indicating a de novo mutation) nor observed in the 278 control subjects. The sequencing chromatograms showing the heterozygous c.290G>A mutation in SOX4 along with its wild-type control are presented in Figure 4A. The schemas describing the main structural domains of the wild-type and Trp97*-mutant human SOX4 proteins are provided in Figure 4B.

A representative electrocardiogram showing AF from the AF patient carrying the SOX4 c.290G>A mutation was given in Figure 5.

Neither of the detected two SOX4 nutations accountable for AF was released from the databases of gnomAD and dbSNP, indicating their novelty.

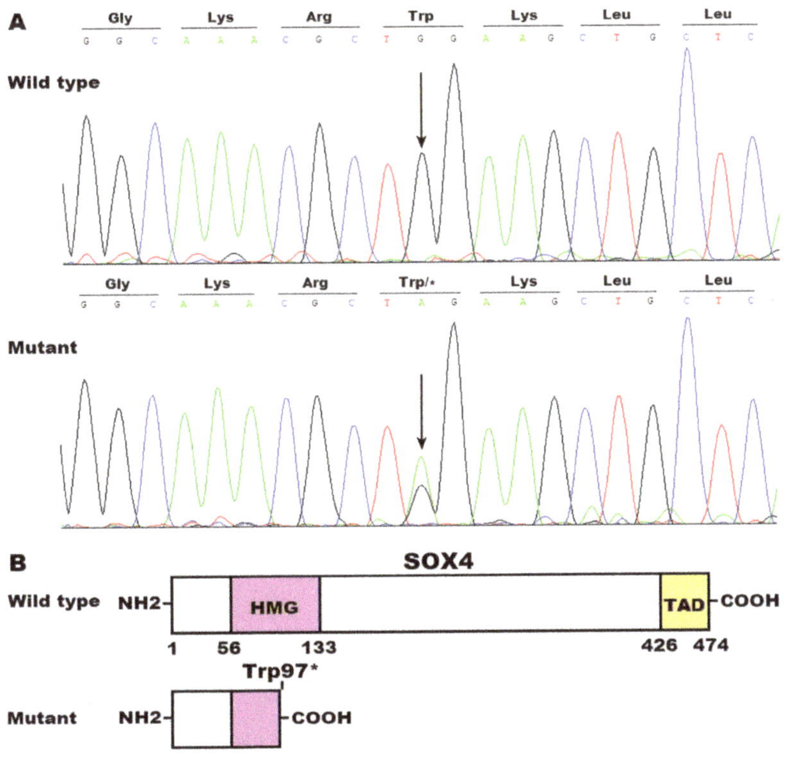

Figure 4. A novel SOX4 mutation contributing to atrial fibrillation. (**A**) Sequencing electropherograms displaying the heterogeneous SOX4 mutation detected in one of the cohort patients with atrial fibrillation (mutant) and its wild-type control detected in a control individual (wild type). An arrow directs where the mutation occurs. (**B**) Schematic diagrams illustrating the main structural domains of SOX4. HMG: high mobility group; TAD: transcriptional activation domain.

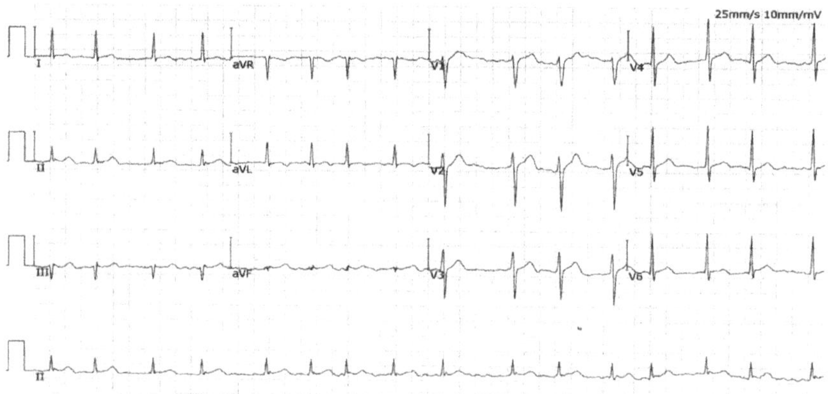

Figure 5. A representative electrocardiogram from the patient harboring the SOX4 c.290G>A mutation. The electrocardiogram documents atrial fibrillation.

3.3. *Inability of Gln71*- or Trp97*-Mutant SOX4 to Transcriptionally Activate GJA1*

As given in Figure 6A, in maintained COS-7 cells expressing multiple vectors, including empty pCI as an external negative control (−), wild-type human SOX4-pCI (SOX4), and Gln71*-mutant human SOX4-pCI (Gln71*), singly or in combination, SOX4 (simulating the physiological state of a healthy person) and Gln71* (simulating the pathological condition of an AF patient with a homozygous SOX4 mutation of Gln71*) transactivated the *GJA1* promoter by ~11-fold and ~1-fold, respectively (SOX4 versus Gln71*: t = 12.4897; p = 0.0002). When SOX4 and Gln71* were co-transfected (simulating the pathological condition of an AF patient with a heterozygous SOX4 mutation of Gln71*), the induced transcriptional activity on the *GJA1* promoter was ~5-fold (SOX4 versus Gln71* + SOX4: t = 6.6464; p = 0.0027). Moreover, equivalent statistical results were generated when multiple comparison analyses were conducted (F = 76.607, $p = 1.842 \times 10^{-7}$). Specifically, for SOX4 versus Gln71*, t = 10.3407; p < 0.0001; for SOX4 versus Gln71* + SOX4, t = 6.0200; p < 0.0001; for SOX4 + (−) versus Gln71* + SOX4, t = 0.9867; p = 0.6296. Similarly, as exhibited in Figure 6B, SOX4 (simulating the physiological state of a healthy person) and Trp97* (simulating the pathological condition of an AF patient with a homozygous SOX4 mutation of Trp97*) transactivated the *GJA1* promoter by ~12-fold and ~1-fold, respectively (SOX4 versus Trp97*: t = 10.8912; p = 0.0004). When SOX4 and Trp97* were co-transfected (simulating the pathological condition of an AF patient with a heterozygous *SOX4* mutation of Trp97*), the induced transcriptional activity on the *GJA1* promoter was ~6-fold (SOX4 versus Trp97* + SOX4: t = 5.1208; p = 0.0069). Moreover, equivalent statistical results were given when multiple comparison analyses were carried out (F = 58.717, $p = 6.583 \times 10^{-7}$). Specifically, for SOX4 versus Trp97*, t = 10.9200; p < 0.0001; for SOX4 versus Trp97* + SOX4, t = 5.8667; p = 0.0003; for SOX4 + (−) versus Trp97* + SOX4, t = 0.8733; p = 0.8354.

3.4. *Failure of Gln71*- or Trp97*-Mutant SOX4 to Transactivate SCN5A Alone or in Synergy with TBX5*

As manifested in Figure 7A, in cultivated HEK-293 cells expressing multiple plasmids, encompassing empty pCI (−), wild-type human SOX4-pCI (SOX4), and Gln71*-mutant human SOX4-pCI (Gln71*), separately or both, SOX4 and Gln71* transactivated the *SCN5A* promoter by ~5-fold and ~1-fold, respectively (SOX4 versus Gln71*: t = 8.3656; p = 0.0011). In the existence of TBX5, SOX4 and Gln71* transactivated the *SCN5A* promoter by ~24-fold and ~8-fold, respectively (SOX4 + TBX5 versus Gln71* + TBX5: t = 8.4538; p = 0.0011). Besides, similar statistical results were produced when multiple comparisons were performed (F = 1047.220, $p = 3.092 \times 10^{-9}$). Specifically, for SOX4 versus Gln71*, t = 4.2333; p = 0.0394; for TBX5 + SOX4 versus TBX5 + Gln71*, t = 15.4567; p < 0.0001. Similarly, as shown in Figure 7B, SOX4 and Trp97* transactivated the *SCN5A* promoter by ~5-fold and ~1-fold, respectively (SOX4 versus Trp97*: t = 7.3407; $p = 0.0018$). In the existence of TBX5, SOX4 and Trp97* transactivated the *SCN5A* promoter by ~22-fold and ~8-fold, respectively (SOX4 + TBX5 versus Trp97* + TBX5: t = 9.3766; $p = 0.0007$). Moreover, similar statistical results were obtained when multiple comparison analyses were conducted ($F = 104.290$, $p = 1.845 \times 10^{-9}$). Specifically, for SOX4 versus Trp97*, t = 3.9567; $p = 0.0259$; for SOX4 + TBX5 versus Trp97* + TBX5, t = 13.9267; $p < 0.0001$.

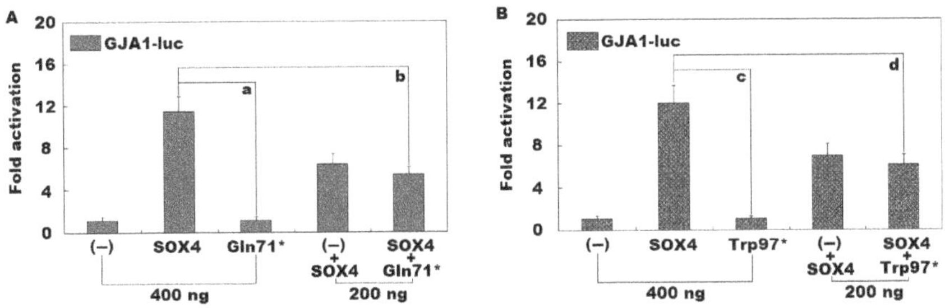

Figure 6. Functional failure of Gln71*- or Trp97*-mutant SOX4. In COS-7 cells cultivated in vitro, dual-reporter measurement of the transactivation of the *GJA1/Cx43* promoter-driven firefly luciferase by wild-type SOX4 or Gln71*-mutant SOX4 (Gln71*) or Trp97*-mutant SOX4 (Trp97*), separately or in combination, unveiled that the Gln71* mutant (**A**) or Trp97* mutant (**B**) failed to transactivate *GJA1/Cx43*. Here "a" and "c" mark $p < 0.001$, and "b" and "d" indicate $p < 0.01$, when compared with wild-type SOX4 (400 ng).

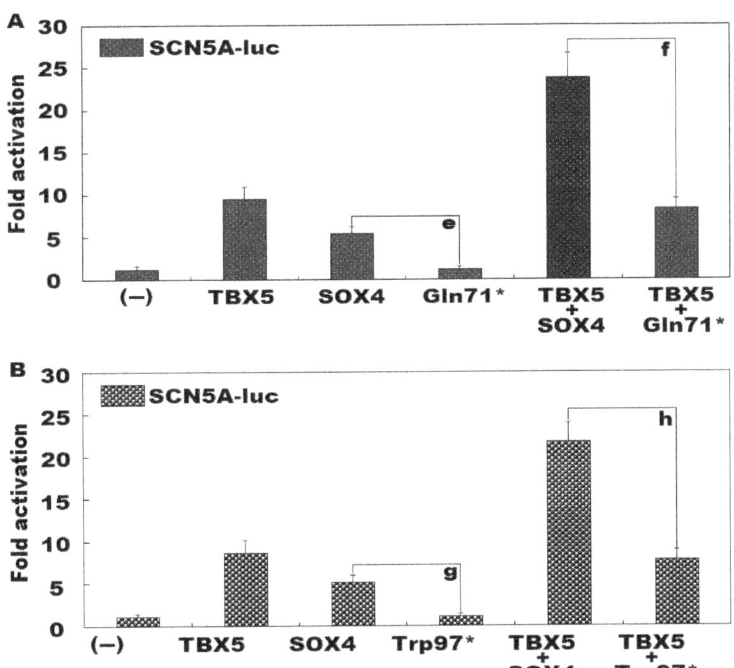

Figure 7. No Synergistic Transactivation of *SCN5A* Between TBX5 and Gln71*- or Trp97*-Mutant SOX4. In cultured HEK293 cells, dual-reporter gene assessment of the synergistic transactivation of the *SCN5A* promoter by TBX5 and SOX4 demonstrated that the synergy was disrupted by the Gln71* (**A**) or Trp97* (**B**) mutation. Herein "e", "f" and "g" denote $p < 0.005$, and "h" means $p < 0.001$ when compared with the corresponding wild-type counterparts.

In addition to multiple SOX4-binding target sites, the congruous DNA sequences of the 5′-WWCAAW-3′ (5′-A/TA/TCAAA/T-3′) motifs within the promoter of *GJA1/Cx43* (GenBank accession version: NC_000006.12) were located and exhibited in Figure 8.

```
-2000 gacctaacaa aacataacag aaataaacac aatttttttt gctttaaaaa tcctattata -1941
-1940 aaacagtcaa tgacaacttt tctttcatgt aacttattct gaattagcat catggtaatg -1881
-1880 ttttttataa ttgtagttga taaataattt tacaaaaata aattgttatg taaattaaat -1821
-1820 gaagcatact tttttgaaaa gaagacatgt attttcttac gataaaaaga ataaatcatt -1761
-1760 ttcatgaata tgttctaagt ccaatgaaat aatttttagt taagacaaaa caatagattt -1701
-1700 atttaaagaa caatataaat acaagaccaa aacaacattt ataggttgct gaacctgaag -1641
-1640 actcgaaagt caggttacaa tttctacttc cagaagattg tttttcccct ttcttaattt -1581
                                  //
-1340 cgaaactccg tctcagaaaa gaaaaaaaaa aacctgcaac atttcaaaag atgttatatt -1281
                                  //
 -680 aacaaaacat gctatatatg taaaaaatgt atatacacac acagcctttt gctaaactga -621
 -620 aagaaagtaa aaattcaaag cagtcaaggg gttgaaagat ggattgggta tggacttttc -561
 -560 ttcatctatg tcaatttttg aaatctctct cccatctcca ccattccctt tgttaattgt -501
 -500 ccttgcttac atttaaagga agggaattag atcatcaaac aggtcactca cacctacctc -441
                                  //
 -320 ccttttttccc ccattatccc tggtcttctg gtgaagcaag cagtgtaatc aatgattctt -261
                                  //
 -140 ccccccagcc cttccacatt taccagcctc ctccctacct cctcccagcc tttccctttg -81
  -80 ccctcccctt tcttctagcc cctcctccca gttgagtcag tggcttgaaa cttttaaaag -21
  -20 ctctgtgctc caagttacaa aaaagctttt acgaggtatc agcacttttc tttcattagg +40
```

Figure 8. Multiple SOX4-binding sites located within the promoter of *GJA1/Cx43*. Red color was used to mark the SOX4-binding sites; green color: the first exon mRNA; underlined bases in bold: primer sequences.

Multiple SOX4- and TBX5-binding sites in the promoter of *SCN5A* (GenBank accession version: NC_000003.12) were located and displayed in Figure 9.

```
-3000 atgtgttcaa caacagaaaa ccctttcaag tgaacactta ggatttgtgc agttttccat -2941
-2940 tatgtaggtt caatggaagt aaaacataaa aagaaaagct ttgggctgtt agaagagaaa -2881
-2880 atgcatatgc acaaacaaca atactacctg ttttacaagg atgtgcacat aaagacctac -2821
-2820 tttagctcct agagggggtgc ctgtgtggag aaggagatgg cagtggagtg atgatggggg -2761
-2760 aagaaataaa acaaaaacac aagagaggag taatgatatg gttaacccag ctttgaacct -2701
                                 //
-2460 actaaaaata caaaaaatta gctgggcatg gtgacgagca cctgcaatcc cagagactca -2401
-2400 ggaggctgag gcaggaggat tgcttgaact ggggaggcgg aggttgcagt gagccgagat -2341
-2340 cacgccactg cacttcagcc ctggaggcaa tgcaagactc cgtctcaaac aaaacaaaac -2281
-2280 aaaacaaaac aaaaaattta aatagggtct ggcaaagaga agggagtagc caagatcttg -2221
-2220 gcaagggctc tttgtaggaa gtgcctgtct ccagacacct gttgcctgtc tctgcactct -2161
-2160 gaggcctctc taggaccaca aggtgtgctt ttatggatga acttctacag gttcagagct -2101
                                 //
-1800 ctactgtact ctcctgcctg gtaggtgtga ctctgggaag gcggctgcac gtgggttgaa -1741
                                 //
-1020 ctgtgaccat gtctctgagt gagaaacagt gggtgtgtaa gtatgtgact ccaggctctg -961
                                 //
 -840 tctctgtgtt ggttgcctgc ttgcttctgg gtctgggtgt gtgtttggtt ctgtcagcat -781
 -780 gtgcgtccgt gtgttcaggt agatgtgtac cattgtgcag atttctgtgg catcactaag -721
                                 //
  -60 cgcccccgac cccgcccccca gcccgagccc gcgccgcctg cccagcccgg gagcccgaac -1
   +1 agagccgcgg agccgagacg gcggcggcgc ccgtaggatg cagggatcgc tcccccgggg +60
```

Figure 9. Multiple SOX4- and TBX5-binding sites mapped within the promoter of *SCN5A*. Red color was used to mark the SOX4-binding sites; blue color: the TBX5-binding sites; green color: the first exon mRNA; underlined bases in bold: primer sequences.

4. Discussion

The SOX family of transcription factor proteins is characteristic of a DNA-binding domain with high homology to the so-called high-mobility group (HMG) box of the sex-determining region Y (*SRY*) gene [87]. To date, there have been 20 members of the SOX family discovered in all vertebrates, which are categorized into 8 groups ranging from SOXA to SOXH [87]. It has been substantiated that the SOX family proteins play critical roles in the development as well as pathological processes of most organs and tissues from the ectoderm, mesoderm, and endoderm, including the heart, central nervous system, pancreas, bone, retina, cartilage, vasculature, hematopoietic system, and lymphatic system, and a large number of genetic diseases caused by genetically compromised SOX proteins, or termed 'SOXopathies,' have been reported, which are involved in the cardiovascular system, reproductive system, muscular system, nervous systems, ocular and auditory systems, urinary system, skin and hair, and skeleton [87,88]. SOX4, together with SOX12 and SOX11, comprises the SOXC group, and SOXC proteins, which share almost identical DNA-binding domains and highly conserved transactivation domains, are amply expressed in many types of progenitor cells and exert redundant functions in regulating cellular survival, differentiation, and fatal determination in response to a great variety of signaling pathways [87]. *SOX4*, a single-exon gene mapped on human chromosome 6p22.3, coding for a 474-amino acid protein [89]. The SOX4 protein possesses 2 functionally key structural domains, including an HMG box and a transcriptional activation domain (TAD) at the C-terminus. The HMG box is responsible for DNA binding, bending, subcellular trafficking, and nuclear distribution, while TAD functions to activate the expression of target genes and mediate the interaction with other transcription factors or cofactors [87]. Previous experiments have revealed that SOX4 is abundantly expressed in the cardiovascular system in both mice and humans, playing critical roles in the embryonic development and postnatal remodeling of the hearts and vessels [90–96]. Noticeably, SOX4 has been validated to transcriptionally regulate the expression of *GJA1*-encoded Cx43, alone or synergistically with the T-box family of transcription factors, including TBX5 [97]. Additionally, TBX5 has been verified to drive the expression of multiple genes, including *SCN5A*, *PITX2*, *GJA1*, and *RyR2* [98–100], and mutations in *SCN5A*, *PITX2*, *GJA1*, and *RyR2* have been reported to cause AF [2,49]. In the current human research, the Gln71* and Trp97* mutations in SOX4 uncovered in patients affected with AF were predicted to produce truncated SOX4 proteins losing TAD along with about halves of the HMG domains, hence anticipated to fail to bind target DNA and activate transcription, and quantitative biochemical measurements demonstrated that Gln71*- or Trp97*-mutant SOX4 failed to transactivate the expression of *GJA1* as well as *SCN5A* in synergy with TBX5. These findings indicate that *SOX4* haploinsufficiency contributes to the pathogenesis of AF probably by reducing the expression of such genes as *GJA1* and *SCN5A*.

The key roles of SOX4 in cardiovascular development and remodeling have been demonstrated in animals and humans [90–96]. In mice, SOX4 is predominantly expressed in the endocardial tissue, including both the outlet tract and atrioventricular canal, and deletion of *Sox4* led to embryonic death because of circulatory failure at embryonic day 14, which was a result of abnormal development of the endocardial ridges into the semilunar valves as well as the outflow tract of the muscular ventricular septum [90,91]. Additionally, in mice with conditional null of *Sox4* alleles ($Sox4^{fl-/fl-}$), embryos died from the same cardiovascular developmental defects as observed in the *Sox4*-knockout mice ($Sox4^{-/-}$) [92]. Both in mice with acute myocardial infarction and in neonatal murine cardiomyocytes treated with hydrogen peroxide (H_2O_2), the expression of SOX4 was markedly increased, and overexpression of SOX4 induced cardiomyocyte apoptosis with or without H_2O_2, whereas knockdown of *Sox4* could alleviate H_2O_2-induced myocardial apoptosis [93]. Furthermore, silencing *Sox4* diminished the area of cardiac infarction, improved heart function, and reversed *Sox4*-induced cardiomyocyte apoptosis in mice with myocardial infarction [93]. Liu and colleagues [94] performed RNA-sequencing analysis of cardiac tissue specimens from patients with heart failure caused by dilated or hypertrophic cardiomyopathy and healthy donors to determine the RNA expression levels of 20 SOX genes and found that only the *SOX4* and *SOX8* RNA levels were significantly upregulated in the heart failure groups, irrespective of four SOX genes whose RNA expression levels were remarkably increased in the dilated or hypertrophic cardiomyopathy group compared to the healthy donor group [94]. Additionally, a moderate to strong association of the RNA level of *SOX4/SOX8* with fibrotic genes was observed, and via a meta-analysis of epigenetics and whole-genome association data, several genomic variants accountable for heart failure were linked to *SOX4* or *SOX8* [94]. Cheng and coworkers [95] identified SOX4 as an atherosclerotic marker in mouse aortic tissues and distinct human arteries (human coronary arteries and human renal arteries), in addition to a cancer marker. Endothelial cell-specific overexpression of SOX4 promoted atherogenesis and increased endothelial-to-mesenchymal transition [95]. In addition, suppression of circHDAC9, a circular RNA, could dramatically lower the level of SOX4 and significantly attenuate myocardial ischemia/reperfusion injury by mediating the miR-671-5p/SOX4 signaling pathway [96]. Importantly, rapidly evolving strong evidence has demonstrated that cardiovascular developmental abnormalities, especially the myocardial sleeves clothing the systemic and pulmonary veins at their junctions with the atria, underly ectopic pace-making electrical activities and are common anatomic matrixes underpinning AF [101–103]. Taken collectively, these research results highlight the fundamental roles of SOX4 in proper cardiovascular development and indicate that genetically defective SOX4 predisposes to AF very likely by creating an abnormal electrophysiological substrate in favor of AF.

Previously in humans, pathogenic *SOX4* variants encompassing causative missense and truncating variants were involved in the etiopathogenesis of Coffin–Siris syndrome, a rare genetic disease characterized by multi-systemic developmental anomalies with a phenotypic spectrum varying from mild cognitive delays to intellectual disabilities with various congenital deformities, including craniofacial dysmorphism, hypoplasia or absence of the fifth distal phalanges/nails, and dental abnormalities [87,104]. Furthermore, copy number variants resulting in entire gene deletion or duplication of *SOX4* were associated with skeletal and mesomelic dysplasia as well as cardiac developmental defects but were not causally linked to any neurodevelopmental disorder [87,105–108]. In the present investigation, two novel *SOX4* mutations were implicated with AF as a notable clinical feature of Coffin–Siris syndrome, thus expanding the *SOX4*-linked phenotypic spectrum. Considering that a significant proportion of AF is paroxysmal or subclinical without symptoms [1], this study indicates that a long-term electrocardiographic screening of the patients with Coffin–Siris syndrome caused by *SOX4* mutations is necessary for the discovery of AF as a component of broader syndromic conditions.

It has been demonstrated that atrial structural remodeling plays a crucial role in the occurrence and maintenance of AF [109]. Furthermore, atrial structural remodeling (characterized by changes in both the size and shape of the left atrium) can be caused by AF itself, which, in turn, results in further progression of AF, a phenomenon referred to as 'AF begets AF' [110,111]. Left atrial enlargement has been an extensively used marker of atrial remodeling, though the left atrium is unsymmetrical, and left atrial dilatation frequently occurs not uniformly in all directions [112]. In the present AF family, significant atrial remodeling as reflected by the enlarged left atrial diameter was observed in the eldest living family member (II-2 from Family AF-01). However, significant atrial remodeling was not observed in the other living family members, which might be explained at least in part by the early echocardiographic detection at the initial diagnosis of AF. Given that left atrial changes represent an advanced stage of AF [111], long-term follow-up with close echocardiographic surveillance will probably discover significant atrial remodeling in the other living family members.

There are some limitations to the current investigation. First, the limited ethnic diversity of the study participants (from a single ethnicity, Han Chinese) and the small sample size may weaken the generalizability of the findings. Second, due to the intrinsic limitations of the WES analysis, we could not rule out the possibility that other genetic mutations might also have a role in the pathogenesis of AF. Specifically, regarding the WES analysis conducted, there are doubts about the robustness of the filtering process used to identify the *SOX4* mutation. The study's reliance on rare, non-synonymous variations with a minor allele frequency of <0.1% might have missed other potential mutations or overemphasized the identified mutation. Third, the study's findings that the *SOX4* mutations are novel and not found in databases like gnomAD and dbSNP are intriguing, but the study lacks broader validation. Fourthly, the mutation was found in all affected family members and none of the controls, yet the possibility of incomplete penetrance or variable expressivity could affect the interpretation of the results. Fifth, the study emphasizes the role of *SOX4* in AF, but the causative link between the mutation and AF is not definitively established, particularly without functional in vivo studies or replication in other ethnic populations. Sixth, we have demonstrated that SOX4 can transactivate the expression of GJA1/Cx43 and SCN5A, alone or in synergy with TBX5, but the complex interaction between SOX4 and the other different players, such as LMNA, PITX2, RyR2, and others for the development of AF, remains to be elucidated. Finally, there is a need for caution in making broad claims about the clinical implications of these findings without further validation.

5. Conclusions

In the current study, two novel *SOX4* loss-of-function mutations are identified to be accountable for idiopathic AF. The findings indicate *SOX4* as a new gene predisposing to AF and unravel a new molecular pathogenesis underlying AF, which provides a novel target for potential prophylactic and therapeutic intervention of AF in a subset of patients.

Author Contributions: Conceptualization, W.-F.J., Y.-Q.Y. and T.-B.J.; methodology, W.-F.J., Y.-Q.Y. and T.-B.J.; validation, W.-F.J., Y.-Q.Y. and T.-B.J.; formal analysis, W.-F.J., Y.-M.S., X.-B.Q., S.-H.W., Y.-Y.D., N.L., C.-X.Y., Y.-J.X., Y.-Q.Y. and T.-B.J.; performed the experiment, W.-F.J., Y.-M.S., N.L., C.-X.Y., Y.-Q.Y. and T.-B.J.; collecting patients, W.-F.J., Y.-M.S., X.-B.Q., S.-H.W., N.L., C.-X.Y., Y.-J.X., Y.-Q.Y. and T.-B.J.; data curation, W.-F.J., Y.-Q.Y. and T.-B.J.; writing—original draft preparation, W.-F.J., Y.-M.S., X.-B.Q., S.-H.W., Y.-Y.D., N.L., C.-X.Y., Y.-J.X., Y.-Q.Y. and T.-B.J.; writing—review and editing, W.-F.J., Y.-Q.Y. and T.-B.J.; visualization W.-F.J., Y.-Q.Y. and T.-B.J.; review the final draft, W.-F.J., Y.-Q.Y. and T.-B.J.; supervision, Y.-Q.Y. and T.-B.J.; project administration, Y.-Q.Y. and T.-B.J.; funding acquisition, Y.-M.S. and Y.-J.X. All authors have read and agreed to the published version of the manuscript.

Funding: This human investigation was funded by the Natural Science Foundation of Shanghai, China (22ZR1454100 and 23ZR1450000).

Institutional Review Board Statement: The present investigation conformed to the tenets of Declaration of Helsinki and approved by the ethical committee of Shanghai Chest Hospital, Shanghai Jiao Tong University School of Medicine, Shanghai, China (ethics approval number: KS1101; ethical approval date: 12 April 2011).

Informed Consent Statement: Informed consent was signed by the subjects participating in the clinical study or their legal guardians.

Data Availability Statement: All the data related to this study are included in the manuscript.

Acknowledgments: The authors cordially thank the research participants for participation in the current clinical investigation.

Conflicts of Interest: The authors declare no conflicts of interest.

References

1. Joglar, J.A.; Chung, M.K.; Armbruster, A.L.; Benjamin, E.J.; Chyou, J.Y.; Cronin, E.M.; Deswal, A.; Eckhardt, L.L.; Goldberger, Z.D.; Gopinathannair, R.; et al. 2023 ACC/AHA/ACCP/HRS Guideline for the Diagnosis and Management of Atrial Fibrillation: A Report of the American College of Cardiology/American Heart Association Joint Committee on Clinical Practice Guidelines. *Circulation* **2024**, *149*, e1–e156. [CrossRef] [PubMed]

2. Lee, D.S.M.; Damrauer, S.M.; Levin, M.G. Genetics of atrial fibrillation. *Curr. Opin. Cardiol.* **2023**, *38*, 162–168. [CrossRef] [PubMed]
3. January, C.T.; Wann, L.S.; Alpert, J.S.; Calkins, H.; Cigarroa, J.E.; Cleveland, J.C., Jr.; Conti, J.B.; Ellinor, P.T.; Ezekowitz, M.D.; Field, M.E.; et al. 2014 AHA/ACC/HRS guideline for the management of patients with atrial fibrillation: A report of the American College of Cardiology/American Heart Association Task Force on practice guidelines and the Heart Rhythm Society. *Circulation* **2014**, *130*, e199–e267. [CrossRef]
4. Hindricks, G.; Potpara, T.; Dagres, N.; Arbelo, E.; Bax, J.J.; Blomström-Lundqvist, C.; Boriani, G.; Castella, M.; Dan, G.A.; Dilaveris, P.E.; et al. 2020 ESC Guidelines for the diagnosis and management of atrial fibrillation developed in collaboration with the European Association for Cardio-Thoracic Surgery (EACTS): The Task Force for the diagnosis and management of atrial fibrillation of the European Society of Cardiology (ESC) Developed with the special contribution of the European Heart Rhythm Association (EHRA) of the ESC. *Eur. Heart J.* **2021**, *42*, 373–498.
5. Imberti, J.F.; Bonini, N.; Tosetti, A.; Mei, D.A.; Gerra, L.; Malavasi, V.L.; Mazza, A.; Lip, G.Y.H.; Boriani, G. Atrial High-Rate Episodes Detected by Cardiac Implantable Electronic Devices: Dynamic Changes in Episodes and Predictors of Incident Atrial Fibrillation. *Biology* **2022**, *11*, 443. [CrossRef]
6. Kim, J.Y.; Park, H.S.; Park, H.W.; Choi, E.K.; Park, J.K.; Kim, J.B.; Kang, K.W.; Shim, J.; Joung, B.; Park, K.M. Clinical Outcomes of Rhythm Control Strategies for Asymptomatic Atrial Fibrillation According to the Quality-of-Life Score: The CODE-AF (Comparison Study of Drugs for Symptom Control and Complication Prevention of Atrial Fibrillation) Registry. *J. Am. Heart Assoc.* **2022**, *11*, e025956. [CrossRef] [PubMed]
7. Lin, J.; Wu, X.Y.; Long, D.Y.; Jiang, C.X.; Sang, C.H.; Tang, R.B.; Li, S.N.; Wang, W.; Guo, X.Y.; Ning, M.; et al. Asymptomatic atrial fibrillation among hospitalized patients: Clinical correlates and in-hospital outcomes in Improving Care for Cardiovascular Disease in China-trial Fibrillation. *Europace* **2023**, *25*, euad272. [CrossRef]
8. Andrade, J.G.; Deyell, M.W.; Bennett, R.; Macle, L. Assessment and management of asymptomatic atrial fibrillation. *Heart* **2024**, *110*, 675–682. [CrossRef]
9. Wazni, O.; Dandamudi, G.; Sood, N.; Hoyt, R.; Tyler, J.; Durrani, S.; Niebauer, M.; Makati, K.; Halperin, B.; Gauri, A.; et al. Quality of life after the initial treatment of atrial fibrillation with cryoablation versus drug therapy. *Heart Rhythm* **2022**, *19*, 197–205. [CrossRef]
10. Seki, Y.; Fujisawa, T.; Ikemura, N.; Ibe, S.; Tsuzuki, I.; Hashimoto, K.; Yamashita, T.; Miyama, H.; Niimi, N.; Suzuki, M.; et al. Catheter ablation improves outcomes and quality of life in Japanese patients with early-stage atrial fibrillation: A retrospective cohort study. *Heart Rhythm* **2022**, *19*, 1076–1083. [CrossRef]
11. Walfridsson, U.; Hassel Jönsson, A.; Karlsson, L.O.; Liuba, I.; Almroth, H.; Sandgren, E.; Walfridsson, H.; Charitakis, E. Symptoms and health-related quality of life 5 years after catheter ablation of atrial fibrillation. *Clin. Cardiol.* **2022**, *45*, 42–50. [CrossRef] [PubMed]
12. Zeitler, E.P.; Li, Y.; Silverstein, A.P.; Russo, A.M.; Poole, J.E.; Daniels, M.R.; Al-Khalidi, H.R.; Lee, K.L.; Bahnson, T.D.; Anstrom, K.J.; et al. Effects of Ablation Versus Drug Therapy on Quality of Life by Sex in Atrial Fibrillation: Results from the CABANA Trial. *J. Am. Heart Assoc.* **2023**, *12*, e027871. [CrossRef] [PubMed]
13. Särnholm, J.; Skúladóttir, H.; Rück, C.; Axelsson, E.; Bonnert, M.; Bragesjö, M.; Venkateshvaran, A.; Ólafsdóttir, E.; Pedersen, S.S.; Ljótsson, B.; et al. Cognitive Behavioral Therapy Improves Quality of Life in Patients with Symptomatic Paroxysmal Atrial Fibrillation. *J. Am. Coll. Cardiol.* **2023**, *82*, 46–56. [CrossRef]
14. Teppo, K.; Airaksinen, K.E.J.; Jaakkola, J.; Halminen, O.; Salmela, B.; Kouki, E.; Haukka, J.; Putaala, J.; Linna, M.; Aro, A.L.; et al. Ischaemic stroke in women with atrial fibrillation: Temporal trends and clinical implications. *Eur. Heart J.* **2024**, *45*, 1819–1827. [CrossRef]
15. Chao, T.F.; Potpara, T.S.; Lip, G.Y.H. Atrial fibrillation: Stroke prevention. *Lancet Reg. Health Eur.* **2024**, *37*, 100797. [CrossRef] [PubMed]
16. Suomalainen, O.P.; Martinez-Majander, N.; Broman, J.; Mannismäki, L.; Aro, A.; Curtze, S.; Pakarinen, S.; Lehto, M.; Putaala, J. Stroke in Patients with Atrial Fibrillation: Epidemiology, Screening, and Prognosis. *J. Clin. Med.* **2023**, *13*, 30. [CrossRef]
17. Lip, G.Y.H.; Proietti, M.; Potpara, T.; Mansour, M.; Savelieva, I.; Tse, H.F.; Goette, A.; Camm, A.J.; Blomstrom-Lundqvist, C.; Gupta, D.; et al. Atrial fibrillation and stroke prevention: 25 years of research at EP Europace journal. *Europace* **2023**, *25*, euad226. [CrossRef]
18. Elsheikh, S.; Hill, A.; Irving, G.; Lip, G.Y.H.; Abdul-Rahim, A.H. Atrial fibrillation and stroke: State-of-the-art and future directions. *Curr. Probl. Cardiol.* **2024**, *49*, 102181. [CrossRef] [PubMed]
19. Kogelschatz, B.; Zenger, B.; Steinberg, B.A.; Ranjan, R.; Jared Bunch, T. Atrial fibrillation and the risk of early-onset dementia and cognitive decline: An updated review. *Trends Cardiovasc. Med.* **2024**, *34*, 236–241. [CrossRef]
20. Giannone, M.E.; Filippini, T.; Whelton, P.K.; Chiari, A.; Vitolo, M.; Boriani, G.; Vinceti, M. Atrial Fibrillation and the Risk of Early-Onset Dementia: A Systematic Review and Meta-Analysis. *J. Am. Heart Assoc.* **2022**, *11*, e025653. [CrossRef]
21. Rivard, L.; Friberg, L.; Conen, D.; Healey, J.S.; Berge, T.; Boriani, G.; Brandes, A.; Calkins, H.; Camm, A.J.; Yee Chen, L.; et al. Atrial Fibrillation and Dementia: A Report From the AF-SCREEN International Collaboration. *Circulation* **2022**, *145*, 392–409. [CrossRef] [PubMed]
22. Blum, S.; Conen, D. Mechanisms and Clinical Manifestations of Cognitive Decline in Atrial Fibrillation Patients: Potential Implications for Preventing Dementia. *Can. J. Cardiol.* **2023**, *39*, 159–171. [CrossRef] [PubMed]

23. Varrias, D.; Saralidze, T.; Borkowski, P.; Pargaonkar, S.; Spanos, M.; Bazoukis, G.; Kokkinidis, D. Atrial Fibrillation and Dementia: Pathophysiological Mechanisms and Clinical Implications. *Biomolecules* **2024**, *14*, 455. [CrossRef] [PubMed]
24. van der Burgh, A.C.; Geurts, S.; Ikram, M.A.; Hoorn, E.J.; Kavousi, M.; Chaker, L. Bidirectional Association Between Kidney Function and Atrial Fibrillation: A Population-Based Cohort Study. *J. Am. Heart Assoc.* **2022**, *11*, e025303. [CrossRef]
25. Öztürk, B.; Göçer, K.; Aksu, E.; Doğan, K. Impaired Renal Vein Flow in Atrial Fibrillation: A Potential Risk for Renal Dysfunction. *Med. Sci. Monit.* **2023**, *29*, e941435. [CrossRef]
26. Venier, S.; Vaxelaire, N.; Jacon, P.; Carabelli, A.; Desbiolles, A.; Garban, F.; Defaye, P. Severe acute kidney injury related to haemolysis after pulsed field ablation for atrial fibrillation. *Europace* **2023**, *26*, euad371. [CrossRef]
27. Yue, B.; Hou, Q.; Bredehorst, J.; Han, Q.; Zhang, B.; Zhang, C.; Zhang, J.; Chen, S.; Wu, S.; Li, K. Atrial fibrillation increases the risk of new-onset myocardial infarction amongst working-age population: A propensity-matched study. *Herz* **2023**, *48*, 408–412. [CrossRef]
28. Wu, J.; Hou, Q.; Han, Q.; Mao, R.; Yue, B.; Yu, J.; Chen, S.; Wu, S.; Li, K. Atrial fibrillation is an independent risk factor for new-onset myocardial infarction: A prospective study. *Acta Cardiol.* **2023**, *78*, 341–348. [CrossRef]
29. Frederiksen, T.C.; Benjamin, E.J.; Trinquart, L.; Lin, H.; Dahm, C.C.; Christiansen, M.K.; Jensen, H.K.; Preis, S.R.; Kornej, J. Bidirectional Association Between Atrial Fibrillation and Myocardial Infarction, and Relation to Mortality in the Framingham Heart Study. *J. Am. Heart Assoc.* **2024**, *13*, e032226. [CrossRef]
30. Gabarin, M.; Hornik-Lurie, T.; Minha, S.; Omelchenko, A.; Barashi, R.; Arow, Z.; Assali, A.; Pereg, D. CHA(2)DS(2)-VASc Score, Mortality and Acute Myocardial Infarction in Patients with Nonvalvular Atrial Fibrillation. *Am. J. Cardiol.* **2022**, *180*, 24–28. [CrossRef]
31. Cismaru, G.; Negru, A.G. Atrial Fibrillation and Heart Failure. *Life* **2024**, *14*, 572. [CrossRef] [PubMed]
32. Ariyaratnam, J.P.; Elliott, A.D.; Mishima, R.S.; Kadhim, K.; McNamee, O.; Kuklik, P.; Emami, M.; Malik, V.; Fitzgerald, J.L.; Gallagher, C.; et al. Identification of Subclinical Heart Failure with Preserved Ejection Fraction in Patients with Symptomatic Atrial Fibrillation. *JACC Heart Fail.* **2023**, *11*, 1626–1638. [CrossRef] [PubMed]
33. Bergonti, M.; Ascione, C.; Marcon, L.; Pambrun, T.; Della Rocca, D.G.; Ferrero, T.G.; Pannone, L.; Kühne, M.; Compagnucci, P.; Bonomi, A.; et al. Left ventricular functional recovery after atrial fibrillation catheter ablation in heart failure: A prediction model. *Eur. Heart J.* **2023**, *44*, 3327–3335. [CrossRef] [PubMed]
34. Fawzy, A.M.; Bisson, A.; Bodin, A.; Herbert, J.; Lip, G.Y.H.; Fauchier, L. Atrial Fibrillation and the Risk of Ventricular Arrhythmias and Cardiac Arrest: A Nationwide Population-Based Study. *J. Clin. Med.* **2023**, *12*, 1075. [CrossRef] [PubMed]
35. Fawzy, A.M.; Bisson, A.; Bentounes, S.A.; Bodin, A.; Herbert, J.; Lip, G.Y.H.; Fauchier, L. Ventricular arrhythmias and cardiac arrest in atrial fibrillation patients with pacemakers and implantable cardioverter-defibrillators. *Eur. J. Intern. Med.* **2023**, *115*, 70–78. [CrossRef] [PubMed]
36. Lenhoff, H.; Järnbert-Petersson, H.; Darpo, B.; Tornvall, P.; Frick, M. Mortality and ventricular arrhythmias in patients on d,l-sotalol for rhythm control of atrial fibrillation: A nationwide cohort study. *Heart Rhythm* **2023**, *20*, 1473–1480. [CrossRef]
37. Kim, Y.G.; Choi, Y.Y.; Han, K.D.; Min, K.; Choi, H.Y.; Shim, J.; Choi, J.I.; Kim, Y.H. Atrial fibrillation is associated with increased risk of lethal ventricular arrhythmias. *Sci. Rep.* **2021**, *11*, 18111. [CrossRef]
38. de Terwangne, C.; Lelubre, C.; Hanotier, P.; de Meester, A.; Descamps, O.; Duray, C.; Pannone, L.; Chierchia, G.B.; de Asmundis, C.; Nokerman, H.; et al. Prevalence and Impact of Atrial Fibrillation on Intra-Hospital Mortality in Patients Aged 75 Years. *Am. J. Cardiol.* **2022**, *177*, 40–47. [CrossRef]
39. Wu, J.; Nadarajah, R.; Nakao, Y.M.; Nakao, K.; Wilkinson, C.; Cowan, J.C.; Camm, A.J.; Gale, C.P. Temporal trends of cause-specific mortality after diagnosis of atrial fibrillation. *Eur. Heart J.* **2023**, *44*, 4422–4431. [CrossRef]
40. Obeid, M.J.; Zhou, J.; Sale, A.J.; Longacre, C.; Zeitler, E.P.; Andrade, J.; Mittal, S.; Piccini, J.P. Early mortality after inpatient versus outpatient catheter ablation in patients with atrial fibrillation. *Heart Rhythm* **2023**, *20*, 833–841. [CrossRef]
41. Zuin, M.; Bertini, M.; Vitali, F.; Turakhia, M.; Boriani, G. Heart Failure-Related Death in Subjects with Atrial Fibrillation in the United States, 1999 to 2020. *J. Am. Heart Assoc.* **2024**, *13*, e033897. [CrossRef] [PubMed]
42. Piccini, J.P.; Hammill, B.G.; Sinner, M.F.; Hernandez, A.F.; Walkey, A.J.; Benjamin, E.J.; Curtis, L.H.; Heckbert, S.R. Clinical course of atrial fibrillation in older adults: The importance of cardiovascular events beyond stroke. *Eur. Heart J.* **2014**, *35*, 250–256. [CrossRef] [PubMed]
43. Chew, D.S.; Li, Y.; Cowper, P.A.; Anstrom, K.J.; Piccini, J.P.; Poole, J.E.; Daniels, M.R.; Monahan, K.H.; Davidson-Ray, L.; Bahnson, T.D.; et al. Cost-Effectiveness of Catheter Ablation Versus Antiarrhythmic Drug Therapy in Atrial Fibrillation: The CABANA Randomized Clinical Trial. *Circulation* **2022**, *146*, 535–547. [CrossRef] [PubMed]
44. Ngo, L.; Denman, R.; Hay, K.; Kaambwa, B.; Ganesan, A.; Ranasinghe, I. Excess Bed Days and Hospitalization Costs Associated with 30-Day Complications Following Catheter Ablation of Atrial Fibrillation. *J. Am. Heart Assoc.* **2023**, *12*, e030236. [CrossRef]
45. Di, J.; Wei, Y.; Zhang, G.; Yue, Y.; Sun, S. Comparison of clinical effects and costs among dabigatran etexilate, rivaroxaban and warfarin in elderly patients with atrial fibrillation. *Am. J. Transl. Res.* **2023**, *15*, 3639–3646.
46. Peigh, G.; Zhou, J.; Rosemas, S.C.; Roberts, A.I.; Longacre, C.; Nayak, T.; Schwab, G.; Soderlund, D.; Passman, R.S. Impact of Atrial Fibrillation Burden on Health Care Costs and Utilization. *JACC Clin. Electrophysiol.* **2024**, *10*, 718–730. [CrossRef]
47. Buja, A.; Rebba, V.; Montecchio, L.; Renzo, G.; Baldo, V.; Cocchio, S.; Ferri, N.; Migliore, F.; Zorzi, A.; Collins, B.; et al. The Cost of Atrial Fibrillation: A Systematic Review. *Value Health* **2024**, *27*, 527–541. [CrossRef]

48. Elliott, A.D.; Middeldorp, M.E.; Van Gelder, I.C.; Albert, C.M.; Sanders, P. Epidemiology and modifiable risk factors for atrial fibrillation. *Nat. Rev. Cardiol.* **2023**, *20*, 404–417. [CrossRef]
49. Vinciguerra, M.; Dobrev, D.; Nattel, S. Atrial fibrillation: Pathophysiology, genetic and epigenetic mechanisms. *Lancet Reg. Health Eur.* **2024**, *37*, 100785. [CrossRef]
50. Young, L.J.; Antwi-Boasiako, S.; Ferrall, J.; Wold, L.E.; Mohler, P.J.; El Refaey, M. Genetic and non-genetic risk factors associated with atrial fibrillation. *Life Sci.* **2022**, *299*, 120529. [CrossRef]
51. Wren, G.; Davies, W. Sex-linked genetic mechanisms and atrial fibrillation risk. *Eur. J. Med. Genet.* **2022**, *65*, 104459. [CrossRef] [PubMed]
52. Giannopoulos, G.; Anagnostopoulos, I.; Kousta, M.; Vergopoulos, S.; Deftereos, S.; Vassilikos, V. Alcohol Consumption and the Risk of Incident Atrial Fibrillation: A Meta-Analysis. *Diagnostics* **2022**, *12*, 479. [CrossRef] [PubMed]
53. Boursiquot, B.C.; Bellettiere, J.; LaMonte, M.J.; LaCroix, A.Z.; Perez, M.V. Sedentary Behavior and Atrial Fibrillation in Older Women: The OPACH Study. *J. Am. Heart Assoc.* **2022**, *11*, e023833. [CrossRef]
54. Ding, M.; Viet, N.N.; Gigante, B.; Lind, V.; Hammar, N.; Modig, K. Elevated Uric Acid Is Associated with New-Onset Atrial Fibrillation: Results from the Swedish AMORIS Cohort. *J. Am. Heart Assoc.* **2023**, *12*, e027089. [CrossRef] [PubMed]
55. Chen, Y.; Wu, Y.; Chu, X.; Wang, M. Meta-analysis of the correlation between recurrence of atrial fibrillation and serum uric acid level after radiofrequency ablation. *Am. J. Transl. Res.* **2022**, *14*, 8793–8799. [PubMed]
56. Xu, Z.; Qian, L.; Zhang, L.; Gao, Y.; Huang, S. Predictive value of NT-proBNP, procalcitonin and CVP in patients with new-onset postoperative atrial fibrillation after cardiac surgery. *Am. J. Transl. Res.* **2022**, *14*, 3481–3487.
57. Hulsmans, M.; Schloss, M.J.; Lee, I.H.; Bapat, A.; Iwamoto, Y.; Vinegoni, C.; Paccalet, A.; Yamazoe, M.; Grune, J.; Pabel, S.; et al. Recruited macrophages elicit atrial fibrillation. *Science* **2023**, *381*, 231–239. [CrossRef]
58. Sun, J.; Xu, J.; Yang, Q. Expression and predictive value of NLRP3 in patients with atrial fibrillation and stroke. *Am. J. Transl. Res.* **2022**, *14*, 3104–3112.
59. Malagù, M.; Marchini, F.; Fiorio, A.; Sirugo, P.; Clò, S.; Mari, E.; Gamberini, M.R.; Rapezzi, C.; Bertini, M. Atrial Fibrillation in β-Thalassemia: Overview of Mechanism, Significance and Clinical Management. *Biology* **2022**, *11*, 148. [CrossRef]
60. Vanchiere, C.; Thirumal, R.; Hendrani, A.; Dherange, P.; Bennett, A.; Shi, R.; Gopinathannair, R.; Olshansky, B.; Smith, D.L.; Dominic, P. Association Between Atrial Fibrillation and Occupational Exposure in Firefighters Based on Self-Reported Survey Data. *J. Am. Heart Assoc.* **2022**, *11*, e022543. [CrossRef]
61. van Wijk, S.W.; Su, W.; Wijdeveld, L.F.J.M.; Ramos, K.S.; Brundel, B.J.J.M. Cytoskeletal Protein Variants Driving Atrial Fibrillation: Potential Mechanisms of Action. *Cells* **2022**, *11*, 416. [CrossRef] [PubMed]
62. Ke, Z.P.; Zhang, G.F.; Guo, Y.H.; Sun, Y.M.; Wang, J.; Li, N.; Qiu, X.B.; Xu, Y.J.; Yang, Y.Q. A novel PRRX1 loss-of-function variation contributing to familial atrial fibrillation and congenital patent ductus arteriosus. *Genet. Mol. Biol.* **2022**, *45*, e20210378. [CrossRef] [PubMed]
63. Vad, O.B.; Yan, Y.; Denti, F.; Ahlberg, G.; Refsgaard, L.; Bomholtz, S.H.; Santos, J.L.; Rasmussen, S.; Haunsø, S.; Svendsen, J.H.; et al. Whole-Exome Sequencing Implicates Neuronal Calcium Channel with Familial Atrial Fibrillation. *Front. Genet.* **2022**, *13*, 806429. [CrossRef]
64. Malakootian, M.; Jalilian, M.; Kalayinia, S.; Hosseini Moghadam, M.; Heidarali, M.; Haghjoo, M. Whole-exome sequencing reveals a rare missense variant in DTNA in an Iranian pedigree with early-onset atrial fibrillation. *BMC Cardiovasc. Disord.* **2022**, *22*, 37. [CrossRef]
65. Guo, Y.H.; Yang, Y.Q. Atrial Fibrillation: Focus on Myocardial Connexins and Gap Junctions. *Biology* **2022**, *11*, 489. [CrossRef] [PubMed]
66. Jameson, H.S.; Hanley, A.; Hill, M.C.; Xiao, L.; Ye, J.; Bapat, A.; Ronzier, E.; Hall, A.W.; Hucker, W.J.; Clauss, S.; et al. Loss of the Atrial Fibrillation-Related Gene, Zfhx3, Results in Atrial Dilation and Arrhythmias. *Circ. Res.* **2023**, *133*, 313–329. [CrossRef]
67. Vad, O.B.; Angeli, E.; Liss, M.; Ahlberg, G.; Andreasen, L.; Christophersen, I.E.; Hansen, C.C.; Møller, S.; Hellsten, Y.; Haunsoe, S.; et al. Loss of Cardiac Splicing Regulator RBM20 Is Associated with Early-Onset Atrial Fibrillation. *JACC Basic Transl. Sci.* **2023**, *9*, 163–180. [CrossRef]
68. Li, N.; Li, Y.J.; Guo, X.J.; Wu, S.H.; Jiang, W.F.; Zhang, D.L.; Wang, K.W.; Li, L.; Sun, Y.M.; Xu, Y.J.; et al. Discovery of TBX20 as a Novel Gene Underlying Atrial Fibrillation. *Biology* **2023**, *12*, 1186. [CrossRef]
69. O'Reilly, M.; Sommerfeld, L.C.; O'Shea, C.; Broadway-Stringer, S.; Andaleeb, S.; Reyat, J.S.; Kabir, S.N.; Stastny, D.; Malinova, A.; Delbue, D.; et al. Familial atrial fibrillation mutation M1875T-SCN5A increases early sodium current and dampens the effect of flecainide. *Europace* **2023**, *25*, 1152–1161. [CrossRef]
70. Chalazan, B.; Freeth, E.; Mohajeri, A.; Ramanathan, K.; Bennett, M.; Walia, J.; Halperin, L.; Roston, T.; Lazarte, J.; Hegele, R.A.; et al. Genetic testing in monogenic early-onset atrial fibrillation. *Eur. J. Hum. Genet.* **2023**, *31*, 769–775. [CrossRef]
71. Huang, Y.; Wang, L.L.; Liu, Z.B.; Chen, C.; Ren, X.; Luo, A.T.; Ma, J.H.; Antzelevitch, C.; Barajas-Martínez, H.; Hu, D. Underlying mechanism of atrial fibrillation-associated Nppa-I137T mutation and cardiac effect of potential drug therapy. *Heart Rhythm* **2024**, *21*, 184–196. [CrossRef] [PubMed]
72. Moreno-Manuel, A.I.; Macías, Á.; Cruz, F.M.; Gutiérrez, L.K.; Martínez, F.; González-Guerra, A.; Martínez Carrascoso, I.; Bermúdez-Jimenez, F.J.; Sánchez-Pérez, P.; Vera-Pedrosa, M.L.; et al. The Kir2.1E299V mutation increases atrial fibrillation vulnerability while protecting the ventricles against arrhythmias in a mouse model of short QT syndrome type 3. *Cardiovasc. Res.* **2024**, *120*, 490–505. [CrossRef] [PubMed]

73. Miyazawa, K.; Ito, K.; Ito, M.; Zou, Z.; Kubota, M.; Nomura, S.; Matsunaga, H.; Koyama, S.; Ieki, H.; Akiyama, M.; et al. Cross-ancestry genome-wide analysis of atrial fibrillation unveils disease biology and enables cardioembolic risk prediction. *Nat. Genet.* **2023**, *55*, 187–197. [CrossRef] [PubMed]
74. Li, Y.J.; Wang, J.; Ye, W.G.; Liu, X.Y.; Li, L.; Qiu, X.B.; Chen, H.; Xu, Y.J.; Yang, Y.Q.; Bai, D.; et al. Discovery of GJC1 (Cx45) as a New Gene Underlying Congenital Heart Disease and Arrhythmias. *Biology* **2023**, *12*, 346. [CrossRef]
75. Gu, J.N.; Yang, C.X.; Ding, Y.Y.; Qiao, Q.; Di, R.M.; Sun, Y.M.; Wang, J.; Yang, L.; Xu, Y.J.; Yang, Y.Q. Identification of BMP10 as a Novel Gene Contributing to Dilated Cardiomyopathy. *Diagnostics* **2023**, *13*, 242. [CrossRef]
76. Huang, R.T.; Guo, Y.H.; Yang, C.X.; Gu, J.N.; Qiu, X.B.; Shi, H.Y.; Xu, Y.J.; Xue, S.; Yang, Y.Q. SOX7 loss-of-function variation as a cause of familial congenital heart disease. *Am. J. Transl. Res.* **2022**, *14*, 1672–1684.
77. Guo, Y.H.; Wang, J.; Guo, X.J.; Gao, R.F.; Yang, C.X.; Li, L.; Sun, Y.M.; Qiu, X.B.; Xu, Y.J.; Yang, Y.Q. KLF13 Loss-of-Function Mutations Underlying Familial Dilated Cardiomyopathy. *J. Am. Heart Assoc.* **2022**, *11*, e027578. [CrossRef] [PubMed]
78. Zhang, X.L.; Qiu, X.B.; Yuan, F.; Wang, J.; Zhao, C.M.; Li, R.G.; Xu, L.; Xu, Y.J.; Shi, H.Y.; Hou, X.M.; et al. TBX5 loss-of-function mutation contributes to familial dilated cardiomyopathy. *Biochem. Biophys. Res. Commun.* **2015**, *459*, 166–171. [CrossRef]
79. Wang, Z.; Qiao, X.H.; Xu, Y.J.; Liu, X.Y.; Huang, R.T.; Xue, S.; Qiu, H.Y.; Yang, Y.Q. SMAD1 Loss-of-Function Variant Responsible for Congenital Heart Disease. *Biomed. Res. Int.* **2022**, *2022*, 9916325. [CrossRef]
80. Wu, S.H.; Wang, X.H.; Xu, Y.J.; Gu, J.N.; Yang, C.X.; Qiao, Q.; Guo, X.J.; Guo, Y.H.; Qiu, X.B.; Jiang, W.F.; et al. ISL1 loss-of-function variation causes familial atrial fibrillation. *Eur. J. Med. Genet.* **2020**, *63*, 104029. [CrossRef]
81. Abhinav, P.; Zhang, G.F.; Zhao, C.M.; Xu, Y.J.; Wang, J.; Yang, Y.Q. A novel KLF13 mutation underlying congenital patent ductus arteriosus and ventricular septal defect, as well as bicuspid aortic valve. *Exp. Ther. Med.* **2022**, *23*, 311. [CrossRef] [PubMed]
82. Shi, H.Y.; Xie, M.S.; Yang, C.X.; Huang, R.T.; Xue, S.; Liu, X.Y.; Xu, Y.J.; Yang, Y.Q. Identification of SOX18 as a New Gene Predisposing to Congenital Heart Disease. *Diagnostics* **2022**, *12*, 1917. [CrossRef] [PubMed]
83. Wang, Y.; Xu, Y.J.; Yang, C.X.; Huang, R.T.; Xue, S.; Yuan, F.; Yang, Y.Q. SMAD4 loss-of-function mutation predisposes to congenital heart disease. *Eur. J. Med. Genet.* **2023**, *66*, 104677. [CrossRef] [PubMed]
84. Dong, B.B.; Li, Y.J.; Liu, X.Y.; Huang, R.T.; Yang, C.X.; Xu, Y.J.; Lv, H.T.; Yang, Y.Q. Discovery of BMP10 as a new gene underpinning congenital heart defects. *Am. J. Transl. Res.* **2024**, *16*, 109–125. [CrossRef]
85. Wang, Z.; Liu, X.Y.; Yang, C.X.; Zhou, H.M.; Li, Y.J.; Qiu, X.B.; Huang, R.T.; Cen, S.S.; Wang, Y.; Xu, Y.J.; et al. Discovery and functional investigation of BMP4 as a new causative gene for human congenital heart disease. *Am. J. Transl. Res.* **2024**, *16*, 2034–2048. [CrossRef]
86. Abhinav, P.; Li, Y.J.; Huang, R.T.; Liu, X.Y.; Gu, J.N.; Yang, C.X.; Xu, Y.J.; Wang, J.; Yang, Y.Q. Somatic GATA4 mutation contributes to tetralogy of Fallot. *Exp. Ther. Med.* **2024**, *27*, 91. [CrossRef]
87. Grippa, M.; Graziano, C. Landscape of Constitutional SOX4 Variation in Human Disorders. *Genes* **2024**, *15*, 158. [CrossRef] [PubMed]
88. Moreno, C.S. SOX4: The unappreciated oncogene. *Semin. Cancer Biol.* **2020**, *67*, 57–64. [CrossRef]
89. Farr, C.J.; Easty, D.J.; Ragoussis, J.; Collignon, J.; Lovell-Badge, R.; Goodfellow, P.N. Characterization and mapping of the human SOX4 gene. *Mamm. Genome* **1993**, *4*, 577–584. [CrossRef] [PubMed]
90. Schilham, M.W.; Oosterwegel, M.A.; Moerer, P.; Ya, J.; de Boer, P.A.; van de Wetering, M.; Verbeek, S.; Lamers, W.H.; Kruisbeek, A.M.; Cumano, A.; et al. Defects in cardiac outflow tract formation and pro-B-lymphocyte expansion in mice lacking Sox-4. *Nature* **1996**, *380*, 711–714. [CrossRef]
91. Ya, J.; Schilham, M.W.; de Boer, P.A.; Moorman, A.F.; Clevers, H.; Lamers, W.H. Sox4-deficiency syndrome in mice is an animal model for common trunk. *Circ. Res.* **1998**, *83*, 986–994. [CrossRef]
92. Penzo-Méndez, A.; Dy, P.; Pallavi, B.; Lefebvre, V. Generation of mice harboring a Sox4 conditional null allele. *Genesis* **2007**, *45*, 776–780. [CrossRef] [PubMed]
93. Zhang, L.; Lv, L.; Zheng, N.; Li, R.; Yang, R.; Li, T.; Li, Y.; Liu, Y.; Luo, H.; Li, X.; et al. Suppression of Sox4 protects against myocardial ischemic injury by reduction of cardiac apoptosis in mice. *J. Cell. Physiol.* **2021**, *236*, 1094–1104. [CrossRef] [PubMed]
94. Liu, C.F.; Ni, Y.; Thachil, V.; Morley, M.; Moravec, C.S.; Tang, W.H.W. Differential expression of members of SOX family of transcription factors in failing human hearts. *Transl. Res.* **2022**, *242*, 66–78. [CrossRef]
95. Cheng, C.K.; Lin, X.; Pu, Y.; Tse, J.K.Y.; Wang, Y.; Zhang, C.L.; Cao, X.; Lau, C.W.; Huang, J.; He, L.; et al. SOX4 is a novel phenotypic regulator of endothelial cells in atherosclerosis revealed by single-cell analysis. *J. Adv. Res.* **2023**, *43*, 187–203. [CrossRef] [PubMed]
96. Liu, Q.; Hu, Y.; Jie, H.; Lu, W.; Chen, Y.; Xing, X.; Tang, B.; Xu, G.; Sun, J.; Liang, Y. CircHDAC9 regulates myocardial ischemia-reperfusion injury via miR-671-5p/SOX4 signaling axis. *Am. J. Med. Sci.* **2024**, *367*, 49–60. [CrossRef]
97. Boogerd, C.J.; Wong, L.Y.; van den Boogaard, M.; Bakker, M.L.; Tessadori, F.; Bakkers, J.; 't Hoen, P.A.; Moorman, A.F.; Christoffels, V.M.; Barnett, P. Sox4 mediates Tbx3 transcriptional regulation of the gap junction protein Cx43. *Cell. Mol. Life Sci.* **2011**, *68*, 3949–3961. [CrossRef]
98. Arnolds, D.E.; Liu, F.; Fahrenbach, J.P.; Kim, G.H.; Schillinger, K.J.; Smemo, S.; McNally, E.M.; Nobrega, M.A.; Patel, V.V.; Moskowitz, I.P. TBX5 drives Scn5a expression to regulate cardiac conduction system function. *J. Clin. Investig.* **2012**, *122*, 2509–2518. [CrossRef]

99. Nieto-Marín, P.; Tinaquero, D.; Utrilla, R.G.; Cebrián, J.; González-Guerra, A.; Crespo-García, T.; Cámara-Checa, A.; Rubio-Alarcón, M.; Dago, M.; Alfayate, S.; et al. Tbx5 variants disrupt Nav1.5 function differently in patients diagnosed with Brugada or Long QT Syndrome. *Cardiovasc. Res.* **2022**, *118*, 1046–1060. [CrossRef]
100. Nadadur, R.D.; Broman, M.T.; Boukens, B.; Mazurek, S.R.; Yang, X.; van den Boogaard, M.; Bekeny, J.; Gadek, M.; Ward, T.; Zhang, M.; et al. Pitx2 modulates a Tbx5-dependent gene regulatory network to maintain atrial rhythm. *Sci. Transl. Med.* **2016**, *8*, 354ra115. [CrossRef]
101. Mommersteeg, M.T.; Christoffels, V.M.; Anderson, R.H.; Moorman, A.F. Atrial fibrillation: A developmental point of view. *Heart Rhythm* **2009**, *6*, 1818–1824. [CrossRef] [PubMed]
102. Postma, A.V.; Dekker, L.R.; Soufan, A.T.; Moorman, A.F. Developmental and genetic aspects of atrial fibrillation. *Trends Cardiovasc. Med.* **2009**, *19*, 123–130. [CrossRef] [PubMed]
103. Kapur, S.; Macrae, C.A. The developmental basis of adult arrhythmia: Atrial fibrillation as a paradigm. *Front. Physiol.* **2013**, *4*, 221. [CrossRef]
104. Zhou, M.; Wang, F.; Dai, Q.; Dou, J.; Wu, Y.; Zhu, Y. Identification of a novel de novo mutation in SOX4 for syndromic tooth agenesis. *Clin. Oral Investig.* **2024**, *28*, 287. [CrossRef]
105. Flöttmann, R.; Wagner, J.; Kobus, K.; Curry, C.J.; Savarirayan, R.; Nishimura, G.; Yasui, N.; Spranger, J.; Van Esch, H.; Lyons, M.J.; et al. Microdeletions on 6p22.3 Are Associated with Mesomelic Dysplasia Savarirayan Type. *J. Med. Genet.* **2015**, *52*, 476–483. [CrossRef] [PubMed]
106. Ladinsky, H.T.; Elizalde, A.; Schickler, R.; Dees, P.B.; Crenshaw, M.L.; Sleasman, J.W. Hypereosinophilic Syndrome and Hemimelia in a Patient with Chromosome 6p22.3 Deletion. *Pediatr. Allergy Immunol.* **2014**, *25*, 500–503. [CrossRef]
107. Trimouille, A.; Barouk-Simonet, E.; Charron, S.; Bouron, J.; Bernhard, J.C.; Lacombe, D.; Fergelot, P.; Rooryck, C. Deletion of the Transcription Factor SOX4 Is Implicated in Syndromic Nephroblastoma. *Clin. Genet.* **2017**, *92*, 449–450. [CrossRef]
108. Gambale, A.; Russo, R.; Andolfo, I.; Quaglietta, L.; De Rosa, G.; Contestabile, V.; De Martino, L.; Genesio, R.; Pignataro, P.; Giglio, S.; et al. Germline Mutations and New Copy Number Variants among 40 Pediatric Cancer Patients Suspected for Genetic Predisposition. *Clin. Genet.* **2019**, *96*, 359–365. [CrossRef] [PubMed]
109. Nattel, S.; Burstein, B.; Dobrev, D. Atrial remodeling and atrial fibrillation. *Circ. Arrhythm. Electrophysiol.* **2008**, *1*, 62–73. [CrossRef]
110. Wijffels, M.C.; Kirchhof, C.J.; Dorland, R.; Allessie, M.A. Atrial fibrillation begets atrial fibrillation. A study in awake chronically instrumented goats. *Circulation* **1995**, *92*, 1954–1968. [CrossRef]
111. Nedios, S.; Dinov, B.; Seewöster, T.; Lindemann, F.; Richter, S.; Arya, A.; Dagres, N.; Husser, D.; Bollmann, A.; Hindricks, G.; et al. Characteristics of left atrial remodeling in patients with atrial fibrillation and hypertrophic cardiomyopathy in comparison to patients without hypertrophy. *Sci. Rep.* **2021**, *11*, 12411. [CrossRef] [PubMed]
112. Mulder, M.J.; Kemme, M.J.B.; Visser, C.L.; Hopman, L.H.G.A.; van Diemen, P.A.; van de Ven, P.M.; Götte, M.J.W.; Danad, I.; Knaapen, P.; van Rossum, A.C.; et al. Left atrial sphericity as a marker of atrial remodeling: Comparison of atrial fibrillation patients and controls. *Int. J. Cardiol.* **2020**, *304*, 69–74. [CrossRef] [PubMed]

diagnostics

Article

Left Ventricular Diastolic Dysfunction Predicts Global Longitudinal Strain Recovery after Surgical Aortic Valve Replacement

Francesca Bonanni [1,2,*], Sabina Caciolli [1], Martina Berteotti [1,3], Andrea Grasso Granchietti [1], Valentina Tozzetti [1], Noemi Cenni [1], Chiara Servoli [1], Marta Bandini [1], Enrico Marchi [1], Stefano Del Pace [1,3,4], Pierluigi Stefano [4] and Niccolò Marchionni [1,3]

1 Division of General Cardiology, Department of Cardiac, Thoracic and Vascular Medicine, Azienda Ospedaliero-Universitaria Careggi, 50134 Florence, Italy; sabinacaciolli@gmail.com (S.C.); martina.berteotti@gmail.com (M.B.); andrea.grasso@unifi.it (A.G.G.); valentina.tozzetti@unifi.it (V.T.); noemi.cenni@unifi.it (N.C.); chiara.servoli@unifi.it (C.S.); marta.bandini@unifi.it (M.B.); enrico.marchi1@gmail.com (E.M.); delpaces@aou-careggi.toscana.it (S.D.P.); niccolo.marchionni@unifi.it (N.M.)
2 Health Science Interdisciplinary Center, Sant'Anna School of Advanced Studies, 56127 Pisa, Italy
3 Department of Experimental and Clinical Medicine, University of Florence, 50134 Florence, Italy
4 Division of Cardiac Surgery, Careggi University Hospital, 50134 Florence, Italy; pierluigi.stefano@unifi.it
* Correspondence: francesca.bonanni@unifi.it

Abstract: Background and Objectives: In patients with severe aortic stenosis (AS), left ventricular systolic dysfunction is one of the main predictors of adverse events after surgical aortic valve replacement (SAVR). However, more patients undergo surgery earlier, often with preserved systolic function. In these cases, global longitudinal strain (GLS) has been proposed as a marker of ventricular remodeling post-surgery. This study aims to evaluate GLS variation in patients undergoing SAVR and explore differences across the diastolic dysfunction classes. Methods: From June 2020 to March 2023, patients with AS and preserved ejection fraction (EF) requiring SAVR were enrolled. Echocardiographic evaluations were conducted preoperatively, seven days post-surgery, and twelve months after surgery. Patients were divided into two groups based on the severity of diastolic dysfunction: Group A (grade I) and Group B (grades II–III). Results: The final analysis included 108 patients (mean age 71.3 ± 7.2 years). Twenty-two patients (20.4%) also underwent coronary artery bypass grafting (CABG). The preoperative EF averaged 61.6 ± 6.03%, with no significant differences between groups. Preoperative GLS was 16 ± 4.3%, decreasing to 12.8 ± 3.4% postoperatively ($p < 0.0001$). GLS was comparable between the groups preoperatively ($p = 0.185$) and postoperatively (0.854). After twelve months, GLS improved in both groups (Group A: 17.7 ± 3.4%, Group B: 15.7 ± 3.2%, $p < 0.0001$), but only Group A showed significant improvement from preoperative values ($p = 0.018$). SAVR improved GLS regardless of CABG intervention. Conclusions: SAVR in patients with preserved LVEF results in an early reduction in GLS, regardless of diastolic dysfunction. After twelve months, GLS improved significantly, with significant recovery only in patients with mild dysfunction.

Keywords: global longitudinal strain; aortic stenosis; cardiac surgery

1. Introduction

Aortic valve stenosis (AS) is the most prevalent acquired heart valve disease in Western countries [1], with a degenerative etiology; in a minority of cases, it is the consequence of rheumatic or infective endocarditis. The prevalence of AS rises with increasing age, affecting nearly 3% of individuals over the age of 75 years, and this valvular disease is characterized by the progressive narrowing of the aortic valve orifice, leading to increased resistance against left ventricular outflow [2,3].

Over time, such a significant pressure overload to the left ventricle (LV) triggers hypertrophic remodeling, which initially helps maintain cardiac output. AS is often associated with left ventricular diastolic dysfunction, while systolic dysfunction appears later during the natural course of the disease; some patients are severely symptomatic, manifesting dyspnea, angina, and syncope, and may have episodes of heart failure even in the presence of preserved LV systolic function [4]. However, defining cases as truly asymptomatic may be challenging, particularly in older and sedentary patients.

Severe, symptomatic AS has a poor prognosis [5], and timely aortic valve replacement (AVR) is mandatory when symptoms or LV systolic dysfunction occur in order to improve long-term prognosis [6]. Without AVR, the survival rate drops dramatically, with a mortality rate exceeding 50% within two years for those with symptomatic AS. Early identification of patients who will benefit from AVR is critical, particularly as the progression of AS may be difficult to predict reliably. Once symptoms manifest, it is often too late to prevent irreversible damage to the myocardium. Current guidelines emphasize early intervention, not only to relieve the mechanical obstruction but also to prevent the downstream effects of prolonged pressure overload on the myocardium, including fibrosis and heart failure. According to guidelines, patients are candidates for surgical (SAVR) or transcatheter (TAVR) AVR based on age, surgical risk, and technical feasibility [6].

LV systolic dysfunction is one of the main predictors of adverse events after AVR, increasing mortality and postoperative complications [7,8]. Subtle LV dysfunction may remain unrevealed by standard routine echocardiography in patients with preserved left ventricular ejection fraction (LVEF). Speckle-tracking echocardiography has been shown to be a promising tool for revealing even initial LV dysfunction [9]. In particular, global longitudinal strain (GLS) proved to be a robust prognostic marker in patients with severe AS and preserved LVEF [5], in whom reduced GLS is usually associated with myocardial fibrosis and ventricular remodeling [10–12]. Conversely, GLS has been shown to improve remarkably after AVR [5,13,14]. However, the correlation between GLS modifications and baseline LV diastolic dysfunction has not yet been investigated. Understanding this correlation may reveal information of crucial clinical relevance because diastolic dysfunction, which is frequently present in AS patients, may have a substantial impact on the prediction of the recovery of myocardial function post-AVR. This study aims to fill this gap in knowledge by exploring how preoperative diastolic function influences the improvement of GLS after SAVR. Exploring this issue might provide valuable insights into the potential benefits of early AVR before the development of advanced diastolic dysfunction, thereby improving long-term outcomes for AS patients

2. Materials and Methods

2.1. Study Population

This observational, prospective, monocentric study enrolled consecutive patients with severe AS (defined as an aortic valve area, AVA, <1.0 cm^2) [6], with an indication for SAVR according to the most recent ESC guidelines [6], who were evaluated at the cardiac surgery department of our hospital from June 2020 to March 2023. The decision to perform SAVR was made after a thorough discussion by the heart team, which included cardiologists, cardiac surgeons, and anesthesiologists, ensuring that all aspects of the patient's health, including their surgical risk and comorbidities, were considered. This interdisciplinary approach permits a balanced evaluation of whether surgical or transcatheter intervention is more appropriate for each patient. SAVR was preferred in patients without any contraindication to open-heart surgery and who were considered to be at an acceptable surgical risk.

Inclusion criteria were (1) severe, symptomatic, high-flow, high-gradient AS with an indication for SAVR, with or without associated coronary artery bypass graft surgery (CABG); (2) LVEF > 50%; (3) age > 18 years. The inclusion of patients undergoing CABG allowed for the investigation of potential confounding effects of coronary artery disease on recovery of myocardial function after SAVR. This was important because coronary artery disease and aortic stenosis often coexist, and addressing both conditions simultaneously could influence the postoperative recovery trajectory.

Exclusion criteria were (1) atrial fibrillation; (2) more than mild associated aortic valve regurgitation; (3) previous heart valve surgery; (4) sub-valvular stenosis or endocarditis; (5) concomitant mitral valve disease requiring surgery; and (6) poor acoustic window.

Clinical and instrumental data were collected immediately before surgery and in the early postoperative period (seven days after surgery) with clinical, electrocardiographic, and echocardiographic data recording. Furthermore, all patients were scheduled for a 12-month follow-up visit, when they were re-evaluated clinically and by echocardiography. The study was approved by the Ethics Committee of Careggi University Hospital, and informed consent for access to individual data was obtained from all patients.

2.2. Echocardiographic Evaluation

Comprehensive transthoracic 2DE and Doppler echocardiographic data were obtained by three experienced cardiologists (SC, MB, SDP) using a commercially available ultrasound system (Vivid 9, General Electric Healthcare, Chicago, IL, USA). The use of such high-end imaging technology allowed for precise measurement and quantification of key echocardiographic parameters, ensuring a high degree of accuracy and reproducibility across different operators. Additionally, the cardiologists' expertise in the study ensured consistency in image acquisition and interpretation, which is essential when evaluating subtle changes in myocardial strain or diastolic parameters that might not be apparent in standard echocardiography. The data were stored and analyzed offline on a dedicated workstation (EchoPAC version 202, General Electric Healthcare, Chicago, IL, USA).

Patients were positioned in left lateral decubitus. Data were acquired in the parasternal (long- and short-axis views) and apical views (two-, three-, and four-chamber apical views).

LV thickness and diameters were measured from the parasternal long-axis view. LV mass was calculated for each exam using Devereux's formula and indexed by body surface area; LV end-diastolic and end-systolic volumes were measured from the apical 2- and 4-chamber views. The LVEF was calculated using Simpson's rule [15].

AS severity was assessed by calculating AVA from the continuity equation. The mean and peak transaortic pressure gradients were calculated by averaging the gradients during systole with the CW Doppler [16]. LV GLS was assessed with speckle-tracking analysis from the three standard 2D apical views. All images were recorded using high-frame-rate loops (50–80 Hz) for accurate software-based analysis. Mean GLS was determined by averaging all values of the 18 segments of the three views. The GLS values, usually reported as negative percentages, are expressed as absolute values in the present work.

LV diastolic function was fully assessed according to the recent updated guidelines [17]: tricuspid regurgitation velocity (TRV) was measured with continuous wave (CW) Doppler; left atrial volume indexed (LAVI) was calculated from apical views; values of early (E wave) and late (A wave) trans-mitral velocities, the E/A ratio, and the E-deceleration time were obtained from pulsed-wave (PW) Doppler recordings; the peak early diastolic velocity (septal and lateral e') and the E/e' ratio were obtained by tissue Doppler.

Following the most recent international guidelines [17], patients were classified by three levels of diastolic dysfunction (I–III), which were used to distinguish two groups: A for grade I (mild) and B for grade II–III (moderate/severe) diastolic dysfunction

2.3. Statistical Analysis

Categorical variables were expressed as absolute and percentage values and compared between groups with a chi-squared test. Continuous variables were expressed as mean ± standard deviation (SD) or median and interquartile range (IQR) in case of non-normal distribution and compared with Student's *t*-test and non-parametric tests as appropriate. Analysis of variance was performed using the ANOVA test and non-parametric tests as appropriate. A $p < 0.05$ was considered statistically significant. Statistical analysis was performed with SPSS for Macintosh (v. 28.0).

3. Results

3.1. Baseline Characteristics

One hundred eight consecutive patients were enrolled in the study. Their mean age was 71.3 ± 7.2 years, and 57 were men (52.8%; Table 1).

Table 1. Baseline characteristics of our study population at preoperative evaluation according to diastolic dysfunction groups.

Variables	Total (n = 108)	Group A (n = 69)	Group B (n = 39)	p-Value
Sex (male), n (%)	57 (52.8)	42 (60.9)	15 (38.5)	0.029
Age, (years), mean ± SD	71.3 ± 7.2	70.5 ± 7.6	72.6 ± 6.3	0.147
BMI (kg/m^2), mean ± SD	26.87 ± 4	26.18 ± 3.7	28.08 ± 4.1	0.017
Hypertension, n (%)	76 (70.4)	46 (66.7)	30 (76.9)	0.283
Diabetes mellitus, n (%)	24 (22.2)	14 (20.3)	10 (25.6)	0.631
Dyslipidemia, n (%)	67 (62)	43 (62.3)	24 (61.5)	1.000
Chronic kidney disease, n (%)	9 (8.3)	3 (4.3)	6 (15.4)	0.069
Smoke, n (%)	51 (47.2)	28 (40.6)	23 (59)	0.074
Indication to CABG, n (%)	22 (20.4)	15 (21.7)	7 (17.9)	0.804
Euroscore II (%), median [IQR]	1.43 [1–2.2]	1.44 [1–2.2]	1.43 [1–2.2]	1.000
Hospital stay (days), median [IQR]	11 [8–14.8]	10 [8–13.5]	12 [10–17]	0.187

Continuous variables are presented as median [IQR] or mean ± SD, whereas categorical variables are presented as n (%). Group A: diastolic dysfunction grade I; Group B: diastolic dysfunction grades II–III. BMI: body mass index; CABG: coronary artery bypass graft surgery.

Ischemic heart disease with an indication for CABG coexisted in 22 patients (20.4%), 6 of whom (27.3%) for acute coronary syndromes and 4 (18.2%) for three-vessel disease. Most patients had mild diastolic dysfunction (69, 63.9%, Group A; Table 1).

Notably, there was a significant difference in sex distribution between the two groups (p = 0.029). In Group A, 42 patients (60.9%) were men, whereas only 15 patients (38.5%) in Group B were men, indicating a higher proportion of women with moderate-to-severe diastolic dysfunction.

The mean BMI was significantly higher in Group B compared with Group A (28.08 ± 4.1 kg/m^2 vs. 26.18 ± 3.7 kg/m^2, p = 0.017), suggesting that patients with more severe diastolic dysfunction tended to have a higher BMI. The presence of comorbidities such as hypertension, diabetes, dyslipidemia, and smoking history did not significantly differ between the two groups. However, chronic kidney disease was more prevalent in Group B (15.4%) compared with Group A (4.3%), although this difference did not attain statistical significance (p = 0.069). The need for CABG during surgery was similar between the two groups, with 15 patients (21.7%) in Group A and 7 patients (17.9%) in Group B undergoing CABG in addition to aortic valve replacement (p = 0.804). Significant coronary artery disease of the left main coronary artery (LMCA) was observed in 4 patients—3 in Group A and 1 in Group B (p = 0.746)—while for the circumflex artery, the prevalence was 11 patients—8 in Group A and 3 in Group B (p = 0.647). For the left anterior descending artery (LAD), it was 14 patients: 9 in Group A and 5 in Group B (p = 0.604). For the right coronary artery (RCA), 13 patients were affected: 8 in Group A and 5 in Group B (p = 0.421).

Hospital stay was also slightly longer in Group B, with a median of 12 days compared with 10 days in Group A, though this difference was not statistically significant (p = 0.187).

The echocardiographic parameters of our study population are summarized in Table 2.

Table 2. Echocardiographic parameters.

Variables	Preoperative Total (n = 108)	Preoperative Group A (n = 69)	Preoperative Group B (n = 39)	p A1 vs. B1	Early Postoperative Total (n = 108)	Early Postoperative Group A (n = 6)	Early Postoperative Group B (n = 39)	p A2 vs. B2	p Total 1 vs. 2	p A1 vs. A2	p B1 vs. B2
LVEF (%)	63 [58–65]	61 [57.5–65]	64 [60–66]	0.562	59 [56–61]	60 [56–60]	60 [57–63]	0.721	<0.001	<0.001	0.002
LV/mass/ BSA (g/m^2)	130.4 ± 30.1	127.8 ± 30.2	134 ± 29.6	0.238	116.6 ± 35.6	111.9 ± 38.3	124.7 ± 29.2	0.074	<0.001	<0.001	0.027
IVS (mm)	13 [12–14]	13 [12–14]	13 [12–14]	0.338	12 [11–13]	12 [11–13]	13 [11–13]	0.292	<0.001	0.032	0.010
PWT (mm)	11 [11–12]	11 [10.25–12]	12 [11–12]	0.010	11 [10–12]	11 [10–12]	11 [10–12]	0.478	0.320	0.683	0.022
LVD (mm)	50 [45–53]	50 [45.5–54]	48 [45–52]	0.253	47 [45–53]	47 [45–52]	48 [45–54]	0.592	0.04	<0.001	0.837
LAVI (mL/m^2)	39.1 ± 12.9	35.9 ± 12.1	44.8 ± 12.4	<0.001	34.4 ± 13	30.9 ± 11.5	40.9 ± 13.2	<0.001	<0.001	0.001	0.036
TAPSE (mm)	22.5 ± 3.6	22.3 ± 3.5	23 ± 3.8	0.310	16 ± 3.1	16.6 ± 3.1	16.5 ± 3.3	0.905	<0.001	<0.001	<0.001
PAPS (mmHg)	29 [25–34]	28.5 [24–32.8]	33 [28–36]	0.001	26 [22–33]	26 [22–31]	28 [25–35]	0.071	0.009	0.263	0.010
RVD2 (mm)	28.4 ± 6.7	28.8 ± 6.6	27.9 ± 6.9	0.530	28.5 ± 5.9	28.8 ± 5.8	27.9 ± 6.2	0.435	0.975	0.988	0.966
Grad-Peak (mmHg)	86 ± 27.9	80.7 ± 26.2	95.3 ± 28.8	0.010	13.3 ± 4.3	13.0 ± 3.9	13.7 ± 4.8	0.630	<0.001	<0.001	<0.001
Grad-Mean (mmHg)	56.7 ± 19.1	54.11 ± 18.1	61.2 ± 20.1	0.065	9.2 ± 3.1	9.4 ± 3.15	9 ± 3.8	0.435	<0.001	<0.001	<0.001
Mean GLS (%)	16 ± 4.3	15.6 ± 4.3	16.7 ± 4.2	0.185	12.8 ± 3.4	12.7 ± 3.3	12.8 ± 3.5	0.854	<0.001	<0.001	<0.001

Continuous variables are presented as median [IQR] or mean ± SD. Group A or A: diastolic dysfunction grade I; Group B or B: diastolic dysfunction grades II–III; 1: preoperative; 2: early postoperative. LVEF: left ventricular ejection fraction; IVS: interventricular septum; PWT: posterior wall thickness; LVD: left ventricular diameter; LAVI: left atrial volume index; TAPSE: tricuspid annular plane systolic excursion; PAPS: systolic pulmonary artery pressure; RVD2: mid-cavity diameter of the right ventricle; Grad-Peak: aortic valve peak gradient; Grad-Mean: aortic valve mean gradient.

The LV mass indexed to body surface area (LV mass/BSA) was slightly higher in Group B preoperatively (134 ± 29.6 g/m^2) compared with Group A (127.8 ± 30.2 g/m^2), though the difference was not statistically significant (p = 0.238). In the early postoperative period, both groups showed a significant reduction in LV mass (p < 0.001 for both). In terms of left atrial volume index (LAVI), Group B had significantly higher values preoperatively (44.8 ± 12.4 mL/m^2 vs. 35.9 ± 12.1 mL/m^2, p < 0.001), and this difference persisted postoperatively (40.9 ± 13.2 mL/m^2 vs. 30.9 ± 11.5 mL/m^2, p < 0.001). Both groups showed a significant reduction in LAVI postoperatively (p < 0.001 for both groups). Systolic pulmonary artery pressure (PAPS) was significantly higher in Group B preoperatively (33 mmHg vs. 28.5 mmHg, p = 0.001). Postoperatively, both groups showed a reduction in PAPS (Group A: 26 mmHg, Group B: 28 mmHg, p = 0.071), though the reduction was statistically significant only for Group B (p = 0.010) but not for Group A (p = 0.0263).

The peak aortic valve gradient was significantly higher in Group B preoperatively (95.3 ± 28.8 mmHg vs. 80.7 ± 26.2 mmHg, p = 0.010). Both groups experienced a significant reduction in peak gradient postoperatively (p < 0.001 for both), with no significant difference between groups in the early postoperative period (p = 0.630).

The aortic valve area (AVA) was 0.73 ± 0.13 cm^2 in Group A and 0.65 ± 0.13 cm^2 in Group B (p = 0.001). The indexed aortic valve area (iAVA) was 0.40 ± 0.09 cm^2/m^2 in Group A and 0.37 ± 0.09 cm^2/m^2 in Group B (p = 0.045).

The prevalence of mild mitral regurgitation was 23 cases (21.3%): 13 in Group A and 10 in Group B (p = 0.407). The prevalence of mild aortic regurgitation was 31 cases (28.7%): 20 in Group A and 11 in Group B (p = 0.931).

The mean GLS values before surgery were 16 ± 4.3%; at early postoperative examination, we observed a reduction to 12.8 ± 3.4%, p < 0.0001. Mean GLS was similar between patients in the two classes of diastolic dysfunction at both preoperative (p = 0.185) and early postoperative (p = 0.854) examination (Table 2).

The analysis was repeated in the 54 and 32 patients undergoing SAVR without CABG in groups A and B to verify whether coronary revascularization might be a confounder. Mean GLS was 16.6 ± 3.8% before surgery and 13.2 ± 2.9% in the early postoperative evaluation, p < 0.0001. Also in this case, values were similar in the two diastolic dysfunction groups: 16.5 ± 3.8% before surgery in Group A and 16.8 ± 4% in Group B (p = 0.718); 13.3 ± 2.8% in the early postoperative in Group A and 13 ± 3% in Group B (p = 0.778).

3.2. Follow-Up

Of the 108 patients included in the study, 86 (80%) completed a follow-up visit, whereas 2 (2%) died from non-cardiological causes, and 20 (18%) refused further cardiological examinations due to geographical distance.

At follow-up examination, no patient had signs or symptoms of heart failure. No patient had been re-hospitalized for heart failure, myocardial infarction, ventricular arrhythmias, or other cardiovascular causes over the follow-up after surgery, while three had had an episode of pneumonia, in one case requiring hospitalization.

A comprehensive echocardiographic examination was obtained in all patients participating in follow-up visits (Table 3).

Table 3. Echocardiographic parameters at follow-up according to diastolic dysfunction groups.

Variables	Total (n = 86)	Group A (n = 55)	Group B (n = 31)	p A3 vs. B3	p Total 1 vs. 3	p A1 vs. A3	p B1 vs. B3
EF (%)	60 [59–65]	60 [60–65]	60 [58–63]	0.034	0.050	0.478	0.025
LV/mass/BSA (g)	95.1 ± 25.7	90.2 ± 26.1	103.4 ± 20.1	0.026	<0.001	<0.001	<0.001
IVS (mm)	11 [11,12]	11 [10–12]	12 [11–13]	0.050	<0.001	<0.001	0.003
PWT (mm)	10 [10–11]	10 [9–11]	11 [10–11]	0.010	<0.001	<0.001	0.827
LVD (mm)	45 [43–48]	45 [43–48]	45 [42–50]	0.825	<0.001	<0.001	0.016
LAVI (mL/m^2)	35.2 ± 12.0	33.4 ± 12.0	38.3 ± 11.5	0.077	0.021	0.289	0.018
TAPSE (mm)	20.1 ± 3.2	19.7 ± 2.8	20.6 ± 3.7	0.208	<0.001	<0.001	0.003
PAPS (mmHg)	25 [20–29]	24 [20–26]	27 [22–33]	0.036	<0.001	0.004	0.031
RVD2 (mm)	30.0 ± 5.3	30.9 ± 5.0	28.7 ± 5.7	0.080	0.011	0.012	0.412
Grad-Peak (mmHg)	20.1 ± 7.4	19.3 ± 6.1	21.5 ± 8.9	0.198	<0.001	<0.001	<0.001
Grad-Mean (mmHg)	11.3 ± 4.5	10.8 ± 3.8	11.0 ± 5.5	0.216	<0.001	<0.001	<0.001
Mean GLS (%)	17.0 ± 3.4	17.7 ± 3.4	15.7 ± 3.2	0.011	0.087	0.018	0.221

Continuous variables are presented as median [IQR] or mean ± SD. Group A or A: diastolic dysfunction grade I; Group B or B: diastolic dysfunction grades II–III; 1: preoperative; 3: follow-up. IVS: interventricular septum; PWT: posterior wall thickness; LVD: left ventricular diameter; LAVI: left atrial volume index; TAPSE: tricuspid annular plane systolic excursion; PAPS: systolic pulmonary artery pressure; RVD2: mid-cavity diameter of the right ventricle; Grad-Peak: aortic valve peak gradient; Grad-Mean: aortic valve mean gradient; GLS: global longitudinal strain.

The median LVEF was marginally but significantly greater, and LV mass was lower, in Group A. Group A exhibited a lower LV mass compared with Group B (90.2 ± 26.1 g/m^2 vs. 103.4 ± 20.1 g/m^2, p = 0.026), and both groups showed significant reductions in LV mass compared with baseline (p < 0.001 for both groups). The mean transvalvular gradient at follow-up was similarly low in the two groups (p = 0.216). From baseline to follow-up, LAVI was significantly reduced only in Group B (p = 0.018). Although Group B had higher LAVI values compared with Group A at follow-up, this difference was not statistically significant (38.3 ± 11.5 mL/m^2 vs. 33.4 ± 12.0 mL/m^2, p = 0.077).

Compared with the early postoperative values, 12-month GLS improved significantly in both groups (Table 3 and Figure 1).

However, the improvement was significantly more considerable in the Group A patients with mild diastolic dysfunction (p = 0.018), in whom 12-month GLS increased over the baseline (Figure 2). The same analysis was repeated in patients undergoing isolated SAVR (without CABG). The mean GLS values were still significantly higher in Group A (18 ± 3.1%) than in B (16.3 ± 3.2%; p = 0.039) and, again, were improved from baseline in Group A only (p = 0.045).

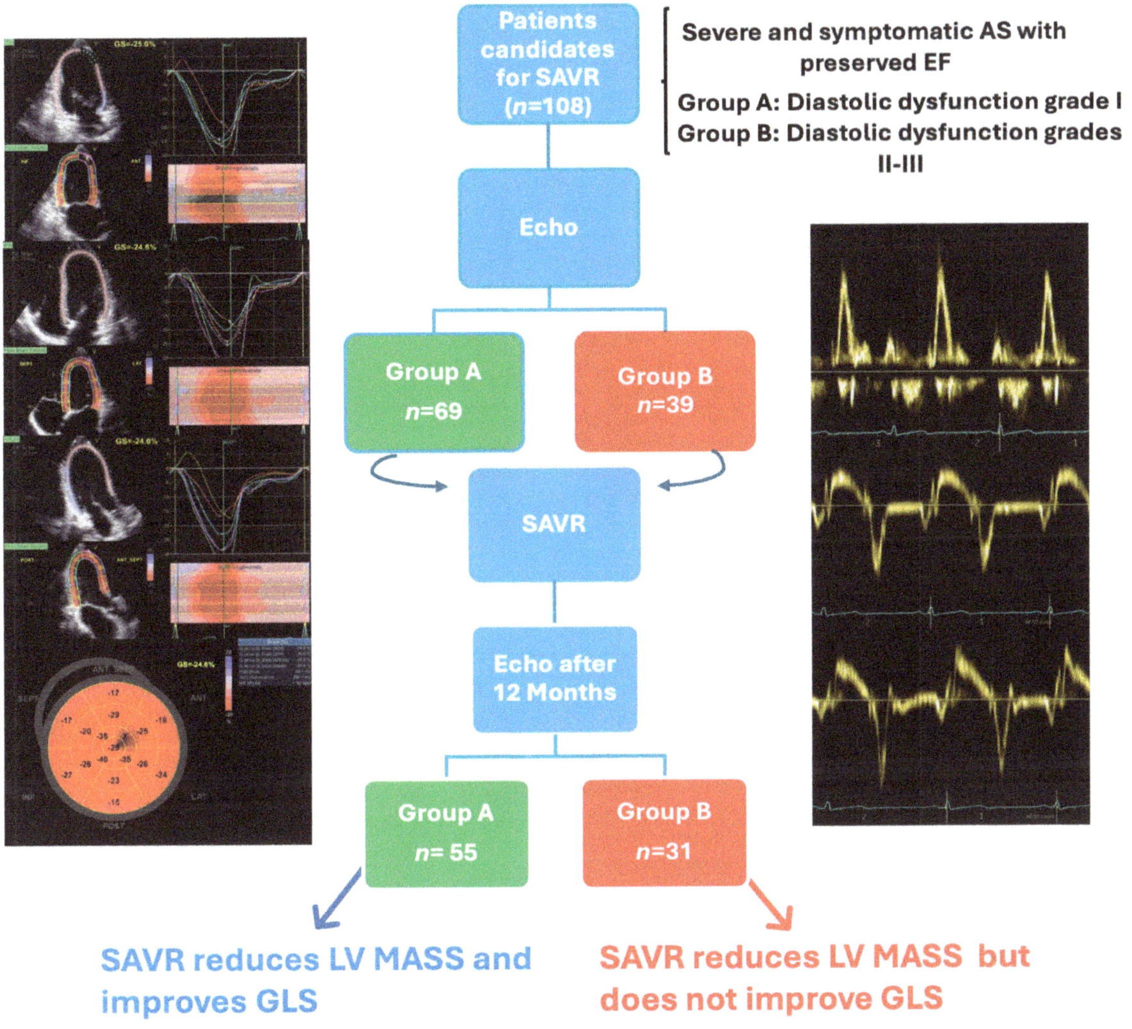

Figure 1. Central illustration: SAVR: surgical aortic valve replacement; LV: left ventricle; GLS: global longitudinal strain. Group A or A: diastolic dysfunction grade I; Group B or B: Diastolic dysfunction grades II–III.

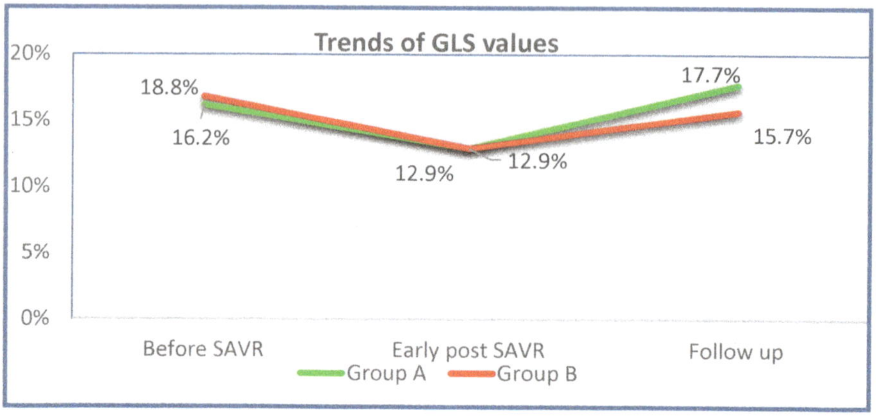

Figure 2. Trends of GLS values. SAVR: surgical aortic valve replacement; GLS: global longitudinal strain. Group A (green): diastolic dysfunction grade I; Group B (red): diastolic dysfunction grades II–III.

4. Discussion

In the present study of patients with severe AS and preserved LVEF, the GLS before SAVR was reduced in most patients and, after an acute decrease in the early postoperative period, improved over the baseline only in patients with mild baseline diastolic dysfunction at 1 year follow-up.

Patients with severe AS and preserved LV function often present a decreased GLS, which suggests subclinical LV systolic dysfunction despite normal LVEF. In fact, the reduction in GLS usually precedes symptom onset [18–20], and it has been associated with a less favorable prognosis [21]. The ability of GLS to detect subclinical myocardial damage before overt symptoms develop underscores its clinical utility. Incorporating GLS into routine diagnostic protocols may allow for more accurate risk stratification, especially in patients who may appear asymptomatic but are already experiencing detrimental cardiac changes.

In some studies, a GLS < 15% or 16% has been taken to indicate initial LV dysfunction, but an optimal cut-off remains to be identified [18,22–24]. Determining an appropriate GLS threshold will be crucial for clinical decision making.

Several studies have demonstrated that SAVR improves GLS, especially in patients with elevated transvalvular gradients [5,14]. Indeed, SAVR causes a dramatic decline in LV pressure overload, leading to reduced wall stress and, over time, reverse remodeling with improved transmural myocardial perfusion and, finally, enhanced LV systolic performance. The timing of intervention remains critical, as earlier relief of pressure overload may enhance the potential for myocardial recovery. Notably, a persistently reduced GLS one year after SAVR has been associated with a less favorable long-term prognosis [20]. In this setting, it would be relevant to identify those patients who may benefit the most from surgery. Persistent GLS reductions suggest ongoing myocardial damage or incomplete recovery. These individuals may require closer postoperative follow-up and possibly adjunctive medical therapies aimed at promoting further reverse remodeling and improving myocardial strain. Targeting these patients with personalized therapeutic strategies could enhance their long-term outcomes and mitigate the risk of adverse events.

The present study found that patients with more severe baseline diastolic dysfunction did not significantly improve GLS at follow-up. Diastolic dysfunction is present in virtually all patients with AS and can affect their outcomes [25]. Some small retrospective studies have already shown that patients with more severe diastolic dysfunction have a greater degree of remodeling and worse outcomes after AVR [26,27]. It can be hypothesized that a higher degree of LV remodeling might hamper GLS improvement or require more time to achieve it.

Beyond GLS, in patients with mild diastolic dysfunction, we also observed that LV mass was reduced to a greater extent after SAVR, confirming a more pronounced reverse remodeling in this group. This supports the concept that early-stage diastolic dysfunction represents a more favorable substrate for recovery following SAVR. The significant reduction in LV mass in these patients indicates that hypertrophy remains relatively plastic and responsive to changes in hemodynamics when the intervention occurs early enough.

It is important to note that the mean values of GLS were reduced to about 12% in the immediate postoperative period, without significant differences in patients with different diastolic dysfunctions. Early postoperative reductions in GLS are likely multifactorial and influenced by the immediate effects of surgery, such as myocardial stunning, altered loading conditions, and inflammatory responses [28,29]. Recognizing that this decline is transient is essential for interpreting early postoperative echocardiographic findings.

Of relevance, the same results were confirmed in patients undergoing isolated SAVR, excluding the coronary revascularization effect on GLS improvement. This consistency across patient groups suggests that the improvement in GLS is primarily related to the correction of AS rather than the concurrent treatment of coronary artery disease. Thus, isolated SAVR has a clear and direct benefit on myocardial strain.

Our study highlights the correlation between diastolic dysfunction and GLS and identifies diastolic dysfunction as a predictor of GLS recovery post-SAVR.

Therefore, we support introducing LV strain as a further indication for surgery in patients with severe AS [30]. Similarly, we claim that diastolic dysfunction should also be considered when deciding the indication for valve replacement before LV systolic dysfunction or overt symptoms occur [18,20]. In fact, we found that SAVR, when patients have not yet developed a marked diastolic dysfunction, is associated with a more extensive—and perhaps an earlier—reverse remodeling, as suggested by higher GLS values at follow-up. This further reinforces the notion that surgical intervention should not be delayed until the later stages of diastolic dysfunction. Instead, early identification of impaired LV strain and mild diastolic dysfunction could serve as key factors for surgery, potentially leading to more favorable long-term outcomes.

The present study has some limitations that need to be acknowledged. First, the study sample was relatively small, with an even smaller number of patients with moderate-to-severe diastolic dysfunction. Moreover, our primary study outcome was represented by changes in GLS at follow-up, while we could not find any association with one-year adverse events, an outcome for which our study was vastly underpowered, also because of the limited follow-up duration.

5. Conclusions

In our study, patients with severe AS and mild diastolic dysfunction undergoing SAVR showed a remarkable improvement in GLS after twelve months, regardless of CABG intervention. The same did not occur in those with moderate-to-severe diastolic dysfunction. Further studies with larger populations and longer follow-ups are required to confirm these findings and explore the association between diastolic dysfunction, GLS recovery, and clinical outcomes.

Author Contributions: Conceptualization, S.C., F.B. and M.B. (Martina Berteotti); writing—original draft preparation, S.C., N.M., M.B. (Martina Berteotti) and F.B.; supervision, N.M, P.S. and S.D.P.; data curation V.T., N.C., M.B. (Marta Bandini), E.M., C.S. and A.G.G.; formal analysis and investigation A.G.G. and F.B. All authors have read and agreed to the published version of the manuscript.

Funding: This research received no external funding.

Institutional Review Board Statement: The study was conducted in accordance with the Declaration of Helsinki and approved by the Ethics Committee of Careggi University Hospital (CEAVC 23627, 31 May 2023).

Informed Consent Statement: Informed consent was obtained from all subjects involved in the study.

Data Availability Statement: Data are contained within the article.

Conflicts of Interest: The authors declare no conflicts of interest.

References

1. Nkomo, V.T.; Gardin, J.M.; Skelton, T.N.; Gottdiener, J.S.; Scott, C.G.; Enriquez-Sarano, M. Burden of valvular heart diseases: A population-based study. *Lancet* **2006**, *368*, 1005–1011. [CrossRef] [PubMed]
2. Joseph, J.; Naqvi, S.Y.; Giri, J.; Goldberg, S. Aortic Stenosis: Pathophysiology, Diagnosis, and Therapy. *Am. J. Med.* **2017**, *130*, 253–263. [CrossRef] [PubMed]
3. Peters, A.S.; Duggan, J.P.; Trachiotis, G.D.; Antevil, J.L. Epidemiology of Valvular Heart Disease. *Surg. Clin. N. Am.* **2022**, *102*, 517–528. [CrossRef] [PubMed]
4. Chahal, N.; Senior, R. Assessing systolic function in aortic stenosis: The earlier the better? *Heart* **2020**, *106*, 1200–1201. [CrossRef]
5. Delgado, V.; Tops, L.F.; Bommel, R.J.; Van Kley, F.; Van Der Marsan, N.A.; Klautz, R.J.; Versteegh, M.I.M.; Holman, E.R.; Schalij, M.J.; Bax, J.J. Strain analysis in patients with severe aortic stenosis and preserved left ventricular ejection fraction undergoing surgical valve replacement. *Eur. Heart J.* **2009**, *30*, 3037–3047. [CrossRef] [PubMed]
6. Vahanian, A.; Beyersdorf, F.; Praz, F.; Milojevic, M.; Baldus, S.; Bauersachs, J.; Capodanno, D.; Conradi, L.; De Bonis, M.; De Paulis, R.; et al. 2021 ESC/EACTS Guidelines for the management of valvular heart diseaseDeveloped by the Task Force for the management of valvular heart disease of the European Society of Cardiology (ESC) and the European Association for Cardio-Thoracic Surgery (EACTS). *Eur. Heart J.* **2022**, *43*, 561–632. [CrossRef]
7. O'Toole, J.D.; Geiser, E.A.; Sudhakar Reddy, P.; Curtiss, E.I.; Landfair, R.M. Effect of Preoperative Ejection Fraction on Survival and Hemodynamic Improvement Following Aortic Valve Replacement. *Circulation* **1978**, *58*, 1175–1184. [CrossRef] [PubMed]
8. Dahl, J.S.; Eleid, M.F.; Michelena, H.I.; Scott, C.G.; Suri, R.M.; Schaff, H.V.; Pellikka, P.A. Effect of left ventricular ejection fraction on postoperative outcome in patients with severe aortic stenosis undergoing aortic valve replacement. *Circ. Cardiovasc. Imaging* **2015**, *8*, e002917. [CrossRef] [PubMed]
9. Januzzi, J.L.; Chandrashekhar, Y. Echocardiography: The New Gold Standard for Imaging Ventricular Function? *J. Am. Coll. Cardiol.* **2017**, *70*, 955–957. [CrossRef]
10. Ng, A.C.T.; Prihadi, E.A.; Antoni, M.L.; Bertini, M.; Ewe, S.H.; Marsan, N.A.; Leung, D.Y.; Delgado, V.; Bax, J.J. Left ventricular global longitudinal strain is predictive of all-cause mortality independent of aortic stenosis severity and ejection fraction. *Eur. Heart J. Cardiovasc. Imaging* **2018**, *19*, 859–867. [CrossRef]
11. Lafitte, S.; Perlant, M.; Reant, P.; Serri, K.; Douard, H.; Demaria, A.; Roudaut, R. Impact of impaired myocardial deformations on exercise tolerance and prognosis in patients with asymptomatic aortic stenosis. *Eur. J. Echocardiogr.* **2009**, *10*, 414–419. [CrossRef] [PubMed]
12. Cosyns, B.; Vannan, M.A. Global longitudinal strain in severe aortic stenosis. *Eur. Heart J. Cardiovasc. Imaging* **2020**, *21*, 1259–1261. [CrossRef] [PubMed]
13. Twing, A.H.; Slostad, B.; Anderson, C.; Konda, S.; Groves, E.M.; Kansal, M.M. Improvements in global longitudinal strain after transcatheter aortic valve replacement according to race. *Am. J. Cardiovasc. Dis.* **2021**, *11*, 203.
14. Grund, F.F.; Myhr, K.A.; Visby, L.; Hassager, C.; Mogelvang, R. Impact of surgical aortic valve replacement on global and regional longitudinal strain across four flow gradient patterns of severe aortic stenosis. *Int. J. Cardiovasc. Imaging* **2021**, *37*, 2175–2187. [CrossRef] [PubMed]
15. Lang, R.M.; Badano, L.P.; Victor, M.A.; Afilalo, J.; Armstrong, A.; Ernande, L.; Flachskampf, F.A.; Foster, E.; Goldstein, S.A.; Kuznetsova, T.; et al. Recommendations for Cardiac Chamber Quantification by Echocardiography in Adults: An Update from the American Society of Echocardiography and the European Association of Cardiovascular Imaging. *J. Am. Soc. Echocardiogr.* **2015**, *28*, 233–271. [CrossRef]
16. Baumgartner, H.; Hung, J.; Bermejo, J.; Chambers, J.B.; Edvardsen, T.; Goldstein, S.; Lancellotti, P.; Lefevre, M.; Miller, F.; Otto, C.M.; et al. Recommendations on the echocardiographic assessment of aortic valve stenosis: A focused update from the European Association of Cardiovascular Imaging and the American Society of Echocardiography. *Eur. Heart J. Cardiovasc. Imaging* **2017**, *18*, 254–275. [CrossRef]
17. Nagueh, S.F.; Smiseth, O.A.; Appleton, C.P.; Byrd, B.F.; Dokainish, H.; Edvardsen, T.; Flachskampf, F.A.; Gillebert, T.C.; Klein, A.L.; Lancellotti, P.; et al. Recommendations for the Evaluation of Left Ventricular Diastolic Function by Echocardiography: An Update from the American Society of Echocardiography and the European Association of Cardiovascular Imaging. *J. Am. Soc. Echocardiogr.* **2016**, *29*, 277–314. [CrossRef] [PubMed]
18. Wang, Y.; Zhang, M.; Chen, H.; Li, H. Prognostic Value of Global Longitudinal Strain in Asymptomatic Aortic Stenosis: A Systematic Review and Meta-Analysis. *Front. Cardiovasc. Med.* **2022**, *9*, 203. [CrossRef] [PubMed]
19. Vollema, E.M.; Sugimoto, T.; Shen, M.; Tastet, L.; Ng, A.C.T.; Abou, R.; Marsan, N.A.; Mertens, B.; Dulgheru, R.; Lancellotti, P.; et al. Association of Left Ventricular Global Longitudinal Strain with Asymptomatic Severe Aortic Stenosis: Natural Course and Prognostic Value. *JAMA Cardiol.* **2018**, *3*, 839–847. [CrossRef]
20. Kafa, R.; Kusunose, K.; Goodman, A.L.; Svensson, L.G.; Sabik, J.F.; Griffin, B.P.; Desai, M.Y. Association of Abnormal Postoperative Left Ventricular Global Longitudinal Strain with Outcomes in Severe Aortic Stenosis Following Aortic Valve Replacement. *JAMA Cardiol* **2016**, *1*, 494–496. [CrossRef] [PubMed]

21. Dahl, J.S.; Videbæk, L.; Poulsen, M.K.; Rudbæk, T.R.; Pellikka, P.A.; Maller, J.E. Global strain in severe aortic valve stenosis relation to clinical outcome after aortic valve replacement. *Circ. Cardiovasc. Imaging* **2012**, *5*, 613–620. [CrossRef] [PubMed]
22. Galli, E.; Lancellotti, P.; Sengupta, P.P.; Donal, E. LV Mechanics in Mitral and Aortic Valve Diseases: Value of Functional Assessment Beyond Ejection Fraction. *JACC Cardiovasc. Imaging* **2014**, *7*, 1151–1166. [CrossRef] [PubMed]
23. Thellier, N.; Altes, A.; Appert, L.; Binda, C.; Leman, B.; Marsou, W.; Debry, N.; Joly, C.; Ennezat, P.V.; Tribouilloy, C.; et al. Prognostic Importance of Left Ventricular Global Longitudinal Strain in Patients with Severe Aortic Stenosis and Preserved Ejection Fraction. *J. Am. Soc. Echocardiogr.* **2020**, *33*, 1454–1464. [CrossRef]
24. Stassen, J.; Pio, S.M.; Ewe, S.H.; Singh, G.K.; Hirasawa, K.; Butcher, S.C.; Cohen, D.J.; Généreux, P.; Leon, M.B.; Marsan, N.A.; et al. Left Ventricular Global Longitudinal Strain in Patients with Moderate Aortic Stenosis. *J. Am. Soc. Echocardiogr.* **2022**, *35*, 791–800.e4. [CrossRef] [PubMed]
25. Kampaktsis, P.N.; Kokkinidis, D.G.; Wong, S.C.; Vavuranakis, M.; Skubas, N.J.; Devereux, R.B. The role and clinical implications of diastolic dysfunction in aortic stenosis. *Heart* **2017**, *103*, 1481–1487. [CrossRef] [PubMed]
26. Chang, S.A.; Park, P.W.; Sung, K.; Lee, S.C.; Park, S.W.; Lee, Y.T.; Oh, J.K. Noninvasive estimate of left ventricular filling pressure correlated with early and midterm postoperative cardiovascular events after isolated aortic valve replacement in patients with severe aortic stenosis. *J. Thorac. Cardiovasc. Surg.* **2010**, *140*, 1361–1366. [CrossRef] [PubMed]
27. Dahl, J.S.; Videbæk, L.; Poulsen, M.K.; Pellikka, P.A.; Veien, K.; Andersen, L.I.; Haghfelt, T.; Møller, J.E. Noninvasive assessment of filling pressure and left atrial pressure overload in severe aortic valve stenosis: Relation to ventricular remodeling and clinical outcome after aortic valve replacement. *J. Thorac. Cardiovasc. Surg.* **2011**, *142*, e77–e83. [CrossRef] [PubMed]
28. Chen, Y.; Zhang, Z.; Cheng, L.; Fan, L.; Wang, C.; Shu, X. The Early Variation of Left Ventricular Strain after Aortic Valve Replacement by Three-Dimensional Echocardiography. *PLoS ONE* **2015**, *10*, e0140469. [CrossRef] [PubMed]
29. Leclercq, C.; Rouge, J.; Burtin, P.; Roussiaux, A.; Ducrocq, N.; Halchini, C. Left ventricular strain variations in cardiac surgery; The role of the type of surgery. *J. Cardiothorac. Vasc. Anesth.* **2019**, *33*, S130. [CrossRef]
30. Magne, J.; Cosyns, B.; Popescu, B.A.; Carstensen, H.G.; Dahl, J.; Desai, M.Y.; Kearney, L.; Lancellotti, P.; Marwick, T.H.; Sato, K.; et al. Distribution and Prognostic Significance of Left Ventricular Global Longitudinal Strain in Asymptomatic Significant Aortic Stenosis: An Individual Participant Data Meta-Analysis. *JACC Cardiovasc. Imaging* **2019**, *12*, 84–92. [CrossRef] [PubMed]

diagnostics

MDPI

Review

Atrial Cardiomyopathy in Atrial Fibrillation: A Multimodal Diagnostic Framework

Paschalis Karakasis [1,*], Panayotis K. Vlachakis [2], Panagiotis Theofilis [2], Nikolaos Ktenopoulos [2], Dimitrios Patoulias [3], Barbara Fyntanidou [4], Antonios P. Antoniadis [1] and Nikolaos Fragakis [1,*]

1 Second Department of Cardiology, Hippokration General Hospital, Aristotle University of Thessaloniki, 54642 Thessaloniki, Greece; aantoniadis@gmail.com

2 First Cardiology Department, School of Medicine, Hippokration General Hospital, National and Kapodistrian University of Athens, 11527 Athens, Greece; vlachakispanag@gmail.com (P.K.V.); panos.theofilis@hotmail.com (P.T.); nikosktenop@gmail.com (N.K.)

3 Second Propedeutic Department of Internal Medicine, Faculty of Medicine, School of Health Sciences Aristotle, University of Thessaloniki, 54642 Thessaloniki, Greece; dipatoulias@gmail.com

4 Emergency Department, AHEPA University General Hospital, Aristotle University of Thessaloniki, 54642 Thessaloniki, Greece; bfyntan@yahoo.com

* Correspondence: pakar15@hotmail.com (P.K.); fragakis.nikos@googlemail.com (N.F.)

Abstract: Atrial fibrillation (AF) is increasingly recognized as the clinical manifestation of an underlying atrial disease process rather than a purely electrical disorder. This evolving paradigm has given rise to the concept of atrial cardiomyopathy (AtCM), encompassing structural, electrical, contractile, and molecular remodeling of the atrial myocardium that contributes to AF initiation, maintenance, and progression. Although consensus definitions of AtCM now exist, its integration into clinical practice remains limited, with AF management still largely guided by arrhythmic patterns rather than substrate characterization. This review synthesizes current diagnostic strategies for AtCM within the context of AF, emphasizing a multimodal approach. We outline advances in cardiac imaging—including echocardiography, cardiac magnetic resonance, and computed tomography—for detailed assessment of atrial morphology, function, and fibrosis. Electroanatomic mapping is discussed as a key invasive tool for substrate localization, while electrocardiographic indices such as P-wave morphology and dispersion serve as accessible surrogates of electrical remodeling. In parallel, we examine the role of circulating biomarkers and emerging genomic, transcriptomic, and epigenomic markers in refining disease phenotyping. Despite promising progress, significant challenges remain. Standardization of imaging protocols, validation of biomarker thresholds, and integration of artificial intelligence tools are needed to enhance clinical utility. A diagnostic framework informed by atrial substrate assessment may support more tailored therapeutic decision-making in AF. Future research should prioritize the harmonization of diagnostic criteria and explore how substrate profiling in AF may refine risk stratification and improve clinical outcomes.

Keywords: atrial cardiomyopathy; atrial fibrillation; cardiac imaging; electrocardiography; electroanatomic mapping; biomarkers; atrial substrate

1. Introduction

Atrial fibrillation (AF) is the most prevalent sustained arrhythmia encountered in clinical practice and is associated with increased morbidity, mortality, and healthcare burden [1–7]. AF has long been conceptualized as an isolated electrical disturbance; however, growing evidence underscores the primacy of atrial structural and functional remodeling in its pathogenesis and progression [8]. This reconceptualization has culminated in the emergence of atrial cardiomyopathy (AtCM)—a term denoting a constellation of architectural, contractile, and electrophysiologic alterations of the atrial myocardium that may exist independently of, or in conjunction with, AF [8,9]. Recognizing AtCM as the arrhythmogenic substrate underlying AF reframes the disorder not merely as an arrhythmia but as the clinical expression of a more diffuse and heterogeneous atrial disease process [10].

Consensus definitions now describe AtCM as any complex of atrial abnormalities capable of producing clinically relevant manifestations, including atrial arrhythmias, thromboembolism, or atrial failure [8]. Recent expert consensus from the European Association of Cardiovascular Imaging (EACVI) and the European Heart Rhythm Association (EHRA) has further emphasized the central role of left atrial structure and function in this context, advocating for comprehensive multimodality imaging to guide clinical assessment and management strategies [11]. Importantly, AF may result from, exacerbate, or even mask the underlying substrate of AtCM, contributing to a bidirectional and progressive disease process. Histopathologic studies, translational investigations, and clinical imaging have each identified diverse mechanisms—fibrosis, inflammation, oxidative stress, adipose infiltration, and aging—that converge to produce electrophysiological instability and mechanical dysfunction in the atria [12–15].

A staging framework for AtCM was recently proposed in a clinical consensus statement by the EHRA, together with HRS, APHRS, and LAHRS, integrating structural, functional, and electrophysiological criteria [8]:

1. Early (subclinical) stage: characterized by the presence of subtle atrial abnormalities identifiable through advanced imaging or electrophysiologic assessment, in the absence of overt arrhythmia or mechanical dysfunction. These changes are often asymptomatic and represent early atrial remodeling;
2. Intermediate (clinically overt) stage: defined by measurable structural alterations (e.g., atrial enlargement), impaired atrial contractility, or elevated circulating biomarkers such as B-type natriuretic peptide (BNP) or atrial natriuretic peptide (ANP). Atrial fibrillation may be present at this stage, indicating clinically significant substrate dysfunction;
3. Advanced (severe) stage: reflects substantial deterioration of atrial function, including marked reduction in left atrial ejection fraction ($\leq$35%), low atrial appendage flow velocities ($\leq$20 cm/s), or severely impaired strain parameters. This stage may also involve extensive structural remodeling such as dense interstitial fibrosis ($\geq$35% of atrial wall volume), fatty or amyloid infiltration, or active inflammation. Severe left atrial enlargement (diameter $\geq$ 5.0 cm or volume index $\geq$ 50 mL/m^2) and the presence of long-standing persistent or permanent atrial fibrillation are common.

Despite this conceptual shift, the clinical integration of AtCM remains underdeveloped and has yet to be fully translated into routine practice [16,17]. Current AF management strategies still largely rely on rhythm classification, without explicit characterization of atrial pathology. Yet, advances in cardiac imaging, electrocardiography, electroanatomic mapping, and circulating biomarkers now offer an unprecedented opportunity to non-invasively assess atrial structure and function (Figure 1). The synthesis of these diagnostic modalities may enable early identification of AtCM, more precise risk stratification for AF-related complications, and refined targeting of therapeutic strategies—including ablation, anticoagulation, and upstream disease modification.

In this review, we summarize current approaches to diagnosing AtCM in the setting of AF, drawing on advances in imaging, electroanatomic mapping, electrocardiography, biomarkers, and omics.

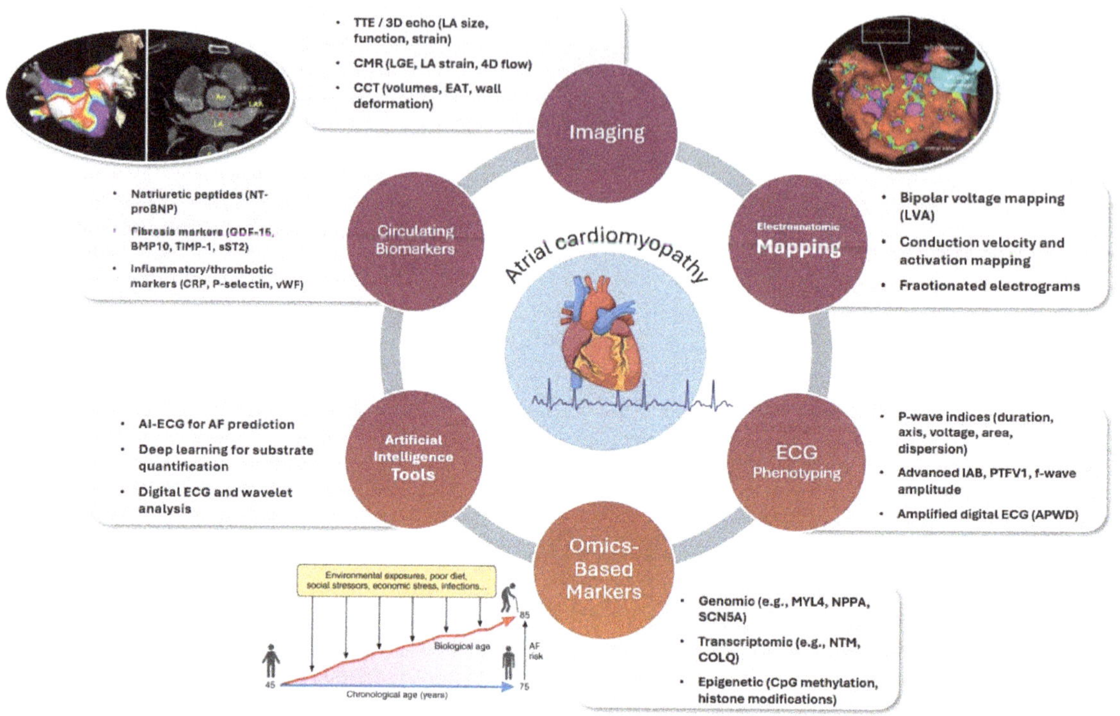

Figure 1. Integrated multimodal diagnostic framework for atrial cardiomyopathy in atrial fibrillation. This schematic illustrates a comprehensive approach to diagnosing atrial cardiomyopathy (AtCM) in the context of atrial fibrillation (AF), emphasizing the convergence of structural, electrical, functional, and molecular assessments. Central to the diagram is AtCM, surrounded by six key diagnostic domains: (1) imaging modalities (echocardiography, cardiac magnetic resonance, and computed tomography) provide detailed assessment of atrial size, function, fibrosis, and wall deformation; (2) electroanatomic mapping offers invasive evaluation of low-voltage areas, conduction velocity, and electrogram complexity; (3) ECG phenotyping captures electrical remodeling through P-wave indices, interatrial block, and f-wave morphology; (4) circulating biomarkers reflect hemodynamic stress, inflammation, and fibrosis; (5) omics-based markers—including genomic, transcriptomic, and epigenetic alterations—offer molecular insight into atrial remodeling; and (6) artificial intelligence tools enhance diagnostic precision through advanced ECG interpretation and substrate quantification. Together, these modalities contribute to a substrate-guided framework for risk stratification, therapy selection, and progression monitoring in AF. Abbreviations: TTE—transthoracic echocardiography; 3D—three-dimensional; CMR—cardiac magnetic resonance; LGE—late gadolinium enhancement; LA—left atrium; CCT—cardiac computed tomography; EAT—epicardial adipose tissue; LVA—low-voltage area; ECG—electrocardiogram; IAB—interatrial block; PTFV1—P-wave terminal force in lead V1; APWD—amplified P-wave duration; AI—artificial intelligence; AF—atrial fibrillation; NT-proBNP—N-terminal pro-B-type natriuretic peptide; GDF-15—growth differentiation factor-15; BMP10—bone morphogenetic protein 10; TIMP-1—tissue inhibitor of metalloproteinase-1; sST2—soluble suppression of tumorigenicity-2; CRP—C-reactive protein; vWF—von Willebrand factor; MYL4—myosin light chain 4; NPPA—natriuretic peptide A; SCN5A—sodium voltage-gated channel alpha subunit 5; NTM—neurotrimin; COLQ—collagen-like tail subunit of asymmetric acetylcholinesterase; CpG—cytosine–phosphate–guanine.

2. Advancements in Diagnostic Imaging

A comprehensive, multimodality assessment of atrial morphology and function through advanced cardiac imaging techniques is essential for the accurate diagnosis of atrial cardiomyopathy and for delineating the extent of atrial structural and functional impairment [18].

2.1. Advanced Atrial Imaging in the Diagnostic Evaluation of AtCM

AtCM can be comprehensively assessed using a range of imaging modalities, including transthoracic echocardiography (TTE), cardiac magnetic resonance (CMR), and cardiac computed tomography (CCT). Both echocardiography and CMR facilitate dynamic quantification of atrial volumes throughout the cardiac cycle, whereas CCT provides precise volumetric measurements at a fixed point in time [19]. Among echocardiographic indices, the left atrial volume index (LAVI) is a particularly well-established marker of atrial remodeling. Easily obtainable via TTE, LAVI is reproducible and widely adopted in clinical practice [20]. A LAVI $\leq$ 34 mL/m^2 is considered normal for both sexes, while values of 35–41 mL/m^2, 42–48 mL/m^2, and >48 mL/m^2 correspond to mild, moderate, and severe atrial enlargement, respectively [20]. It is strongly associated with the presence, progression, and recurrence of AF, and remains central to echocardiographic risk stratification [21]. Beyond static volumetric measures, recent studies have introduced the left atrioventricular coupling index (LACI) as a dynamic marker of atrial–ventricular interaction [22,23]. Defined as the ratio of LA to LV end-diastolic volume, LACI reflects atrioventricular mechanical coupling and has demonstrated strong prognostic value. In the Multi-Ethnic Study of Atherosclerosis (MESA), both an elevated LACI (>30%) and its annualized increase (>1.5% per year) were independently associated with incident AF, offering superior predictive discrimination compared to conventional structural indices [23]. These findings suggest that LACI may provide complementary insights into atrial remodeling and functional integration within the left heart.

Three-dimensional TTE mitigates the limitations of atrial foreshortening inherent to two-dimensional approaches, thereby enhancing anatomical fidelity [19]. The left and right atrial appendages are more accurately delineated using transesophageal echocardiography (TEE) and CCT, which offer superior spatial resolution for these anatomically complex structures. In addition, CCT has been validated for the assessment of epicardial adipose tissue (EAT) and is now increasingly augmented by deep learning algorithms to improve image interpretation and prognostic insights [24]. Notably, alterations in the morphology of the left atrial (LA) roof, as visualized on CT, have been associated with the emergence of non–pulmonary vein (non-PV) arrhythmogenic substrates in patients undergoing atrial fibrillation (AF) ablation [25,26]. Furthermore, pre-procedural assessment of regional LA wall deformation via CCT has demonstrated superior predictive accuracy for low-voltage areas compared to traditional remodeling indices, underscoring its potential value in procedural planning and risk stratification [27]. CMR has been instrumental in delineating atrial fibrotic remodeling through late gadolinium enhancement imaging, while also facilitating the assessment of intra-atrial flow dynamics via 4D flow-derived velocity mapping [28,29].

Moreover, although not widely implemented in routine clinical practice, molecular positron emission tomography (PET) imaging represents a novel approach for characterizing atrial inflammation and fibrosis. Increased atrial uptake of 18F-fluorodeoxyglucose (18F-FDG) has been observed in patients with AF and in those with cardiac sarcoidosis, where it may predict subsequent arrhythmia onset [30,31]. More recently, 68Ga–fibroblast activation protein inhibitor (68Ga-FAPI) PET imaging has been employed to detect elevated fibroblast activation in the atria of individuals with AF or following pulmonary vein isolation, suggesting its potential utility in substrate characterization [32,33]. Although atrial activity is infrequently reported in oncologic PET/CT imaging, incidental atrial uptake has been associated with a heightened risk of AF and thromboembolic complications, underscoring its possible relevance in cardio-oncology [30]. These findings have been substantiated by a recent prospective case-control study demonstrating that right atrial 18F-FDG uptake was significantly greater in patients with persistent AF and independently associated with inflammatory burden [34]. Complementary evidence from a systematic review and meta-analysis further confirmed a strong association between atrial FDG uptake and AF, with markedly elevated odds of both left and right atrial involvement in affected patients [35]. Moreover, hybrid PET-MRI imaging using FAPI tracers has enabled visualization of atrial fibroblast activation that correlates with histological fibrosis, particularly in persistent AF [36]. Preclinical work utilizing 11C-hydroxyephedrine PET has also revealed sympathetic denervation accompanying fibrosis, emphasizing the multifaceted remodeling processes underpinning AF [37]. While these advances position PET imaging as a powerful research tool for substrate phenotyping, its clinical utility remains investigational and awaits further validation. A comparative overview of the technical capabilities and diagnostic applications of each modality in the context of atrial evaluation is presented in Table 1.

Table 1. Diagnostic utility of multimodality imaging in atrial structural and functional assessment.

Atrial Parameters	TTE	TEE	Cardiac MRI	Cardiac CT	Electroanatomic Mapping [a]
Atrial size	++ (more accurate in 3D TTE)	−	++	+++	+
Atrial function	+++ (more accurate in 3D TTE)	−	+++	+	−
Atrial appendage size	−	+++	+	+++	−
Thrombus	+	+++	++	+++ (delayed acquisition) [b]	−
Epicardial adipose tissue	+	+	+	+++	−
Atrial fibrosis	−	−	+ (LGE) [b]	−	+
Atrial velocity	−	++ (exclusively in atrial appendage)	++ (4D flow imaging)	−	−
Atrial activation/conduction time	−	−	−	−	+ (P-wave duration)

Abbreviations: 3D, three-dimensional; CT, computed tomography; LGE, late gadolinium enhancement; MRI, magnetic resonance imaging; TEE, transoesophageal echocardiography; TTE, transthoracic echocardiography. [a] Electroanatomic mapping comprises a comprehensive array of electrophysiological techniques, including voltage mapping, activation sequence reconstruction, and the analysis of conduction velocity and fractionated electrograms, collectively aimed at delineating atrial substrate complexity. [b] Administration of contrast agents is required to enhance image quality and enable accurate tissue characterization within the respective imaging protocols.

2.2. Quantitative Assessment of Atrial Chamber Dimensions and Functional Dynamics

The atria perform a series of dynamic and physiologically distinct roles throughout the cardiac cycle, functioning sequentially as a reservoir for pulmonary or systemic venous return during ventricular systole, as a passive conduit during early diastole, and as an active booster pump during atrial systole [38,39]. These three functional phases are quantitatively represented by specific atrial volumetric parameters: (i) the maximum atrial volume, measured at end-systole and corresponding to the reservoir phase; (ii) the pre-atrial contraction volume, assessed in mid-diastole during the conduit phase; and (iii) the minimum atrial volume, determined in late diastole following atrial contraction, indicative of booster–pump function [38,39].

Derived indices of atrial function integrate these volume measurements to characterize global atrial performance [40,41]. The total atrial emptying fraction and expansion index are calculated based on the difference between maximal and minimal atrial volumes, reflecting overall contractile reserve and compliance [41]. The passive emptying fraction is derived from the difference between maximal and pre-atrial contraction volumes, corresponding to conduit function, while the active emptying fraction quantifies booster pump activity by comparing pre-atrial contraction and minimal volumes [42].

Beyond volumetric assessment, atrial strain analysis, predominantly performed using TTE and CMR, provides a sensitive measure of myocardial deformation during the reservoir, conduit, and contractile phases [43]. In this context, reservoir strain, conduit strain, and contractile strain offer mechanistic insights into atrial compliance, passive filling, and active contraction, respectively [43].

CCT has recently emerged as a promising modality for the comprehensive evaluation of atrial structure and function in AtCM. Owing to its superior spatial resolution relative to other imaging techniques, CCT enables detailed anatomical and functional characterization, facilitating advanced phenotyping of atrial remodeling [44]. Normative data for left and right atrial size and function across the cardiac cycle have been established using TTE, CMR, and CCT [20,45–51]. Notably, recent evidence suggests that a left atrial reservoir strain below 18% may serve as a surrogate marker of elevated left ventricular filling pressures in the context of heart failure. However, definitive thresholds to demarcate irreversible atrial dysfunction in AtCM have not yet been established (Table 2).

Table 2. Multimodal characterization of cardiac chamber structure and function in atrial fibrillation: prognostic utility of left atrial strain in stroke risk stratification.

Author, Year	Population	Methods of Evaluating Parameters	Results
Leung et al., 2017 [44]	1361 patients with first-diagnosed AF (mean age 65 ± 12 years; 74% male) followed over 7.9 years for stroke/TIA occurrence	TTE with 2D speckle tracking for LA reservoir, conduit, and booster pump strain; PA-TDI for total atrial conduction time; LV GLS and standard parameters	Reduced LA reservoir and conduit strain and prolonged PA-TDI were independently associated with stroke risk beyond CHA2DS2-VASc score, age, and anticoagulant use. LA volume and LVEF were not predictive.
Obokata et al., 2014 [28]	286 patients with nonvalvular AF: 82 with acute embolism (stroke or systemic embolism); 204 controls; prospectively followed	Speckle-tracking TTE to assess global peak LA longitudinal strain (LAS) during AF rhythm; LAVI, LA emptying fraction, and Doppler echocardiography; CHA2DS2-VASc score; outcome follow-up for mortality	Global LAS was significantly lower in patients with acute embolism. LAS < 15.4% identified embolism with AUC 0.83 and predicted mortality post-embolism. LAS was independently associated with embolism risk and added incremental value over CHA2DS2-VASc. LA volumes were not predictive.
Azemi et al., 2012 [29]	57 patients with AF, stroke/TIA, and CHADS2 ≤ 1; compared with matched controls without stroke or TIA	TTE with velocity vector imaging to assess peak positive and negative LA strain and strain rate; LAVI, LA dimension, and LVEF; binary logistic regression for outcome prediction	Both peak positive and negative LA strain were significantly reduced in patients with stroke/TIA. These strain indices were independently associated with stroke/TIA risk, outperforming conventional metrics such as LAVI and LVEF.

Table 2. *Cont.*

Author, Year	Population	Methods of Evaluating Parameters	Results
Shih et al., 2011 [24]	66 patients with permanent AF, 20 with previous ischemic stroke, and 46 without stroke	TTE with 2D speckle tracking to assess peak positive LA strain (LASp), reservoir strain rate (LASRr), and conduit strain rate (LASRc); standard echocardiographic measures (LAVI, LATEF, and E/E′); multivariate logistic regression	.LASp and LASRr were significantly lower in stroke group and independently associated with stroke after adjusting for age, LAVI, and LVEF. LASp < 13.5% had 80% sensitivity and 63% specificity for stroke. LAVI and LATEF were not predictive.

Abbreviations: AF, atrial fibrillation; AUC, area under the curve; CI, confidence interval; E/E′, ratio of early mitral inflow velocity to early diastolic mitral annular velocity; GLS, global longitudinal strain; LA, left atrium; LAVI, left atrial volume index; LAS, left atrial strain; LASp, peak positive left atrial strain; LASRc, left atrial strain rate during the conduit phase; LASRr, left atrial strain rate during the reservoir phase; LATEF, left atrial total emptying fraction; LVEF, left ventricular ejection fraction; OR, odds ratio; PA-TDI, total atrial conduction time by tissue Doppler imaging; TIA, transient ischemic attack; TTE, transthoracic echocardiography.

2.3. *Late Gadolinium-Enhanced Cardiac Magnetic Resonance for Left Atrial Tissue Characterization*

Late gadolinium-enhanced cardiac magnetic resonance (LGE-CMR) has become a cornerstone in the diagnostic and prognostic evaluation of ventricular cardiomyopathies, primarily by identifying myocardial regions with expanded interstitial space or delayed gadolinium washout [52]. Traditionally, LGE serves as a surrogate for myocardial fibrosis in both chronic ischemic injury and a spectrum of non-ischemic cardiomyopathies [52]. However, it may also reflect other pathological processes such as interstitial edema, inflammatory infiltration—as seen in acute myocarditis or cardiac sarcoidosis—or amyloid protein deposition, all of which alter the tissue composition and gadolinium kinetics [52,53].

In recent years, efforts have been made to adapt LGE-CMR techniques for the evaluation of the left atrium, particularly to visualize and quantify atrial fibrotic remodeling [27]. Despite these advances, the application of LGE-CMR to atrial tissue remains fraught with methodological challenges [54,55]. These include the absence of universally accepted imaging protocols, heterogeneity in post-processing algorithms, and inherent limitations related to the thin atrial wall, which compromises spatial resolution [54,55]. Furthermore, access to high-quality LA LGE imaging remains largely restricted to specialized research centers.

Comparative analyses of existing LGE-CMR methodologies have demonstrated marked variability in the reported burden and spatial distribution of fibrosis, both globally and regionally [56,57]. Importantly, conventional LGE imaging is optimized to highlight focal areas of fibrosis by nulling normal myocardium [58]. Consequently, it lacks the sensitivity to detect diffuse interstitial fibrosis—a significant drawback, particularly given autopsy evidence suggesting that atrial fibrosis in AF is often a widespread, rather than localized, process [59–61]. Thus, LGE-positive regions in the atria may correspond to advanced fibrotic lesions or potentially represent nonfibrotic pathological changes.

Despite these limitations, LGE-CMR has yielded important mechanistic insights. Studies have shown that electrical rotors—key drivers of AF—tend to anchor at fibrotic–nonfibrotic interfaces, underscoring the arrhythmogenic potential of structurally heterogeneous tissue [62]. Moreover, the extent of LA fibrosis identified by LGE has been linked to higher rates of AF recurrence following catheter ablation [63] and an elevated risk of stroke [64]. Nonetheless, recent randomized controlled trials investigating LGE-guided ablation strategies have failed to demonstrate superior outcomes in arrhythmia suppression when targeting fibrotic regions in addition to pulmonary vein isolation, suggesting that the clinical utility of LGE-CMR–guided substrate modification remains uncertain and may require further refinement [65,66].

2.4. Electroanatomic Mapping

Electroanatomic mapping has become an essential tool for substrate characterization in AtCM, particularly during catheter ablation procedures. This technique enables the acquisition of detailed voltage maps, activation sequence patterns, conduction velocity profiles, and the identification of fractionated electrograms, thereby facilitating a comprehensive electrophysiological assessment of the atrial substrate [57,67–69].

One of the principal applications of electroanatomic mapping in AtCM involves the delineation of low-voltage areas (LVAs), typically defined as regions exhibiting bipolar endocardial voltage amplitudes below 0.5 mV [68]. These regions are commonly interpreted as surrogates for underlying atrial fibrosis. Studies have demonstrated that the extent and distribution of LVAs are correlated with atrial structural remodeling, including increased left atrial volume and the presence of cardiovascular comorbidities. Importantly, the voltage attenuation observed in these maps often represents a diffuse electrophysiological phenomenon, with LVAs reflecting localized manifestations within a broader context of global voltage reduction [68]. Currently, no consensus exists regarding definitive cutoff values for atrial fibrosis, and the extent of LVA exhibits only weak correlation with clinical AF history [70–74]. Although combined use of electroanatomic mapping and advanced imaging may enhance substrate identification in atrial arrhythmias [75], substantial variability persists. The development of non-invasive atrial mapping techniques to detect and quantify AtCM remains an area of active and promising investigation [76].

Histopathological investigations have supported the link between reduced atrial voltage and fibrotic remodeling; however, recent biopsy-based evidence indicates that voltage diminution is not exclusively attributable to fibrosis [69]. Other structural alterations—including expanded extracellular matrix volume, loss of myofibrils, diminished cardiomyocyte nuclear density, and amyloid infiltration—have also been implicated in voltage reduction [69]. These findings highlight the multifactorial and heterogeneous nature of atrial substrate pathology in AtCM and suggest that LVA mapping captures a complex spectrum of tissue alterations beyond fibrosis alone.

In addition to catheter-based electroanatomic mapping, emerging non-invasive current-based techniques such as magnetocardiography (MCG) offer novel avenues for atrial substrate assessment [77]. MCG reconstructs pseudo-current distributions derived from cardiac magnetic field measurements and provides contact-free characterization of atrial electrophysiology [77]. In a study by Her et al. [78], MCG was able to detect impaired left atrial (LA) pseudo-current responses during exercise in patients with paroxysmal atrial fibrillation (PAF), compared to healthy controls and endurance athletes [78]. Notably, the exercise-induced change in the LA pseudo-current demonstrated high diagnostic accuracy for PAF, with a sensitivity of 77%, specificity of 92%, and an area under the receiver-operating curve of 0.896. These findings suggest that MCG may serve as a valuable adjunctive tool for identifying functional atrial abnormalities in patients with early-stage AF or at risk for atrial cardiomyopathy.

2.5. Imaging Predictors of AF Onset

To date, no AF management strategy is guided exclusively by cardiac imaging findings. However, accumulating evidence from large-scale cohort studies employing TTE and CMR has demonstrated that structural and functional remodeling of the LA—manifested as increased LA volumes and reduced LA functional indices—is significantly associated with the future development of AF [79–83]. In addition to LA parameters, right atrial (RA) enlargement has emerged as an independent predictor of incident AF. Data from a diverse, multi-ethnic population cohort revealed that elevated RA volume indices remained significantly associated with new-onset AF even after adjusting for cardiovascular comorbidities and concomitant LA characteristics [84].

CCT-based quantification of epicardial adipose tissue (EAT) has further expanded the imaging toolkit for AF risk prediction. Increased EAT mass, as measured by CCT, has been independently linked to a heightened risk of incident AF [85,86]. Moreover, EAT volume demonstrates a strong association with both paroxysmal and persistent AF phenotypes, independent of conventional risk factors [87]. Functional impairment and dilation of the LA have also been correlated with subclinical atrial ectopy. Specifically, both reduced LA function and LA enlargement are associated with an increased frequency of premature atrial contractions on prolonged ambulatory electrocardiographic monitoring—an established harbinger of AF [88,89].

2.6. Recurrent AF

Morphological remodeling of the left atrial (LA) roof, as characterized by CT, has been implicated in the emergence of non–pulmonary vein (non-PV) arrhythmogenic substrates among patients undergoing AF ablation [25,26]. Additionally, pre-procedural assessment of regional LA wall deformation using cardiac CT has demonstrated superior predictive accuracy for the presence of low-voltage areas (LVAs) compared to conventional structural remodeling markers, underscoring its potential utility in procedural planning [27].

LGE-CMR–derived quantification of LA fibrosis has also been shown to independently predict arrhythmia recurrence following AF ablation, reinforcing its relevance in substrate characterization [63]. Nevertheless, two recent randomized controlled trials evaluating fibrosis-guided ablation strategies failed to demonstrate a significant reduction in AF recurrence when LA fibrosis-targeted ablation was performed in addition to standard pulmonary vein isolation (PVI), compared with PVI alone [65,66].

These findings may be attributable to several key factors. One trial postulated that the absence of clinical benefit reflected a relatively low overall fibrotic burden within the study cohort, limiting the efficacy of adjunctive substrate modification [65]. The other highlighted ongoing technical limitations, particularly the lack of standardized imaging protocols, reproducible fibrosis quantification methods, and validated ablation endpoints specific to fibrotic tissue [66]. Furthermore, the authors hypothesized that atrial fibrosis is not a homogeneous entity; rather, distinct histopathological subtypes—such as interstitial versus reparative fibrosis—may exhibit differential contributions to the arrhythmogenic substrate, complicating efforts to tailor ablation strategies solely on the basis of LGE distribution [66].

2.7. Atrial Remodeling and Reverse Remodeling

Cardiac imaging offers valuable prognostic insights into the potential for reverse atrial remodeling following therapeutic interventions for AF [90]. It enables detailed evaluation of both structural and functional recovery, including improvements in atrioventricular valve competence—particularly in cases of functional mitral and tricuspid regurgitation—after restoration of sinus rhythm [91]. In a mechanistic study using CCT, Huang et al. [92] reported that atrial remodeling may commence rapidly, with measurable increases in left and right atrial volumes occurring after just 6 min of atrial high-rate episodes (AHREs). When the duration of AHRE exceeded 6 h, biatrial contractile dysfunction became evident. These findings suggest a time-dependent progression of atrial remodeling, where structural dilation precedes functional decline, and underscore the potential of imaging to detect early changes and guide therapeutic timing.

2.8. Imaging and Risk Stratification

CCT-based evaluations of atrial remodeling in the context of atrial fibrillation (AF)-related stroke have demonstrated that advanced age and diminished atrial booster–pump function are independently associated with stroke occurrence [93]. These findings support the potential utility of substrate-oriented imaging parameters in refining thromboembolic risk stratification among patients with AF [93]. In parallel, the assessment of LA strain has been shown to enhance the predictive accuracy for thromboembolic events in AF and may aid in identifying the presence of LA or left atrial appendage (LAA) thrombus, as detected by transesophageal echocardiography [28,94,95]. Furthermore, variations in LAA morphology and increased LAA volume have been correlated with a prior history of ischemic stroke or transient ischemic attack, reinforcing the link between anatomical features and thromboembolic vulnerability [96,97]. AF itself is associated with globally impaired flow dynamics within both the LA and LAA, contributing to stasis and thrombus formation [98].

Irrespective of AF history, individuals at intermediate-to-high thromboembolic risk demonstrate altered intra-atrial flow dynamics on CMR, indicating the presence of functional atrial abnormalities prior to clinically manifested arrhythmia [98,99]. Furthermore, elevated atrial FDG uptake on PET imaging has been associated with increased stroke prevalence among patients with AF, implicating atrial metabolic or inflammatory activity in the pathophysiology of thromboembolic events [100].

Electroanatomic mapping findings have reinforced these associations, with the presence of LVA being linked to both clinical stroke history and subclinical SCI [101]. Additionally, LVA has shown significant associations with coexisting heart failure with reduced systolic function in individuals with non-valvular AF, suggesting its utility as a marker of more advanced atrial remodeling [102].

In a contemporary observational cohort of 1488 patients undergoing AF ablation, both the presence and burden of LVA were independently predictive of adverse long-term outcomes, including all-cause mortality, heart failure, and stroke, independent of AF recurrence or other confounding clinical variables [103]. These data collectively underscore the potential of cardiac imaging to detect subclinical atrial dysfunction and support its integration into the diagnostic framework of AtCM. Its emerging prognostic value further positions it as a promising tool for refined risk stratification and therapeutic decision-making, warranting continued investigation [100].

3. Biomarkers, Omics, and ECG Parameters

A diverse array of biomarkers from blood, urine, imaging, and ECG have been proposed as surrogates for identifying an underlying AtCM phenotype predictive of AF onset and progression. When combined with ECG, electrophysiological, or echocardiographic data, blood-based markers offer a multidimensional strategy for defining AtCM. Spanning electrophysiological, structural, hemodynamic, serological, and genetic domains, these biomarkers have all been linked to thromboembolic risk—even in the absence of manifest AF.

Consequently, such biomarkers have been adopted as diagnostic criteria in studies exploring the relationship between AtCM and cerebrovascular events. For instance, the ARCADIA trial [104] enrolled patients aged $\geq$ 45 years with embolic stroke of undetermined source (ESUS) and AtCM, as defined by the presence of at least one of the following: P-wave terminal force > 5000 μV·ms in ECG lead V1, NT-proBNP > 250 pg/mL, or an indexed left atrial diameter $\geq$ 3 cm/m^2 on echocardiography. Notably, the trial did not establish the efficacy of biomarker-guided identification of AtCM in selecting patients with cryptogenic stroke who would benefit from oral anticoagulation [105].

The ARIC study further examined echocardiographic parameters of LA function—namely reservoir, conduit, and contractile strain—as well as the LA volume index in relation to incident ischemic stroke [106]. Subsequent analysis from the same cohort revealed that the presence of AtCM, defined by the P-wave terminal force, NT-proBNP, and LA volume index, was also significantly associated with an elevated risk of dementia, independent of AF or prior stroke [107].

3.1. Blood Biomarkers

Although biomarkers associated with AF have been extensively investigated [108,109], comparatively limited attention has been given to those specifically indicative of AtCM in the absence of AF [110]. Notably, a substantial number of biomarkers have been evaluated in both contexts, underscoring their limited specificity for distinguishing between AF-driven and substrate-driven pathology. In certain cases, the transition from paroxysmal to persistent AF is interpreted as a clinical marker of progressive atrial remodeling, and several biomarkers have been explored in this setting to elucidate underlying substrate alterations.

Natriuretic peptides, traditionally linked to heart failure pathophysiology [111], have also been independently associated with the risk of incident AF and its clinical sequelae, including thromboembolic events [112,113]. Similarly, cardiac troponin T (cTnT)—a well-established biomarker of myocardial injury—was identified as a predictor of cardioembolic stroke in the ARIC cohort [114].

Beyond markers of myocardial stress and necrosis, biomarkers of extracellular matrix remodeling have also demonstrated relevance in the context of AF [115]. Elevated levels of soluble suppression of tumorigenicity-2 (sST2) and tissue inhibitor of matrix metalloproteinase-1 (TIMP-1) have been associated with AF progression, independent of conventional clinical risk factors and other circulating biomarkers [115]. Fibrosis-related markers have additionally shown prognostic utility in predicting AF recurrence following electrical cardioversion [110].

More recently, bone morphogenetic protein 10 (BMP10)—a protein predominantly expressed in atrial myocytes—has emerged as a potential biomarker of arrhythmic vulnerability. Elevated serum levels of BMP10 have been linked to AF recurrence following catheter ablation [116]. Furthermore, observational cohort studies in AF populations have demonstrated associations between biomarker elevations and increased risks of ischemic stroke [117], all-cause mortality, and major adverse cerebrovascular and cardiovascular events [118].

C-reactive protein (CRP) has been consistently associated with both the presence and burden of AF in nonsurgical populations, implicating systemic inflammation as a key driver of atrial remodeling [119]. Additionally, elevated levels of von Willebrand factor and soluble P-selectin—markers of endothelial dysfunction and platelet activation—have been linked to increased stroke risk in AF, reinforcing the contribution of prothrombotic and endothelial injury pathways to thromboembolic complications in this population [120].

Limitations of Blood Biomarkers

Several biomarkers lack disease specificity and may simply reflect systemic illness or advanced cardiovascular pathology [121,122]. Growth differentiation factor 15 (GDF-15), for example, is a stress-inducible cytokine implicated in a range of cardiovascular conditions and has been proposed as a prognostic marker for bleeding risk in patients with AF [123]. However, elevated GDF-15 levels are also associated with increased mortality and major adverse cardiovascular events (MACEs) in individuals without AF, highlighting its limited specificity [124].

Practical barriers such as assay variability, diurnal fluctuations in biomarker concentrations, and cost-related constraints further challenge the routine clinical implementation of both blood- and urine-based biomarkers [125]. As such, a careful balance must be struck between the added prognostic insight offered by these markers and the need for operational simplicity in high-volume clinical environments. To date, no biomarker has been prospectively validated to guide clinical decision-making in the management of AtCM.

3.2. Mult-Omics, AF, and AtCM

Primary AtCM may originate from pathogenic variants in genes with either atrial-specific expression or broader systemic functions that manifest within the atrial myocardium [126,127]. Among atrial-selective genetic determinants, MYL4, encoding the essential myosin light chain specific to atrial tissue, has been linked to profound electromechanical abnormalities and atrial contractile dysfunction [128,129]. Similarly, mutations in NPPA, which encodes atrial natriuretic peptide (ANP), have been associated with marked biatrial dilation, heightened thromboembolic risk, and atrial standstill syndromes [126,130]. In addition, several gene variants traditionally associated with ventricular or multisystem disorders have been implicated in atrial dysfunction. These include HCN4, responsible for the cardiac pacemaker channel [131]; SCN5A, which encodes the primary voltage-gated sodium channel in cardiomyocytes [132]; SCN1B, encoding the $\beta 1/\beta 1B$ subunits of the sodium channel complex [133]; and LMNA, which encodes the nuclear envelope protein Lamin A/C [134].

Current understanding of the genomic, transcriptomic, and epigenomic landscape of AtCM is largely extrapolated from investigations focused on AF, a condition that itself contributes to atrial structural and functional impairment. Large-scale genome-wide association studies (GWASs) have identified numerous chromosomal loci containing common genetic variants associated with atrial electrophysiological traits, including heart rate, P-wave duration, and the PR interval [135]. Additional variants have been linked to cardiac anatomical and functional characteristics, such as chamber size, myocardial structure, and contractility [136]. Collectively, GWASs have uncovered more than 100 genetic loci associated with increased susceptibility to AF [137,138]. While it remains unclear whether these variants directly drive the development of primary AtCM, their established association with AF suggests a potential role in promoting secondary atrial remodeling.

Transcriptomic analyses of human atrial tissue—employing both microarray and RNA sequencing technologies—have uncovered a range of genes implicated in atrial remodeling processes [139]. Among these are NPPA and NPPB, which encode natriuretic peptides and are well-established markers of myocardial stress [140], as well as GPR22 and RGS6, involved in G-protein–mediated signaling pathways. Other notable findings include NTM, encoding neurotrimin, a novel biomarker associated with heart failure [141], and COLQ, expressed in both atria, which encodes a specialized collagen responsible for anchoring acetylcholinesterase to the basal lamina, suggesting neuromodulatory involvement in atrial pathology [142].

A recent mRNA sequencing study of atrial tissue obtained during cardiac surgery identified 35 genes significantly associated with AF, independent of coexisting heart failure [140]. These genes were functionally enriched in pathways regulating cardiomyocyte architecture, electrical conduction, fibrosis, inflammation, and endothelial dysfunction, highlighting the multifaceted nature of the arrhythmogenic substrate [143]. Furthermore, in a large-animal model of AF progression, substantial transcriptomic and proteomic alterations were observed early in the transition from paroxysmal to persistent AF, indicating that molecular remodeling precedes sustained arrhythmogenesis [144].

It is now well established that AF arises from the complex interplay between genetic predisposition and environmental influences [145]. Notably, the majority of AF-associated loci identified through GWAS are located within non-coding regions of the genome [146], where they exert regulatory effects on gene expression by modulating transcription factor binding and altering chromatin accessibility through epigenetic mechanisms. Epigenetics, defined as heritable changes in gene expression that do not involve alterations to the DNA sequence itself, has emerged as a key regulatory layer in AF pathophysiology [146].

Among the epigenetic modifications implicated in AF, DNA methylation of CpG islands—mediated by DNA methyltransferases (DNMTs)—has been observed in both paroxysmal and persistent forms of the arrhythmia [147,148]. A recent study employed four DNA methylation-based biological age estimators to provide an epigenetic rationale for the established relationship between chronological aging and AF susceptibility, further bridging aging-related processes with atrial remodeling [149,150].

Beyond DNA methylation, histone modifications have also been implicated in AF pathogenesis [151,152]. Enzymes such as histone deacetylases (HDACs) and histone acetyltransferases (HATs) have been shown to modulate transcriptional activity in atrial tissue, with dysregulation contributing to electrical and structural remodeling [151]. In patients with AF, increased expression of EZH2, which encodes a histone methyltransferase responsible for catalyzing the repressive H3K27me3 epigenetic mark, has been documented in both atrial cardiomyocytes and fibroblasts [152].

3.3. ECG Abnormalities in AtCM

In AtCM, the underlying structural, architectural, contractile, and electrophysiological remodeling of the atria is reflected in distinct abnormalities on surface electrocardiography, serving as a non-invasive manifestation of atrial pathology [153,154].

Atrial remodeling is frequently accompanied by electrocardiographic manifestations, particularly alterations in P-wave duration and morphology, which reflect underlying conduction disturbances. A P-wave duration $\geq$ 120 ms is indicative of partial interatrial block (IAB), while a biphasic P-wave morphology in the inferior leads (II, III, and aVF) signifies advanced IAB, typically caused by complete conduction block at Bachmann's bundle [153–155]. IAB, particularly in its advanced form, is closely associated with LA enlargement [156] and has been linked to the presence of atrial fibrosis, a substrate conducive to reentrant activity, ectopic triggering, and ultimately the initiation and perpetuation of AF [157,158].

Patients exhibiting prolonged P-wave duration or advanced IAB are at heightened risk for AF recurrence following catheter ablation [159,160]. Interestingly, abnormally short P-wave durations have also been identified as predictors of both incident [161] and recurrent AF [162], suggesting that deviations in atrial conduction time at either extreme may reflect distinct forms of electrical remodeling.

High-resolution digital analysis of P-wave duration—typically using amplification settings of 150–200 mm/sec and 80–100 mm/mV—has been proposed as a valuable non-invasive marker for stratifying the severity of AtCM. Specific thresholds of amplified P-wave duration (APWD) correspond to discrete AtCM stages: 140–150 ms for early-stage disease, 150–180 ms for moderate AtCM, and >180 ms or the presence of biphasic P-waves in $\geq$2 inferior leads for advanced stages [163–165]. These electrocardiographic phenotypes have been associated with reduced LAA flow velocities, increased thrombogenic potential, and elevated risk for major adverse cardiovascular and cerebrovascular events, underscoring their clinical relevance [163–165].

In addition to P-wave duration, P-wave terminal force in lead V1 (PTFV1) has emerged as a widely utilized electrocardiographic marker of AtCM. Calculated as the product of the terminal duration (PTDV1) and the absolute amplitude (PTAV1) of the terminal negative P-wave deflection in V1, a PTFV1 exceeding 4 mV·ms is considered pathological [166]. Abnormal PTFV1 has been linked to impaired LAA contractile function, as evidenced by reduced ejection velocities on TEE [167]. While initially interpreted as a marker of LA enlargement and posterior anatomical displacement, more recent evidence suggests that abnormal PTFV1 reflects altered conduction patterns, particularly delayed and posteriorly directed terminal activation of the LA [168,169].

The presence of an abnormal PTFV1 has also been associated with reduced LA strain on speckle-tracking echocardiography, indicating a strong relationship with functional atrial remodeling [169]. A recent meta-analysis confirmed that pathological PTFV1 is predictive of AF occurrence in both general and cardiovascular populations and serves as a prognostic marker for arrhythmia recurrence following CA [170,171]. Furthermore, advanced ECG analysis using amplified digital 12-lead recordings has shown that prolonged P-wave duration is significantly associated with low-voltage substrate in the LA, providing a non-invasive surrogate for fibrotic remodeling [172]. This approach has demonstrated utility in identifying patients with persistent AF at high risk for recurrence after CA and has also been linked to the development of new-onset AF in individuals with HFpEF [173].

The frontal P-wave axis, normally ranging between 0° and +75°, serves as an indicator of the overall direction of atrial depolarization [153]. Deviations from this range have been associated with increased susceptibility to AF. A recent meta-analysis demonstrated that an abnormal P-wave axis is predictive of future AF detection, reinforcing its utility as a non-invasive marker of atrial electrical remodeling [166].

Similarly, reduced P-wave voltage—specifically a P-wave amplitude ≤ 0.1 mV in lead I—is considered pathological [153]. Incorporation of low P-wave voltage into predictive scoring systems has shown utility in identifying individuals at risk for developing AF [174]. Moreover, attenuated P-wave amplitude in lead I has been independently associated with arrhythmia recurrence following radiofrequency ablation, suggesting that diminished atrial electrical signal strength may reflect a pro-arrhythmic substrate [175].

The P-wave area, calculated in lead II as ½ × P-wave duration × P-wave voltage [153], provides an integrative measure of atrial depolarization magnitude. A value ≥ 4 ms·mV is considered abnormal and has been associated with LA enlargement, reflecting structural atrial remodeling [176]. In specific clinical contexts, such as mitral stenosis, the P-wave area measured in lead V3 has demonstrated predictive value for new-onset AF, suggesting its applicability in disease-specific risk stratification [177]. P-wave dispersion, defined as the difference between the maximum and minimum P-wave durations across the 12-lead ECG, serves as a surrogate for intra-atrial conduction heterogeneity [153]. Increased P-wave dispersion has been linked to a higher risk of both incident AF and arrhythmia recurrence following cardioversion [178]. Moreover, in patients with cryptogenic stroke monitored by implantable loop recorders, a P-wave dispersion exceeding 40 ms has been identified as a significant predictor of subsequent AF detection, underscoring its relevance in subclinical AF surveillance [179].

Fibrillatory wave (f-wave) characteristics on surface ECG offer valuable insights into the underlying atrial substrate in AF. Diminished f-wave amplitude has been consistently associated with increased LA size and reduced LAA flow velocity, suggesting a correlation between attenuated atrial electrical activity and structural remodeling [180,181]. Distinct f-wave patterns also demonstrate diagnostic and prognostic relevance. Specifically, low-amplitude f-waves in lead II combined with elevated dominant frequencies in lead V1 have been shown to differentiate long-standing persistent AF from paroxysmal or short-duration persistent forms, indicating a more advanced electrophysiological phenotype [182]. In community-based cohorts, the presence of fine rather than coarse f-waves has been observed more frequently in individuals with persistent AF or greater LA enlargement and independently correlates with elevated risk of incident HF [183]. Moreover, lower pre-ablation f-wave amplitudes have been identified as predictors of post-procedural AF recurrence [184,185], reinforcing the utility of f-wave analysis as a non-invasive marker of advanced atrial disease and procedural outcome.

Advancements in ECG signal processing, particularly through signal-averaged P-wave analysis, have significantly enhanced the resolution and prognostic utility of surface electrocardiography in the context of atrial cardiomyopathy. Unlike standard 10 s ECGs, this approach involves the continuous recording and averaging of P-wave morphology over extended periods, thereby yielding more precise and reproducible data on atrial activation patterns. Prolonged signal-averaged P-wave durations have been consistently observed in individuals with paroxysmal AF compared to controls without a history of arrhythmia [186], and are also predictive of arrhythmia recurrence following catheter ablation [187]. Furthermore, parameters derived from prolonged ECG monitoring have demonstrated superior discriminatory capacity in differentiating paroxysmal from persistent AF relative to conventional short-duration recordings [188], underscoring their potential value in refining atrial substrate characterization and risk stratification strategies.

Finally, premature atrial contractions (PACs) have emerged as clinically relevant electrophysiological markers associated with atrial cardiomyopathy, heightened susceptibility to AF, and elevated risk of thromboembolic events [189,190]. Multiple studies have demonstrated that an increased PAC burden reflects underlying atrial electrical instability and may serve as an early indicator of pathological atrial remodeling [189,190].

Artificial Intelligence-Based ECG Analysis in AtCM

Artificial intelligence (AI)-driven analysis of P-wave morphology, including wavelet-based techniques, represents a promising frontier in the electrocardiographic assessment of atrial cardiomyopathy [153,191]. Emerging deep learning models have demonstrated the capacity to identify individuals at elevated risk for AF with high precision, even when ECGs are recorded in sinus rhythm [192], suggesting that traditional P-wave metrics may insufficiently capture the full spectrum of atrial electrical abnormalities.

In a pragmatic trial evaluating the utility of AI-guided ECG screening, individuals classified as high-risk by the AI algorithm exhibited a fivefold increase in AF detection rates compared to those in the low-risk group [193], underscoring the potential of AI to enhance the sensitivity and efficiency of population-based AF screening strategies. Beyond risk prediction, AI methodologies have also been applied successfully to quantify the atrial substrate and to identify electrophysiological targets for catheter ablation [194,195], highlighting their expanding role in both diagnostic and therapeutic dimensions of atrial disease management.

4. Conclusions and Future Directions

The concept of AtCM has redefined AF as a consequence of progressive structural, electrical, contractile, and molecular remodeling of the atria. This paradigm shift necessitates diagnostic approaches that extend beyond rhythm classification to capture the underlying substrate driving arrhythmogenesis. As reviewed herein, advances in multimodality imaging, electroanatomic mapping, biomarker profiling, ECG analysis, and omics technologies have enabled a more comprehensive assessment of AtCM, offering novel opportunities for early detection, refined risk stratification, and individualized therapy.

Despite these advances, substantial gaps remain. Imaging modalities such as CMR and CCT have improved visualization of atrial fibrosis and morphology, yet standardization of acquisition protocols, quantification thresholds, and clinical integration remain limited.

ECG-derived metrics—including P-wave duration, morphology, dispersion, and f-wave characteristics—hold promise as accessible markers of electrical remodeling but require further validation in prospective studies. Similarly, biomarker-based definitions of AtCM remain challenged by limited specificity and variability across populations. The integration of omics approaches, while conceptually powerful, has yet to yield clinically actionable phenotypes distinct from AF-driven remodeling.

The future of AtCM characterization lies in harmonizing these diverse diagnostic platforms to enable early, precise, and clinically meaningful identification of atrial disease. Key priorities include the following: (i) establishing standardized, reproducible criteria for AtCM across imaging, ECG, and biomarker domains; (ii) elucidating substrate-specific predictors of stroke, heart failure, and arrhythmia recurrence; (iii) validating AI-enhanced tools for non-invasive mapping and risk prediction; and (iv) designing interventional trials that incorporate AtCM phenotypes to guide therapy selection.

Ultimately, embedding substrate characterization into clinical decision-making may transform the contemporary management of AF—from reactive treatment of arrhythmia to proactive modification of its underlying disease substrate.

Author Contributions: P.K.: Conceptualization, Methodology, Investigation, Visualization, Project administration, Writing—original draft, Writing—review and editing; P.K.V.: Writing—review and editing; P.T.: Writing—review and editing; N.K.: Writing—review and editing; D.P.: Writing—review and editing; B.F.: Writing—review and editing; A.P.A.: Writing—review and editing; N.F.: Conceptualization, Writing—review and editing, Supervision. All authors have read and agreed to the published version of the manuscript.

Funding: This research did not receive any specific grant from funding agencies in the public, commercial, or not-for-profit sectors for its design or conduction.

Institutional Review Board Statement: Not applicable.

Informed Consent Statement: Not applicable.

Data Availability Statement: All data generated in this research are included within this article.

Conflicts of Interest: The authors declare no conflicts of interest.

References

1. Tan, S.; Zhou, J.; Veang, T.; Lin, Q.; Liu, Q. Global, Regional, and National Burden of Atrial Fibrillation and Atrial Flutter from 1990 to 2021: Sex Differences and Global Burden Projections to 2046—A Systematic Analysis of the Global Burden of Disease Study 2021. *Europace* **2025**, *27*, euaf027. [CrossRef] [PubMed]
2. Kornej, J.; Börschel, C.S.; Benjamin, E.J.; Schnabel, R.B. Epidemiology of Atrial Fibrillation in the 21st Century: Novel Methods and New Insights. *Circ. Res.* **2020**, *127*, 4–20. [CrossRef]
3. Linz, D.; Gawalko, M.; Betz, K.; Hendriks, J.M.; Lip, G.Y.H.; Vinter, N.; Guo, Y.; Johnsen, S. Atrial Fibrillation: Epidemiology, Screening and Digital Health. *Lancet Reg. Health Eur.* **2024**, *37*, 100786. [CrossRef] [PubMed]
4. Alonso, A.; Bengtson, L.G.S. A Rising Tide: The Global Epidemic of Atrial Fibrillation. *Circulation* **2014**, *129*, 829–830. [CrossRef] [PubMed]
5. Linz, D.; Andrade, J.G.; Arbelo, E.; Boriani, G.; Breithardt, G.; Camm, A.J.; Caso, V.; Nielsen, J.C.; De Melis, M.; De Potter, T.; et al. Longer and Better Lives for Patients with Atrial Fibrillation: The 9th AFNET/EHRA Consensus Conference. *Europace* **2024**, *26*, euae070. [CrossRef]
6. Karakasis, P.; Pamporis, K.; Siontis, K.C.; Theofilis, P.; Samaras, A.; Patoulias, D.; Stachteas, P.; Karagiannidis, E.; Stavropoulos, G.; Tzikas, A.; et al. Major Clinical Outcomes in Symptomatic vs. Asymptomatic Atrial Fibrillation: A Meta-Analysis. *Eur. Heart J.* **2024**, *46*, 1189–1202. [CrossRef]
7. Pamporis, K.; Karakasis, P.; Sagris, M.; Theofilis, P.; Milaras, N.; Pantelidaki, A.; Mourouzis, I.; Fragakis, N.; Vlachos, K.; Kordalis, A.; et al. Prevalence of Asymptomatic Atrial Fibrillation and Risk Factors Associated with Asymptomatic Status: A Systematic Review and Meta-Analysis. *Eur. J. Prev. Cardiol.* **2025**, zwaf138. [CrossRef]

8. Goette, A.; Corradi, D.; Dobrev, D.; Aguinaga, L.; Cabrera, J.-A.; Chugh, S.S.; de Groot, J.R.; Soulat-Dufour, L.; Fenelon, G.; Hatem, S.N.; et al. Atrial Cardiomyopathy Revisited-Evolution of a Concept: A Clinical Consensus Statement of the European Heart Rhythm Association (EHRA) of the ESC, the Heart Rhythm Society (HRS), the Asian Pacific Heart Rhythm Society (APHRS), and the Latin American Heart Rhythm Society (LAHRS). *Europace* **2024**, *26*, euae204. [CrossRef]
9. Shen, M.J.; Arora, R.; Jalife, J. Atrial Myopathy. *JACC Basic Transl. Sci.* **2019**, *4*, 640–654. [CrossRef]
10. Boriani, G.; Gerra, L.; Mantovani, M.; Tartaglia, E.; Mei, D.A.; Imberti, J.F.; Vitolo, M.; Bonini, N. Atrial Cardiomyopathy: An Entity of Emerging Interest in the Clinical Setting. *Eur. J. Intern. Med.* **2023**, *118*, 14–21. [CrossRef]
11. Sade, L.E.; Faletra, F.F.; Pontone, G.; Gerber, B.L.M.; Muraru, D.; Edvardsen, T.; Cosyns, B.; Popescu, B.A.; Klein, A.; Marwick, T.H.; et al. The Role of Multi-Modality Imaging for the Assessment of Left Atrium and Left Atrial Appendage: A Clinical Consensus Statement of the European Association of Cardiovascular Imaging (EACVI), European Heart Rhythm Association (EHRA) of the European Society. *Eur. Heart J. Cardiovasc. Imaging* **2025**, *26*, 385–413. [CrossRef]
12. Nattel, S. Molecular and Cellular Mechanisms of Atrial Fibrosis in Atrial Fibrillation. *JACC Clin. Electrophysiol.* **2017**, *3*, 425–435. [CrossRef] [PubMed]
13. Heijman, J.; Muna, A.P.; Veleva, T.; Molina, C.E.; Sutanto, H.; Tekook, M.; Wang, Q.; Abu-Taha, I.H.; Gorka, M.; Künzel, S.; et al. Atrial Myocyte NLRP3/CaMKII Nexus Forms a Substrate for Postoperative Atrial Fibrillation. *Circ. Res.* **2020**, *127*, 1036–1055. [CrossRef]
14. Yao, C.; Veleva, T.; Scott, L.J.; Cao, S.; Li, L.; Chen, G.; Jeyabal, P.; Pan, X.; Alsina, K.M.; Abu-Taha, I.D.; et al. Enhanced Cardiomyocyte NLRP3 Inflammasome Signaling Promotes Atrial Fibrillation. *Circulation* **2018**, *138*, 2227–2242. [CrossRef] [PubMed]
15. Iacobellis, G. Epicardial Adipose Tissue in Contemporary Cardiology. *Nat. Rev. Cardiol.* **2022**, *19*, 593–606. [CrossRef]
16. Van Gelder, I.C.; Rienstra, M.; Bunting, K.V.; Casado-Arroyo, R.; Caso, V.; Crijns, H.J.G.M.; De Potter, T.J.R.; Dwight, J.; Guasti, L.; Hanke, T.; et al. 2024 ESC Guidelines for the Management of Atrial Fibrillation Developed in Collaboration with the European Association for Cardio-Thoracic Surgery (EACTS). *Eur. Heart J.* **2024**, *45*, 3314–3414. [CrossRef] [PubMed]
17. Joglar, J.A.; Chung, M.K.; Armbruster, A.L.; Benjamin, E.J.; Chyou, J.Y.; Cronin, E.M.; Deswal, A.; Eckhardt, L.L.; Goldberger, Z.D.; Gopinathannair, R.; et al. 2023 ACC/AHA/ACCP/HRS Guideline for the Diagnosis and Management of Atrial Fibrillation: A Report of the American College of Cardiology/American Heart Association Joint Committee on Clinical Practice Guidelines. *Circulation* **2024**, *149*, e1–e156. [CrossRef]
18. Inoue, K.; Khan, F.H.; Remme, E.W.; Ohte, N.; García-Izquierdo, E.; Chetrit, M.; Moñivas-Palomero, V.; Mingo-Santos, S.; Andersen, Ø.S.; Gude, E.; et al. Determinants of Left Atrial Reservoir and Pump Strain and Use of Atrial Strain for Evaluation of Left Ventricular Filling Pressure. *Eur. Heart J. Cardiovasc. Imaging* **2021**, *23*, 61–70. [CrossRef]
19. Smiseth, O.A.; Morris, D.A.; Cardim, N.; Cikes, M.; Delgado, V.; Donal, E.; Flachskampf, F.A.; Galderisi, M.; Gerber, B.L.; Gimelli, A.; et al. Multimodality Imaging in Patients with Heart Failure and Preserved Ejection Fraction: An Expert Consensus Document of the European Association of Cardiovascular Imaging. *Eur. Heart J. Cardiovasc. Imaging* **2022**, *23*, e34–e61. [CrossRef]
20. Lang, R.M.; Badano, L.P.; Mor-Avi, V.; Afilalo, J.; Armstrong, A.; Ernande, L.; Flachskampf, F.A.; Foster, E.; Goldstein, S.A.; Kuznetsova, T.; et al. Recommendations for Cardiac Chamber Quantification by Echocardiography in Adults: An Update from the American Society of Echocardiography and the European Association of Cardiovascular Imaging. *Eur. Heart J. Cardiovasc. Imaging* **2015**, *16*, 233–270. [CrossRef]
21. Raniga, D.; Goda, M.; Hattingh, L.; Thorning, S.; Rowe, M.; Howes, L. Left Atrial Volume Index: A Predictor of Atrial Fibrillation Recurrence Following Direct Current Cardioversion—A Systematic Review and Meta-Analysis. *Int. J. Cardiol. Heart Vasc.* **2024**, *51*, 101364. [CrossRef] [PubMed]
22. Meucci, M.C.; Fortuni, F.; Galloo, X.; Bootsma, M.; Crea, F.; Bax, J.J.; Marsan, N.A.; Delgado, V. Left Atrioventricular Coupling Index in Hypertrophic Cardiomyopathy and Risk of New-Onset Atrial Fibrillation. *Int. J. Cardiol.* **2022**, *363*, 87–93. [CrossRef]
23. Pezel, T.; Ambale-Venkatesh, B.; Quinaglia, T.; Heckbert, S.R.; Kato, Y.; de Vasconcellos, H.D.; Wu, C.O.; Post, W.S.; Henry, P.; Bluemke, D.A.; et al. Change in Left Atrioventricular Coupling Index to Predict Incident Atrial Fibrillation: The Multi-Ethnic Study of Atherosclerosis (MESA). *Radiology* **2022**, *303*, 317–326. [CrossRef] [PubMed]
24. Shih, J.-Y.; Tsai, W.-C.; Huang, Y.-Y.; Liu, Y.-W.; Lin, C.-C.; Huang, Y.-S.; Tsai, L.-M.; Lin, L.-J. Association of Decreased Left Atrial Strain and Strain Rate with Stroke in Chronic Atrial Fibrillation. *J. Am. Soc. Echocardiogr.* **2011**, *24*, 513–519. [CrossRef] [PubMed]
25. Kurotobi, T.; Iwakura, K.; Inoue, K.; Kimura, R.; Toyoshima, Y.; Ito, N.; Mizuno, H.; Shimada, Y.; Fujii, K.; Nanto, S.; et al. The Significance of the Shape of the Left Atrial Roof as a Novel Index for Determining the Electrophysiological and Structural Characteristics in Patients with Atrial Fibrillation. *Europace* **2011**, *13*, 803–808. [CrossRef]
26. Mangiafico, V.; Saberwal, B.; Lavalle, C.; Raharja, A.; Ahmed, Z.; Papageorgiou, N.; Ahsan, S. The Role of CT in Detecting AF Substrate. *Trends Cardiovasc. Med.* **2021**, *31*, 457–466. [CrossRef]
27. Nedios, S.; Sanatkhani, S.; Oladosu, M.; Seewöster, T.; Richter, S.; Arya, A.; Heijman, J.; Crijns, H.J.; Hindricks, G.; Bollmann, A.; et al. Association of Low-Voltage Areas with the Regional Wall Deformation and the Left Atrial Shape in Patients with Atrial Fibrillation: A Proof of Concept Study. *Int. J. Cardiol. Heart Vasc.* **2021**, *33*, 100730. [CrossRef]

28. Obokata, M.; Negishi, K.; Kurosawa, K.; Tateno, R.; Tange, S.; Arai, M.; Amano, M.; Kurabayashi, M. Left Atrial Strain Provides Incremental Value for Embolism Risk Stratification over CHA_2DS_2-VASc Score and Indicates Prognostic Impact in Patients with Atrial Fibrillation. *J. Am. Soc. Echocardiogr.* **2014**, *27*, 709–716.e4. [CrossRef]
29. Azemi, T.; Rabdiya, V.M.; Ayirala, S.R.; McCullough, L.D.; Silverman, D.I. Left Atrial Strain Is Reduced in Patients with Atrial Fibrillation, Stroke or TIA, and Low Risk CHADS(2) Scores. *J. Am. Soc. Echocardiogr.* **2012**, *25*, 1327–1332. [CrossRef]
30. Watanabe, E.; Miyagawa, M.; Uetani, T.; Kinoshita, M.; Kitazawa, R.; Kurata, M.; Ishimura, H.; Matsuda, T.; Tanabe, Y.; Kido, T.; et al. Positron Emission Tomography/Computed Tomography Detection of Increased (18)F-Fluorodeoxyglucose Uptake in the Cardiac Atria of Patients with Atrial Fibrillation. *Int. J. Cardiol.* **2019**, *283*, 171–177. [CrossRef]
31. Niemelä, M.; Uusitalo, V.; Pöyhönen, P.; Schildt, J.; Lehtonen, J.; Kupari, M. Incidence and Predictors of Atrial Fibrillation in Cardiac Sarcoidosis: A Multimodality Imaging Study. *JACC Cardiovasc. Imaging* **2022**, *15*, 1622–1631. [CrossRef] [PubMed]
32. Kupusovic, J.; Kessler, L.; Bruns, F.; Bohnen, J.-E.; Nekolla, S.G.; Weber, M.M.; Lauenroth, A.; Rattka, M.; Hermann, K.; Dobrev, D.; et al. Visualization of Fibroblast Activation Using (68)Ga-FAPI PET/CT after Pulmonary Vein Isolation with Pulsed Field Compared with Cryoballoon Ablation. *J. Nucl. Cardiol.* **2023**, *30*, 2018–2028. [CrossRef] [PubMed]
33. Li, L.; Gao, J.; Chen, B.-X.; Liu, X.; Shi, L.; Wang, Y.; Wang, L.; Wang, Y.; Su, P.; Yang, M.-F.; et al. Fibroblast Activation Protein Imaging in Atrial Fibrillation: A Proof-of-Concept Study. *J. Nucl. Cardiol.* **2023**, *30*, 2712–2720. [CrossRef] [PubMed]
34. Wan, P.; Wang, B.; Yu, W.; Zhai, L.S.; Qian, B.; Zhang, F.; Liu, B.; Wang, J.; Shao, X.; Shi, Y.; et al. Right Atrial Wall Inflammation Detected by (18)F-FDG PET/CT May Be Significantly Associated with Persistent Atrial Fibrillation: A Prospective Case-Control Study. *BMC Cardiovasc. Disord.* **2023**, *23*, 587. [CrossRef]
35. Kassar, A.; Chamoun, N.; Haykal, R.; Chahine, Y.; Babb, M.; Al Yasiri, H.; Hensley, T.; Andrikopoulou, E.; Akoum, N. Atrial FDG Uptake and Atrial Fibrillation: A Systematic Review and Meta-Analysis. *Heart Rhythm O2* **2025**, *6*, 417–423. [CrossRef]
36. Wang, H.; Po, S.; Zhang, Y. PET-MRI Imaging to Assess Severity and Distribution of Atrial Fibrosis in Atrial Fibrillation. *Europace* **2024**, *26*, euae102–euae682. [CrossRef]
37. Wilk, B.; Lim, H.; Hicks, J.; Sullivan, R.; Thiessen, J.D.; Kovacs, M.; Sykes, J.; Biernaski, H.; Duffett, S.; Prato, F.S.; et al. Complexities of Atrial Fibrosis Imaging: Exploring Relationships Between Endocardial Voltage, Extracellular Volume, and Sympathetic Innervation. *JACC Clin. Electrophysiol.* **2025**, *11*, 735–748. [CrossRef]
38. Bäck, S.; Lantz, J.; Skoda, I.; Henriksson, L.; Persson, A.; Karlsson, L.O.; Carlhäll, C.-J.; Ebbers, T. Comprehensive Left Atrial Flow Component Analysis Reveals Abnormal Flow Patterns in Paroxysmal Atrial Fibrillation. *Am. J. Physiol. Heart Circ. Physiol.* **2024**, *326*, H511–H521. [CrossRef] [PubMed]
39. Henriksen, E.; Selmeryd, J.; Leppert, J.; Hedberg, P. Echocardiographic Assessment of Maximum and Minimum Left Atrial Volumes: A Population-Based Study of Middle-Aged and Older Subjects without Apparent Cardiovascular Disease. *Int. J. Cardiovasc. Imaging* **2015**, *31*, 57–64. [CrossRef]
40. Higuchi, S.; Voskoboinik, A.; Im, S.I.; Lee, A.; Olgin, J.; Arbil, A.; Afzal, J.; Marcus, G.M.; Stillson, C.; Bibby, D.; et al. Frequent Premature Atrial Contractions Lead to Adverse Atrial Remodeling and Atrial Fibrillation in a Swine Model. *Circulation* **2024**, *149*, 463–474. [CrossRef]
41. Marino, P.N.; Degiovanni, A.; Zanaboni, J. Complex Interaction between the Atrium and the Ventricular Filling Process: The Role of Conduit. *Open Heart* **2019**, *6*, e001042. [CrossRef]
42. Aronson, D.; Sliman, H.; Abadi, S.; Maiorov, I.; Perlow, D.; Mutlak, D.; Lessick, J. Conduit Flow Compensates for Impaired Left Atrial Passive and Booster Functions in Advanced Diastolic Dysfunction. *Circ. Cardiovasc. Imaging* **2024**, *17*, e016276. [CrossRef]
43. Hoit, B.D. Left Atrial Reservoir Strain: Its Time Has Come. *JACC Cardiovasc. Imaging* **2022**, *15*, 392–394. [CrossRef]
44. Leung, M.; van Rosendael, P.J.; Abou, R.; Ajmone Marsan, N.; Leung, D.Y.; Delgado, V.; Bax, J.J. Left Atrial Function to Identify Patients with Atrial Fibrillation at High Risk of Stroke: New Insights from a Large Registry. *Eur. Heart J.* **2018**, *39*, 1416–1425. [CrossRef] [PubMed]
45. Stojanovska, J.; Cronin, P.; Patel, S.; Gross, B.H.; Oral, H.; Chughtai, K.; Kazerooni, E.A. Reference Normal Absolute and Indexed Values from ECG-Gated MDCT: Left Atrial Volume, Function, and Diameter. *AJR Am. J. Roentgenol.* **2011**, *197*, 631–637. [CrossRef] [PubMed]
46. Doria de Vasconcellos, H.; Win, T.T.; Chamera, E.; Hong, S.Y.; Venkatesh, B.A.; Young, P.; Yang, X.; Ciuffo, L.; Sharma, R.K.; Imai, M.; et al. References Values for Left Atrial Volumes, Emptying Fractions, Strains, and Strain Rates and Their Determinants by Age, Gender, and Ethnicity: The Multiethnic Study of Atherosclerosis (MESA). *Acad. Radiol.* **2021**, *28*, 356–363. [CrossRef] [PubMed]
47. Kawel-Boehm, N.; Hetzel, S.J.; Ambale-Venkatesh, B.; Captur, G.; Francois, C.J.; Jerosch-Herold, M.; Salerno, M.; Teague, S.D.; Valsangiacomo-Buechel, E.; van der Geest, R.J.; et al. Reference Ranges ("normal Values") for Cardiovascular Magnetic Resonance (CMR) in Adults and Children: 2020 Update. *J. Cardiovasc. Magn. Reson.* **2020**, *22*, 87. [CrossRef]
48. Krittanawong, C.; Maitra, N.S.; Hassan Virk, H.U.; Farrell, A.; Hamzeh, I.; Arya, B.; Pressman, G.S.; Wang, Z.; Marwick, T.H. Normal Ranges of Right Atrial Strain: A Systematic Review and Meta-Analysis. *JACC Cardiovasc. Imaging* **2023**, *16*, 282–294. [CrossRef]

49. Soulat-Dufour, L.; Addetia, K.; Miyoshi, T.; Citro, R.; Daimon, M.; Fajardo, P.G.; Kasliwal, R.R.; Kirkpatrick, J.N.; Monaghan, M.J.; Muraru, D.; et al. Normal Values of Right Atrial Size and Function According to Age, Sex, and Ethnicity: Results of the World Alliance Societies of Echocardiography Study. *J. Am. Soc. Echocardiogr.* **2021**, *34*, 286–300. [CrossRef]
50. Kou, S.; Caballero, L.; Dulgheru, R.; Voilliot, D.; De Sousa, C.; Kacharava, G.; Athanassopoulos, G.D.; Barone, D.; Baroni, M.; Cardim, N.; et al. Echocardiographic Reference Ranges for Normal Cardiac Chamber Size: Results from the NORRE Study. *Eur. Heart J. Cardiovasc. Imaging* **2014**, *15*, 680–690. [CrossRef]
51. Singh, A.; Carvalho Singulane, C.; Miyoshi, T.; Prado, A.D.; Addetia, K.; Bellino, M.; Daimon, M.; Gutierrez Fajardo, P.; Kasliwal, R.R.; Kirkpatrick, J.N.; et al. Normal Values of Left Atrial Size and Function and the Impact of Age: Results of the World Alliance Societies of Echocardiography Study. *J. Am. Soc. Echocardiogr.* **2022**, *35*, 154–164.e3. [CrossRef]
52. Aquaro, G.D.; De Gori, C.; Faggioni, L.; Parisella, M.L.; Cioni, D.; Lencioni, R.; Neri, E. Diagnostic and Prognostic Role of Late Gadolinium Enhancement in Cardiomyopathies. *Eur. Heart J. Suppl.* **2023**, *25*, C130–C136. [CrossRef]
53. Okasha, O.; Kazmirczak, F.; Chen, K.-H.A.; Farzaneh-Far, A.; Shenoy, C. Myocardial Involvement in Patients with Histologically Diagnosed Cardiac Sarcoidosis: A Systematic Review and Meta-Analysis of Gross Pathological Images From Autopsy or Cardiac Transplantation Cases. *J. Am. Heart Assoc.* **2019**, *8*, e011253. [CrossRef] [PubMed]
54. Donal, E.; Lip, G.Y.H.; Galderisi, M.; Goette, A.; Shah, D.; Marwan, M.; Lederlin, M.; Mondillo, S.; Edvardsen, T.; Sitges, M.; et al. EACVI/EHRA Expert Consensus Document on the Role of Multi-Modality Imaging for the Evaluation of Patients with Atrial Fibrillation. *Eur. Heart J. Cardiovasc. Imaging* **2016**, *17*, 355–383. [CrossRef] [PubMed]
55. Goette, A.; Kalman, J.M.; Aguinaga, L.; Akar, J.; Cabrera, J.A.; Chen, S.A.; Chugh, S.S.; Corradi, D.; D'Avila, A.; Dobrev, D.; et al. EHRA/HRS/APHRS/SOLAECE Expert Consensus on Atrial Cardiomyopathies: Definition, Characterization, and Clinical Implication. *Europace* **2016**, *18*, 1455–1490. [CrossRef] [PubMed]
56. Boyle, P.M.; Sarairah, S.; Kwan, K.T.; Scott, G.D.; Mohamedali, F.; Anderson, C.A.; Bifulco, S.F.; Ordovas, K.G.; Prutkin, J.; Robinson, M.; et al. Elevated Fibrosis Burden as Assessed by MRI Predicts Cryoballoon Ablation Failure. *J. Cardiovasc. Electrophysiol.* **2023**, *34*, 302–312. [CrossRef]
57. Eichenlaub, M.; Mueller-Edenborn, B.; Minners, J.; Figueras i Ventura, R.M.; Forcada, B.R.; Colomer, A.V.; Hein, M.; Ruile, P.; Lehrmann, H.; Schoechlin, S.; et al. Comparison of Various Late Gadolinium Enhancement Magnetic Resonance Imaging Methods with High-Definition Voltage and Activation Mapping for Detection of Atrial Cardiomyopathy. *EP Eur.* **2022**, *24*, 1102–1111. [CrossRef]
58. Flett, A.S.; Hayward, M.P.; Ashworth, M.T.; Hansen, M.S.; Taylor, A.M.; Elliott, P.M.; McGregor, C.; Moon, J.C. Equilibrium Contrast Cardiovascular Magnetic Resonance for the Measurement of Diffuse Myocardial Fibrosis: Preliminary Validation in Humans. *Circulation* **2010**, *122*, 138–144. [CrossRef]
59. Platonov, P.G.; Mitrofanova, L.B.; Orshanskaya, V.; Ho, S.Y. Structural Abnormalities in Atrial Walls Are Associated with Presence and Persistency of Atrial Fibrillation but Not with Age. *J. Am. Coll. Cardiol.* **2011**, *58*, 2225–2232. [CrossRef]
60. Karakasis, P.; Theofilis, P.; Vlachakis, P.K.; Korantzopoulos, P.; Patoulias, D.; Antoniadis, A.P.; Fragakis, N. Atrial Fibrosis in Atrial Fibrillation: Mechanistic Insights, Diagnostic Challenges, and Emerging Therapeutic Targets. *Int. J. Mol. Sci.* **2024**, *26*, 209. [CrossRef]
61. Karakasis, P.; Theofilis, P.; Lefkou, E.; Antoniadis, A.P.; Patoulias, D.; Korantzopoulos, P.; Fragakis, N. Clonal Hematopoiesis of Indeterminate Potential and Atrial Fibrillation: Insights into Pathophysiology and Clinical Implications. *Int. J. Mol. Sci.* **2025**, *26*, 2739. [CrossRef] [PubMed]
62. Zahid, S.; Cochet, H.; Boyle, P.M.; Schwarz, E.L.; Whyte, K.N.; Vigmond, E.J.; Dubois, R.; Hocini, M.; Haïssaguerre, M.; Jaïs, P.; et al. Patient-Derived Models Link Re-Entrant Driver Localization in Atrial Fibrillation to Fibrosis Spatial Pattern. *Cardiovasc. Res.* **2016**, *110*, 443–454. [CrossRef]
63. Marrouche, N.F.; Wilber, D.; Hindricks, G.; Jais, P.; Akoum, N.; Marchlinski, F.; Kholmovski, E.; Burgon, N.; Hu, N.; Mont, L.; et al. Association of Atrial Tissue Fibrosis Identified by Delayed Enhancement MRI and Atrial Fibrillation Catheter Ablation: The DECAAF Study. *JAMA* **2014**, *311*, 498–506. [CrossRef] [PubMed]
64. King, J.B.; Azadani, P.N.; Suksaranjit, P.; Bress, A.P.; Witt, D.M.; Han, F.T.; Chelu, M.G.; Silver, M.A.; Biskupiak, J.; Wilson, B.D.; et al. Left Atrial Fibrosis and Risk of Cerebrovascular and Cardiovascular Events in Patients with Atrial Fibrillation. *J. Am. Coll. Cardiol.* **2017**, *70*, 1311–1321. [CrossRef] [PubMed]
65. Bisbal, F.; Benito, E.; Teis, A.; Alarcón, F.; Sarrias, A.; Caixal, G.; Villuendas, R.; Garre, P.; Soto, N.; Cozzari, J.; et al. Magnetic Resonance Imaging-Guided Fibrosis Ablation for the Treatment of Atrial Fibrillation: The ALICIA Trial. *Circ. Arrhythm. Electrophysiol.* **2020**, *13*, e008707. [CrossRef]
66. Marrouche, N.F.; Wazni, O.; McGann, C.; Greene, T.; Dean, J.M.; Dagher, L.; Kholmovski, E.; Mansour, M.; Marchlinski, F.; Wilber, D.; et al. Effect of MRI-Guided Fibrosis Ablation vs Conventional Catheter Ablation on Atrial Arrhythmia Recurrence in Patients with Persistent Atrial Fibrillation: The DECAAF II Randomized Clinical Trial. *JAMA* **2022**, *327*, 2296–2305. [CrossRef]

67. Chen, J.; Arentz, T.; Cochet, H.; Müller-Edenborn, B.; Kim, S.; Moreno-Weidmann, Z.; Minners, J.; Kohl, P.; Lehrmann, H.; Allgeier, J.; et al. Extent and Spatial Distribution of Left Atrial Arrhythmogenic Sites, Late Gadolinium Enhancement at Magnetic Resonance Imaging, and Low-Voltage Areas in Patients with Persistent Atrial Fibrillation: Comparison of Imaging vs. Electrical Parameters of Fibr. *EP Eur.* **2019**, *21*, 1484–1493. [CrossRef]
68. Yamaguchi, T.; Otsubo, T.; Takahashi, Y.; Nakashima, K.; Fukui, A.; Hirota, K.; Ishii, Y.; Shinzato, K.; Osako, R.; Tahara, M.; et al. Atrial Structural Remodeling in Patients with Atrial Fibrillation Is a Diffuse Fibrotic Process: Evidence from High-Density Voltage Mapping and Atrial Biopsy. *J. Am. Heart Assoc.* **2022**, *11*, e024521. [CrossRef]
69. Takahashi, Y.; Yamaguchi, T.; Otsubo, T.; Nakashima, K.; Shinzato, K.; Osako, R.; Shichida, S.; Kawano, Y.; Fukui, A.; Kawaguchi, A.; et al. Histological Validation of Atrial Structural Remodelling in Patients with Atrial Fibrillation. *Eur. Heart J.* **2023**, *44*, 3339–3353. [CrossRef]
70. Xintarakou, A.; Tzeis, S.; Psarras, S.; Asvestas, D.; Vardas, P. Atrial Fibrosis as a Dominant Factor for the Development of Atrial Fibrillation: Facts and Gaps. *EP Eur.* **2020**, *22*, 342–351. [CrossRef]
71. Kottkamp, H.; Bender, R.; Berg, J. Catheter Ablation of Atrial Fibrillation: How to Modify the Substrate? *J. Am. Coll. Cardiol.* **2015**, *65*, 196–206. [CrossRef] [PubMed]
72. Kapa, S.; Desjardins, B.; Callans, D.J.; Marchlinski, F.E.; Dixit, S. Contact Electroanatomic Mapping Derived Voltage Criteria for Characterizing Left Atrial Scar in Patients Undergoing Ablation for Atrial Fibrillation. *J. Cardiovasc. Electrophysiol.* **2014**, *25*, 1044–1052. [CrossRef] [PubMed]
73. Spragg, D.D.; Khurram, I.; Zimmerman, S.L.; Yarmohammadi, H.; Barcelon, B.; Needleman, M.; Edwards, D.; Marine, J.E.; Calkins, H.; Nazarian, S. Initial Experience with Magnetic Resonance Imaging of Atrial Scar and Co-Registration with Electroanatomic Voltage Mapping during Atrial Fibrillation: Success and Limitations. *Heart Rhythm.* **2012**, *9*, 2003–2009. [CrossRef]
74. Jadidi, A.S.; Cochet, H.; Shah, A.J.; Kim, S.J.; Duncan, E.; Miyazaki, S.; Sermesant, M.; Lehrmann, H.; Lederlin, M.; Linton, N.; et al. Inverse Relationship between Fractionated Electrograms and Atrial Fibrosis in Persistent Atrial Fibrillation: Combined Magnetic Resonance Imaging and High-Density Mapping. *J. Am. Coll. Cardiol.* **2013**, *62*, 802–812. [CrossRef]
75. Zghaib, T.; Keramati, A.; Chrispin, J.; Huang, D.; Balouch, M.A.; Ciuffo, L.; Berger, R.D.; Marine, J.E.; Ashikaga, H.; Calkins, H.; et al. Multimodal Examination of Atrial Fibrillation Substrate: Correlation of Left Atrial Bipolar Voltage Using Multi-Electrode Fast Automated Mapping, Point-by-Point Mapping, and Magnetic Resonance Image Intensity Ratio. *JACC Clin. Electrophysiol.* **2018**, *4*, 59–68. [CrossRef] [PubMed]
76. Eichenlaub, M.; Mueller-Edenborn, B.; Lehrmann, H.; Minners, J.; Nairn, D.; Loewe, A.; Allgeier, J.; Jander, N.; Allgeier, M.; Ruile, P.; et al. Non-Invasive Body Surface Electrocardiographic Imaging for Diagnosis of Atrial Cardiomyopathy. *EP Eur.* **2021**, *23*, 2010–2019. [CrossRef]
77. Brisinda, D.; Fenici, P.; Fenici, R. Clinical Magnetocardiography: The Unshielded Bet-Past, Present, and Future. *Front. Cardiovasc. Med.* **2023**, *10*, 1232882. [CrossRef]
78. Her, A.-Y.; Shin, E.-S.; Zhou, Q.; Wierzbinski, J.; Vidal-Lopez, S.; Saleh, A.; Kim, Y.H.; Garg, S.; Jung, F.; Brachmann, J. Magnetocardiography Detects Left Atrial Dysfunction in Paroxysmal Atrial Fibrillation. *Clin. Hemorheol. Microcirc.* **2019**, *72*, 353–363. [CrossRef]
79. Lim, D.J.; Ambale-Ventakesh, B.; Ostovaneh, M.R.; Zghaib, T.; Ashikaga, H.; Wu, C.; Watson, K.E.; Hughes, T.; Shea, S.; Heckbert, S.R.; et al. Change in Left Atrial Function Predicts Incident Atrial Fibrillation: The Multi-Ethnic Study of Atherosclerosis. *Eur. Heart J. Cardiovasc. Imaging* **2019**, *20*, 979–987. [CrossRef]
80. Habibi, M.; Samiei, S.; Ambale Venkatesh, B.; Opdahl, A.; Helle-Valle, T.M.; Zareian, M.; Almeida, A.L.C.; Choi, E.-Y.; Wu, C.; Alonso, A.; et al. Cardiac Magnetic Resonance-Measured Left Atrial Volume and Function and Incident Atrial Fibrillation: Results From MESA (Multi-Ethnic Study of Atherosclerosis). *Circ. Cardiovasc. Imaging* **2016**, *9*, e004299. [CrossRef]
81. Nielsen, A.B.; Skaarup, K.G.; Hauser, R.; Johansen, N.D.; Lassen, M.C.H.; Jensen, G.B.; Schnohr, P.; Møgelvang, R.; Biering-Sørensen, T. Normal Values and Reference Ranges for Left Atrial Strain by Speckle-Tracking Echocardiography: The Copenhagen City Heart Study. *Eur. Heart J. Cardiovasc. Imaging* **2021**, *23*, 42–51. [CrossRef]
82. Hauser, R.; Nielsen, A.B.; Skaarup, K.G.; Lassen, M.C.H.; Duus, L.S.; Johansen, N.D.; Sengeløv, M.; Marott, J.L.; Jensen, G.; Schnohr, P.; et al. Left Atrial Strain Predicts Incident Atrial Fibrillation in the General Population: The Copenhagen City Heart Study. *Eur. Heart J. Cardiovasc. Imaging* **2021**, *23*, 52–60. [CrossRef]
83. Hirose, T.; Kawasaki, M.; Tanaka, R.; Ono, K.; Watanabe, T.; Iwama, M.; Noda, T.; Watanabe, S.; Takemura, G.; Minatoguchi, S. Left Atrial Function Assessed by Speckle Tracking Echocardiography as a Predictor of New-Onset Non-Valvular Atrial Fibrillation: Results from a Prospective Study in 580 Adults. *Eur. Heart J. Cardiovasc. Imaging* **2012**, *13*, 243–250. [CrossRef] [PubMed]
84. Xie, E.; Yu, R.; Ambale-Venkatesh, B.; Bakhshi, H.; Heckbert, S.R.; Soliman, E.Z.; Bluemke, D.A.; Kawut, S.M.; Wu, C.O.; Nazarian, S.; et al. Association of Right Atrial Structure with Incident Atrial Fibrillation: A Longitudinal Cohort Cardiovascular Magnetic Resonance Study from the Multi-Ethnic Study of Atherosclerosis (MESA). *J. Cardiovasc. Magn. Reson.* **2020**, *22*, 36. [CrossRef]

85. West, H.W.; Siddique, M.; Williams, M.C.; Volpe, L.; Desai, R.; Lyasheva, M.; Thomas, S.; Dangas, K.; Kotanidis, C.P.; Tomlins, P.; et al. Deep-Learning for Epicardial Adipose Tissue Assessment with Computed Tomography: Implications for Cardiovascular Risk Prediction. *JACC Cardiovasc. Imaging* **2023**, *16*, 800–816. [CrossRef] [PubMed]
86. Chong, B.; Jayabaskaran, J.; Ruban, J.; Goh, R.; Chin, Y.H.; Kong, G.; Ng, C.H.; Lin, C.; Loong, S.; Muthiah, M.D.; et al. Epicardial Adipose Tissue Assessed by Computed Tomography and Echocardiography Are Associated with Adverse Cardiovascular Outcomes: A Systematic Review and Meta-Analysis. *Circ. Cardiovasc. Imaging* **2023**, *16*, e015159. [CrossRef] [PubMed]
87. Al Chekakie, M.O.; Welles, C.C.; Metoyer, R.; Ibrahim, A.; Shapira, A.R.; Cytron, J.; Santucci, P.; Wilber, D.J.; Akar, J.G. Pericardial Fat Is Independently Associated with Human Atrial Fibrillation. *J. Am. Coll. Cardiol.* **2010**, *56*, 784–788. [CrossRef]
88. Huber, M.P.; Pandit, J.A.; Jensen, P.N.; Wiggins, K.L.; Patel, R.B.; Freed, B.H.; Bertoni, A.G.; Shah, S.J.; Heckbert, S.R.; Floyd, J.S. Left Atrial Strain and the Risk of Atrial Arrhythmias from Extended Ambulatory Cardiac Monitoring: MESA. *J. Am. Heart Assoc.* **2022**, *11*, e026875. [CrossRef]
89. Heckbert, S.R.; Jensen, P.N.; Austin, T.R.; Chen, L.Y.; Post, W.S.; Ambale Venkatesh, B.; Soliman, E.Z.; Floyd, J.S.; Sotoodehnia, N.; Kronmal, R.A.; et al. Associations of Left Atrial Function and Structure with Supraventricular Ectopy: The Multi-Ethnic Study of Atherosclerosis. *J. Am. Heart Assoc.* **2021**, *10*, e018093. [CrossRef]
90. Tops, L.F.; Delgado, V.; Bertini, M.; Marsan, N.A.; Den Uijl, D.W.; Trines, S.A.I.P.; Zeppenfeld, K.; Holman, E.; Schalij, M.J.; Bax, J.J. Left Atrial Strain Predicts Reverse Remodeling after Catheter Ablation for Atrial Fibrillation. *J. Am. Coll. Cardiol.* **2011**, *57*, 324–331. [CrossRef]
91. Soulat-Dufour, L.; Lang, S.; Addetia, K.; Ederhy, S.; Adavane-Scheuble, S.; Chauvet-Droit, M.; Jean, M.-L.; Nhan, P.; Ben Said, R.; Kamami, I.; et al. Restoring Sinus Rhythm Reverses Cardiac Remodeling and Reduces Valvular Regurgitation in Patients with Atrial Fibrillation. *J. Am. Coll. Cardiol.* **2022**, *79*, 951–961. [CrossRef] [PubMed]
92. Huang, S.-H.; Tsao, H.-M.; Liao, C.-F.; Chen, Z.-Y.; Chao, T.-F.; Chen, S.-A. Multimodality Imaging Assessment of the Biatrial Remodeling of the Burden of Atrial High-Rate Episodes in Patients with Cardiac Implanted Electronic Devices. *Int. J. Cardiol.* **2023**, *371*, 175–183. [CrossRef] [PubMed]
93. Tsao, H.-M.; Hu, W.-C.; Tsai, P.-H.; Lee, C.-L.; Wang, H.-H.; Chang, S.-L.; Chao, T.-F.; Chen, S.-A. Functional Remodeling of Both Atria Is Associated with Occurrence of Stroke in Patients with Paroxysmal and Persistent Atrial Fibrillation. *Acta Cardiol. Sin.* **2017**, *33*, 50–57. [CrossRef] [PubMed]
94. Kim, D.; Shim, C.Y.; Hong, G.-R.; Kim, M.-H.; Seo, J.; Cho, I.J.; Kim, Y.D.; Chang, H.-J.; Ha, J.-W.; Heo, J.H.; et al. Clinical Implications and Determinants of Left Atrial Mechanical Dysfunction in Patients with Stroke. *Stroke* **2016**, *47*, 1444–1451. [CrossRef]
95. Huang, T.; Patrick, S.; Mayer, L.K.; Müller-Edenborn, B.; Eichenlaub, M.; Allgeier, M.; Allgeier, J.; Lehrmann, H.; Ahlgrim, C.; Bohnen, M.; et al. Echocardiographic and Electrocardiographic Determinants of Atrial Cardiomyopathy Identify Patients with Atrial Fibrillation at Risk for Left Atrial Thrombogenesis. *J. Clin. Med.* **2022**, *11*, 1332. [CrossRef]
96. Burrell, L.D.; Horne, B.D.; Anderson, J.L.; Muhlestein, J.B.; Whisenant, B.K. Usefulness of Left Atrial Appendage Volume as a Predictor of Embolic Stroke in Patients with Atrial Fibrillation. *Am. J. Cardiol.* **2013**, *112*, 1148–1152. [CrossRef]
97. Di Biase, L.; Santangeli, P.; Anselmino, M.; Mohanty, P.; Salvetti, I.; Gili, S.; Horton, R.; Sanchez, J.E.; Bai, R.; Mohanty, S.; et al. Does the Left Atrial Appendage Morphology Correlate with the Risk of Stroke in Patients with Atrial Fibrillation? Results from a Multicenter Study. *J. Am. Coll. Cardiol.* **2012**, *60*, 531–538. [CrossRef]
98. Markl, M.; Lee, D.C.; Furiasse, N.; Carr, M.; Foucar, C.; Ng, J.; Carr, J.; Goldberger, J.J. Left Atrial and Left Atrial Appendage 4D Blood Flow Dynamics in Atrial Fibrillation. *Circ. Cardiovasc. Imaging* **2016**, *9*, e004984. [CrossRef]
99. Spartera, M.; Stracquadanio, A.; Pessoa-Amorim, G.; Von Ende, A.; Fletcher, A.; Manley, P.; Ferreira, V.M.; Hess, A.T.; Hopewell, J.C.; Neubauer, S.; et al. The Impact of Atrial Fibrillation and Stroke Risk Factors on Left Atrial Blood Flow Characteristics. *Eur. Heart J. Cardiovasc. Imaging* **2021**, *23*, 115–123. [CrossRef]
100. Ayubcha, C.; Amanullah, A.; Patel, K.H.; Teichner, E.; Gokhale, S.; Marquez-Valenzuela, U.; Werner, T.J.; Alavi, A. Stroke and Molecular Imaging: A Focus on FDG-PET. *Am. J. Nucl. Med. Mol. Imaging* **2023**, *13*, 51–63.
101. Müller, P.; Makimoto, H.; Dietrich, J.W.; Fochler, F.; Nentwich, K.; Krug, J.; Duncker, D.; Blockhaus, C.; Kelm, M.; Fürnkranz, A.; et al. Association of Left Atrial Low-Voltage Area and Thromboembolic Risk in Patients with Atrial Fibrillation. *EP Eur.* **2018**, *20*, f359–f365. [CrossRef] [PubMed]
102. Prabhu, S.; Voskoboinik, A.; McLellan, A.J.A.; Peck, K.Y.; Pathik, B.; Nalliah, C.J.; Wong, G.R.; Azzopardi, S.M.; Lee, G.; Mariani, J.; et al. Biatrial Electrical and Structural Atrial Changes in Heart Failure: Electroanatomic Mapping in Persistent Atrial Fibrillation in Humans. *JACC Clin. Electrophysiol.* **2018**, *4*, 87–96. [CrossRef] [PubMed]
103. Masuda, M.; Matsuda, Y.; Uematsu, H.; Sugino, A.; Ooka, H.; Kudo, S.; Fujii, S.; Asai, M.; Okamoto, S.; Ishihara, T.; et al. Prognostic Impact of Atrial Cardiomyopathy: Long-Term Follow-up of Patients with and without Low-Voltage Areas Following Atrial Fibrillation Ablation. *Heart Rhythm* **2024**, *21*, 378–386. [CrossRef]

104. Kamel, H.; Longstreth, W.T.J.; Tirschwell, D.L.; Kronmal, R.A.; Broderick, J.P.; Palesch, Y.Y.; Meinzer, C.; Dillon, C.; Ewing, I.; Spilker, J.A.; et al. The AtRial Cardiopathy and Antithrombotic Drugs In Prevention After Cryptogenic Stroke Randomized Trial: Rationale and Methods. *Int. J. Stroke* **2019**, *14*, 207–214. [CrossRef] [PubMed]
105. Kamel, H.; Longstreth, W.T.J.; Tirschwell, D.L.; Kronmal, R.A.; Marshall, R.S.; Broderick, J.P.; Aragón García, R.; Plummer, P.; Sabagha, N.; Pauls, Q.; et al. Apixaban to Prevent Recurrence After Cryptogenic Stroke in Patients With Atrial Cardiopathy: The ARCADIA Randomized Clinical Trial. *JAMA* **2024**, *331*, 573–581. [CrossRef]
106. Maheshwari, A.; Norby, F.L.; Inciardi, R.M.; Wang, W.; Zhang, M.J.; Soliman, E.Z.; Alonso, A.; Johansen, M.C.; Gottesman, R.F.; Solomon, S.D.; et al. Left Atrial Mechanical Dysfunction and the Risk for Ischemic Stroke in People Without Prevalent Atrial Fibrillation or Stroke: A Prospective Cohort Study. *Ann. Intern. Med.* **2023**, *176*, 39–48. [CrossRef]
107. Johansen, M.C.; Wang, W.; Zhang, M.; Knopman, D.S.; Ndumele, C.; Mosley, T.H.; Selvin, E.; Shah, A.M.; Solomon, S.D.; Gottesman, R.F.; et al. Risk of Dementia Associated With Atrial Cardiopathy: The ARIC Study. *J. Am. Heart Assoc.* **2022**, *11*, e025646. [CrossRef]
108. Khan, A.A.; Lip, G.Y.H. The Prothrombotic State in Atrial Fibrillation: Pathophysiological and Management Implications. *Cardiovasc. Res.* **2019**, *115*, 31–45. [CrossRef]
109. Ding, W.Y.; Gupta, D.; Lip, G.Y.H. Atrial Fibrillation and the Prothrombotic State: Revisiting Virchow's Triad in 2020. *Heart* **2020**, *106*, 1463–1468. [CrossRef]
110. Mukherjee, R.; Akar, J.G.; Wharton, J.M.; Adams, D.K.; McClure, C.D.; Stroud, R.E.; Rice, A.D.; DeSantis, S.M.; Spinale, F.G.; Gold, M.R. Plasma Profiles of Matrix Metalloproteinases and Tissue Inhibitors of the Metalloproteinases Predict Recurrence of Atrial Fibrillation Following Cardioversion. *J. Cardiovasc. Transl. Res.* **2013**, *6*, 528–535. [CrossRef]
111. Lee, K.K.; Doudesis, D.; Anwar, M.; Astengo, F.; Chenevier-Gobeaux, C.; Claessens, Y.-E.; Wussler, D.; Kozhuharov, N.; Strebel, I.; Sabti, Z.; et al. Development and Validation of a Decision Support Tool for the Diagnosis of Acute Heart Failure: Systematic Review, Meta-Analysis, and Modelling Study. *BMJ* **2022**, *377*, e068424. [CrossRef]
112. Mu, X.; Qiu, M.; Li, Y.; Li, Z.; Qi, B.; Jing, Z.; Jing, Q. N-Terminal pro-B-Type Natriuretic Peptide Improves the Predictive Value of CHA(2) DS(2) -VASc Risk Score for Long-Term Cardiovascular Events in Acute Coronary Syndrome Patients with Atrial Fibrillation. *Clin. Cardiol.* **2023**, *46*, 810–817. [CrossRef] [PubMed]
113. Nasab Mehrabi, E.; Toupchi-Khosroshahi, V.; Athari, S.S. Relationship of Atrial Fibrillation and N Terminal pro Brain Natriuretic Peptide in Heart Failure Patients. *ESC Heart Fail.* **2023**, *10*, 3250–3257. [CrossRef] [PubMed]
114. Folsom, A.R.; Nambi, V.; Bell, E.J.; Oluleye, O.W.; Gottesman, R.F.; Lutsey, P.L.; Huxley, R.R.; Ballantyne, C.M. Troponin T, N-Terminal pro-B-Type Natriuretic Peptide, and Incidence of Stroke: The Atherosclerosis Risk in Communities Study. *Stroke* **2013**, *44*, 961–967. [CrossRef]
115. Sun, W.-P.; Du, X.; Chen, J.-J. Biomarkers for Predicting the Occurrence and Progression of Atrial Fibrillation: Soluble Suppression of Tumorigenicity 2 Protein and Tissue Inhibitor of Matrix Metalloproteinase-1. *Int. J. Clin. Pract.* **2022**, *2022*, 6926510. [CrossRef] [PubMed]
116. Reyat, J.S.; Chua, W.; Cardoso, V.R.; Witten, A.; Kastner, P.M.; Kabir, S.N.; Sinner, M.F.; Wesselink, R.; Holmes, A.P.; Pavlovic, D.; et al. Reduced Left Atrial Cardiomyocyte PITX2 and Elevated Circulating BMP10 Predict Atrial Fibrillation after Ablation. *JCI Insight* **2020**, *5*, e139179. [CrossRef]
117. Hijazi, Z.; Benz, A.P.; Lindbäck, J.; Alexander, J.H.; Connolly, S.J.; Eikelboom, J.W.; Granger, C.B.; Kastner, P.; Lopes, R.D.; Ziegler, A.; et al. Bone Morphogenetic Protein 10: A Novel Risk Marker of Ischaemic Stroke in Patients with Atrial Fibrillation. *Eur. Heart J.* **2023**, *44*, 208–218. [CrossRef]
118. Hennings, E.; Blum, S.; Aeschbacher, S.; Coslovsky, M.; Knecht, S.; Eken, C.; Lischer, M.; Paladini, R.E.; Krisai, P.; Reichlin, T.; et al. Bone Morphogenetic Protein 10-A Novel Biomarker to Predict Adverse Outcomes in Patients with Atrial Fibrillation. *J. Am. Heart Assoc.* **2023**, *12*, e028255. [CrossRef]
119. Chung, M.K.; Martin, D.O.; Sprecher, D.; Wazni, O.; Kanderian, A.; Carnes, C.A.; Bauer, J.A.; Tchou, P.J.; Niebauer, M.J.; Natale, A.; et al. C-Reactive Protein Elevation in Patients with Atrial Arrhythmias: Inflammatory Mechanisms and Persistence of Atrial Fibrillation. *Circulation* **2001**, *104*, 2886–2891. [CrossRef]
120. Conway, D.S.G.; Pearce, L.A.; Chin, B.S.P.; Hart, R.G.; Lip, G.Y.H. Prognostic Value of Plasma von Willebrand Factor and Soluble P-Selectin as Indices of Endothelial Damage and Platelet Activation in 994 Patients with Nonvalvular Atrial Fibrillation. *Circulation* **2003**, *107*, 3141–3145. [CrossRef]
121. Esteve-Pastor, M.A.; Roldán, V.; Rivera-Caravaca, J.M.; Ramírez-Macías, I.; Lip, G.Y.H.; Marín, F. The Use of Biomarkers in Clinical Management Guidelines: A Critical Appraisal. *Thromb. Haemost.* **2019**, *119*, 1901–1919. [CrossRef] [PubMed]
122. Camelo-Castillo, A.; Rivera-Caravaca, J.M.; Marín, F.; Vicente, V.; Lip, G.Y.H.; Roldán, V. Predicting Adverse Events beyond Stroke and Bleeding with the ABC-Stroke and ABC-Bleeding Scores in Patients with Atrial Fibrillation: The Murcia AF Project. *Thromb. Haemost.* **2020**, *120*, 1200–1207. [CrossRef] [PubMed]

123. Gorog, D.A.; Gue, Y.X.; Chao, T.-F.; Fauchier, L.; Ferreiro, J.L.; Huber, K.; Konstantinidis, S.V.; Lane, D.A.; Marin, F.; Oldgren, J.; et al. Assessment and Mitigation of Bleeding Risk in Atrial Fibrillation and Venous Thromboembolism: Executive Summary of a European and Asia-Pacific Expert Consensus Paper. *Thromb. Haemost.* **2022**, *122*, 1625–1652. [CrossRef]
124. Xie, S.; Li, Q.; Luk, A.O.Y.; Lan, H.-Y.; Chan, P.K.S.; Bayés-Genís, A.; Chan, F.K.L.; Fung, E. Major Adverse Cardiovascular Events and Mortality Prediction by Circulating GDF-15 in Patients with Type 2 Diabetes: A Systematic Review and Meta-Analysis. *Biomolecules* **2022**, *12*, 934. [CrossRef]
125. Ding, W.Y.; McDowell, G.; Lip, G.Y.H. Utilizing Biomarkers in Atrial Fibrillation: The Pros and Cons. *Cardiovasc. Res.* **2022**, *118*, 2035–2036. [CrossRef] [PubMed]
126. Fatkin, D.; Huttner, I.G.; Johnson, R. Genetics of Atrial Cardiomyopathy. *Curr. Opin. Cardiol.* **2019**, *34*, 275–281. [CrossRef]
127. Marcoux, E.; Sosnowski, D.; Ninni, S.; Mackasey, M.; Cadrin-Tourigny, J.; Roberts, J.D.; Olesen, M.S.; Fatkin, D.; Nattel, S. Genetic Atrial Cardiomyopathies: Common Features, Specific Differences, and Broader Relevance to Understanding Atrial Cardiomyopathy. *Circ. Arrhythm. Electrophysiol.* **2023**, *16*, 675–698. [CrossRef]
128. Orr, N.; Arnaout, R.; Gula, L.J.; Spears, D.A.; Leong-Sit, P.; Li, Q.; Tarhuni, W.; Reischauer, S.; Chauhan, V.S.; Borkovich, M.; et al. A Mutation in the Atrial-Specific Myosin Light Chain Gene (MYL4) Causes Familial Atrial Fibrillation. *Nat. Commun.* **2016**, *7*, 11303. [CrossRef]
129. Gudbjartsson, D.F.; Arnar, D.O.; Helgadottir, A.; Gretarsdottir, S.; Holm, H.; Sigurdsson, A.; Jonasdottir, A.; Baker, A.; Thorleifsson, G.; Kristjansson, K.; et al. Variants Conferring Risk of Atrial Fibrillation on Chromosome 4q25. *Nature* **2007**, *448*, 353–357. [CrossRef]
130. Disertori, M.; Quintarelli, S.; Grasso, M.; Pilotto, A.; Narula, N.; Favalli, V.; Canclini, C.; Diegoli, M.; Mazzola, S.; Marini, M.; et al. Autosomal Recessive Atrial Dilated Cardiomyopathy with Standstill Evolution Associated with Mutation of Natriuretic Peptide Precursor A. *Circ. Cardiovasc. Genet.* **2013**, *6*, 27–36. [CrossRef]
131. Schulze-Bahr, E.; Neu, A.; Friederich, P.; Kaupp, U.B.; Breithardt, G.; Pongs, O.; Isbrandt, D. Pacemaker Channel Dysfunction in a Patient with Sinus Node Disease. *J. Clin. Investig.* **2003**, *111*, 1537–1545. [CrossRef] [PubMed]
132. Tan, R.B.; Gando, I.; Bu, L.; Cecchin, F.; Coetzee, W. A Homozygous SCN5A Mutation Associated with Atrial Standstill and Sudden Death. *Pacing Clin. Electrophysiol.* **2018**, *41*, 1036–1042. [CrossRef] [PubMed]
133. Ramos-Mondragon, R.; Edokobi, N.; Hodges, S.L.; Wang, S.; Bouza, A.A.; Canugovi, C.; Scheuing, C.; Juratli, L.; Abel, W.R.; Noujaim, S.F.; et al. Neonatal Scn1b-Null Mice Have Sinoatrial Node Dysfunction, Altered Atrial Structure, and Atrial Fibrillation. *JCI Insight* **2022**, *7*, e152050. [CrossRef]
134. Duparc, A.; Cintas, P.; Somody, E.; Bieth, E.; Richard, P.; Maury, P.; Delay, M. A Cardio-Neurological Form of Laminopathy: Dilated Cardiomyopathy with Permanent Partial Atrial Standstill and Axonal Neuropathy. *Pacing Clin. Electrophysiol.* **2009**, *32*, 410–415. [CrossRef] [PubMed]
135. van Setten, J.; Brody, J.A.; Jamshidi, Y.; Swenson, B.R.; Butler, A.M.; Campbell, H.; Del Greco, F.M.; Evans, D.S.; Gibson, Q.; Gudbjartsson, D.F.; et al. PR Interval Genome-Wide Association Meta-Analysis Identifies 50 Loci Associated with Atrial and Atrioventricular Electrical Activity. *Nat. Commun.* **2018**, *9*, 2904. [CrossRef]
136. Wild, P.S.; Felix, J.F.; Schillert, A.; Teumer, A.; Chen, M.-H.; Leening, M.J.G.; Völker, U.; Großmann, V.; Brody, J.A.; Irvin, M.R.; et al. Large-Scale Genome-Wide Analysis Identifies Genetic Variants Associated with Cardiac Structure and Function. *J. Clin. Investig.* **2017**, *127*, 1798–1812. [CrossRef]
137. Roselli, C.; Chaffin, M.D.; Weng, L.-C.; Aeschbacher, S.; Ahlberg, G.; Albert, C.M.; Almgren, P.; Alonso, A.; Anderson, C.D.; Aragam, K.G.; et al. Multi-Ethnic Genome-Wide Association Study for Atrial Fibrillation. *Nat. Genet.* **2018**, *50*, 1225–1233. [CrossRef]
138. Nielsen, J.B.; Thorolfsdottir, R.B.; Fritsche, L.G.; Zhou, W.; Skov, M.W.; Graham, S.E.; Herron, T.J.; McCarthy, S.; Schmidt, E.M.; Sveinbjornsson, G.; et al. Biobank-Driven Genomic Discovery Yields New Insight into Atrial Fibrillation Biology. *Nat. Genet.* **2018**, *50*, 1234–1239. [CrossRef]
139. Steenman, M. Insight into Atrial Fibrillation through Analysis of the Coding Transcriptome in Humans. *Biophys. Rev.* **2020**, *12*, 817–826. [CrossRef]
140. Man, J.C.K.; van Duijvenboden, K.; Krijger, P.H.L.; Hooijkaas, I.B.; van der Made, I.; de Gier-de Vries, C.; Wakker, V.; Creemers, E.E.; de Laat, W.; Boukens, B.J.; et al. Genetic Dissection of a Super Enhancer Controlling the Nppa-Nppb Cluster in the Heart. *Circ. Res.* **2021**, *128*, 115–129. [CrossRef]
141. Cao, T.H.; Quinn, P.A.; Sandhu, J.K.; Voors, A.A.; Lang, C.C.; Parry, H.M.; Mohan, M.; Jones, D.J.L.; Ng, L.L. Identification of Novel Biomarkers in Plasma for Prediction of Treatment Response in Patients with Heart Failure. *Lancet* **2015**, *385*, S26. [CrossRef] [PubMed]
142. Çubukçuoğlu Deniz, G.; Durdu, S.; Doğan, Y.; Erdemli, E.; Özdağ, H.; Akar, A.R. Molecular Signatures of Human Chronic Atrial Fibrillation in Primary Mitral Regurgitation. *Cardiovasc. Ther.* **2021**, *2021*, 5516185. [CrossRef]

143. Zeemering, S.; Isaacs, A.; Winters, J.; Maesen, B.; Bidar, E.; Dimopoulou, C.; Guasch, E.; Batlle, M.; Haase, D.; Hatem, S.N.; et al. Atrial Fibrillation in the Presence and Absence of Heart Failure Enhances Expression of Genes Involved in Cardiomyocyte Structure, Conduction Properties, Fibrosis, Inflammation, and Endothelial Dysfunction. *Heart Rhythm* **2022**, *19*, 2115–2124. [CrossRef] [PubMed]
144. Alvarez-Franco, A.; Rouco, R.; Ramirez, R.J.; Guerrero-Serna, G.; Tiana, M.; Cogliati, S.; Kaur, K.; Saeed, M.; Magni, R.; Enriquez, J.A.; et al. Transcriptome and Proteome Mapping in the Sheep Atria Reveal Molecular Featurets of Atrial Fibrillation Progression. *Cardiovasc. Res.* **2021**, *117*, 1760–1775. [CrossRef] [PubMed]
145. Roselli, C.; Rienstra, M.; Ellinor, P.T. Genetics of Atrial Fibrillation in 2020: GWAS, Genome Sequencing, Polygenic Risk, and Beyond. *Circ. Res.* **2020**, *127*, 21–33. [CrossRef]
146. van Ouwerkerk, A.F.; Bosada, F.M.; van Duijvenboden, K.; Hill, M.C.; Montefiori, L.E.; Scholman, K.T.; Liu, J.; de Vries, A.A.F.; Boukens, B.J.; Ellinor, P.T.; et al. Identification of Atrial Fibrillation Associated Genes and Functional Non-Coding Variants. *Nat. Commun.* **2019**, *10*, 4755. [CrossRef]
147. Shen, K.; Tu, T.; Yuan, Z.; Yi, J.; Zhou, Y.; Liao, X.; Liu, Q.; Zhou, X. DNA Methylation Dysregulations in Valvular Atrial Fibrillation. *Clin. Cardiol.* **2017**, *40*, 686–691. [CrossRef]
148. Doñate Puertas, R.; Meugnier, E.; Romestaing, C.; Rey, C.; Morel, E.; Lachuer, J.; Gadot, N.; Scridon, A.; Julien, C.; Tronc, F.; et al. Atrial Fibrillation Is Associated with Hypermethylation in Human Left Atrium, and Treatment with Decitabine Reduces Atrial Tachyarrhythmias in Spontaneously Hypertensive Rats. *Transl. Res.* **2017**, *184*, 57–67.e5. [CrossRef]
149. Ward-Caviness, C.K. Accelerated Epigenetic Aging and Incident Atrial Fibrillation: New Outlook on an Immutable Risk Factor? *Circulation* **2021**, *144*, 1912–1914. [CrossRef]
150. Roberts, J.D.; Vittinghoff, E.; Lu, A.T.; Alonso, A.; Wang, B.; Sitlani, C.M.; Mohammadi-Shemirani, P.; Fornage, M.; Kornej, J.; Brody, J.A.; et al. Epigenetic Age and the Risk of Incident Atrial Fibrillation. *Circulation* **2021**, *144*, 1899–1911. [CrossRef]
151. Brundel, B.J.J.M.; Li, J.; Zhang, D. Role of HDACs in Cardiac Electropathology: Therapeutic Implications for Atrial Fibrillation. *Biochim. Biophys. Acta Mol. Cell Res.* **2020**, *1867*, 118459. [CrossRef] [PubMed]
152. Song, S.; Zhang, R.; Mo, B.; Chen, L.; Liu, L.; Yu, Y.; Cao, W.; Fang, G.; Wan, Y.; Gu, Y.; et al. EZH2 as a Novel Therapeutic Target for Atrial Fibrosis and Atrial Fibrillation. *J. Mol. Cell. Cardiol.* **2019**, *135*, 119–133. [CrossRef]
153. Chen, L.Y.; Ribeiro, A.L.P.; Platonov, P.G.; Cygankiewicz, I.; Soliman, E.Z.; Gorenek, B.; Ikeda, T.; Vassilikos, V.P.; Steinberg, J.S.; Varma, N.; et al. P Wave Parameters and Indices: A Critical Appraisal of Clinical Utility, Challenges, and Future Research-A Consensus Document Endorsed by the International Society of Electrocardiology and the International Society for Holter and Noninvasive Electrocardi. *Circ. Arrhythm. Electrophysiol.* **2022**, *15*, e010435. [CrossRef]
154. Kreimer, F.; Gotzmann, M. Left Atrial Cardiomyopathy—A Challenging Diagnosis. *Front. Cardiovasc. Med.* **2022**, *9*, 942385. [CrossRef]
155. Bayés de Luna, A.; Platonov, P.; Cosio, F.G.; Cygankiewicz, I.; Pastore, C.; Baranowski, R.; Bayés-Genis, A.; Guindo, J.; Viñolas, X.; Garcia-Niebla, J.; et al. Interatrial Blocks. A Separate Entity from Left Atrial Enlargement: A Consensus Report. *J. Electrocardiol.* **2012**, *45*, 445–451. [CrossRef] [PubMed]
156. Tsao, C.W.; Josephson, M.E.; Hauser, T.H.; O'Halloran, T.D.; Agarwal, A.; Manning, W.J.; Yeon, S.B. Accuracy of Electrocardiographic Criteria for Atrial Enlargement: Validation with Cardiovascular Magnetic Resonance. *J. Cardiovasc. Magn. Reson.* **2008**, *10*, 7. [CrossRef] [PubMed]
157. Bayés de Luna, A.; Cladellas, M.; Oter, R.; Torner, P.; Guindo, J.; Martí, V.; Rivera, I.; Iturralde, P. Interatrial Conduction Block and Retrograde Activation of the Left Atrium and Paroxysmal Supraventricular Tachyarrhythmia. *Eur. Heart J.* **1988**, *9*, 1112–1118. [CrossRef]
158. Power, D.A.; Lampert, J.; Camaj, A.; Bienstock, S.W.; Kocovic, N.; Bayes-Genis, A.; Miller, M.A.; Bayés-de-Luna, A.; Fuster, V. Cardiovascular Complications of Interatrial Conduction Block: JACC State-of-the-Art Review. *J. Am. Coll. Cardiol.* **2022**, *79*, 1199–1211. [CrossRef]
159. Intzes, S.; Zagoridis, K.; Symeonidou, M.; Spanoudakis, E.; Arya, A.; Dinov, B.; Dagres, N.; Hindricks, G.; Bollmann, A.; Kanoupakis, E.; et al. P-Wave Duration and Atrial Fibrillation Recurrence after Catheter Ablation: A Systematic Review and Meta-Analysis. *Europace* **2023**, *25*, 450–459. [CrossRef]
160. Wu, J.-T.; Long, D.-Y.; Dong, J.-Z.; Wang, S.-L.; Fan, X.-W.; Yang, H.-T.; Duan, H.-Y.; Yan, L.-J.; Qian, P.; Yang, C.-K. Advanced Interatrial Block Predicts Clinical Recurrence of Atrial Fibrillation after Catheter Ablation. *J. Cardiol.* **2016**, *68*, 352–356. [CrossRef]
161. Nielsen, J.B.; Kühl, J.T.; Pietersen, A.; Graff, C.; Lind, B.; Struijk, J.J.; Olesen, M.S.; Sinner, M.F.; Bachmann, T.N.; Haunsø, S.; et al. P-Wave Duration and the Risk of Atrial Fibrillation: Results from the Copenhagen ECG Study. *Heart Rhythm.* **2015**, *12*, 1887–1895. [CrossRef] [PubMed]
162. Auricchio, A.; Özkartal, T.; Salghetti, F.; Neumann, L.; Pezzuto, S.; Gharaviri, A.; Demarchi, A.; Caputo, M.L.; Regoli, F.; De Asmundis, C.; et al. Short P-Wave Duration Is a Marker of Higher Rate of Atrial Fibrillation Recurrences after Pulmonary Vein Isolation: New Insights into the Pathophysiological Mechanisms Through Computer Simulations. *J. Am. Heart Assoc.* **2021**, *10*, e018572. [CrossRef]

163. Jadidi, A.; Müller-Edenborn, B.; Chen, J.; Keyl, C.; Weber, R.; Allgeier, J.; Moreno-Weidmann, Z.; Trenk, D.; Neumann, F.-J.; Lehrmann, H.; et al. The Duration of the Amplified Sinus-P-Wave Identifies Presence of Left Atrial Low Voltage Substrate and Predicts Outcome After Pulmonary Vein Isolation in Patients With Persistent Atrial Fibrillation. *JACC Clin. Electrophysiol.* **2018**, *4*, 531–543. [CrossRef] [PubMed]
164. Müller-Edenborn, B.; Chen, J.; Allgeier, J.; Didenko, M.; Moreno-Weidmann, Z.; Neumann, F.-J.; Lehrmann, H.; Weber, R.; Arentz, T.; Jadidi, A. Amplified Sinus-P-Wave Reveals Localization and Extent of Left Atrial Low-Voltage Substrate: Implications for Arrhythmia Freedom Following Pulmonary Vein Isolation. *EP Eur.* **2020**, *22*, 240–249. [CrossRef]
165. Müller-Edenborn, B.; Minners, J.; Keyl, C.; Eichenlaub, M.; Jander, N.; Abdelrazek, S.; Ahlgrim, C.; Allgeier, J.; Lehrmann, H.; Neumann, F.-J.; et al. Electrocardiographic Diagnosis of Atrial Cardiomyopathy to Predict Atrial Contractile Dysfunction, Thrombogenesis and Adverse Cardiovascular Outcomes. *Sci. Rep.* **2022**, *12*, 576. [CrossRef]
166. German, D.M.; Kabir, M.M.; Dewland, T.A.; Henrikson, C.A.; Tereshchenko, L.G. Atrial Fibrillation Predictors: Importance of the Electrocardiogram. *Ann. Noninvasive Electrocardiol.* **2016**, *21*, 20–29. [CrossRef]
167. McConkey, N.; Malamas, P.; Norby, F.L.; Plamenac, J.; Park, R.; Weigel, F.; Gonzalez, M.D.; Naccarelli, G.V.; Maheshwari, A. Abnormal P-Wave Terminal Force in Lead V1 Is Associated with Low Left Atrial Appendage Ejection Velocity. *J. Electrocardiol.* **2021**, *67*, 142–147. [CrossRef] [PubMed]
168. Truong, Q.A.; Charipar, E.M.; Ptaszek, L.M.; Taylor, C.; Fontes, J.D.; Kriegel, M.; Irlbeck, T.; Mahabadi, A.A.; Blankstein, R.; Hoffmann, U. Usefulness of Electrocardiographic Parameters as Compared with Computed Tomography Measures of Left Atrial Volume Enlargement: From the ROMICAT Trial. *J. Electrocardiol.* **2011**, *44*, 257–264. [CrossRef]
169. Lebek, S.; Wester, M.; Pec, J.; Poschenrieder, F.; Tafelmeier, M.; Fisser, C.; Provaznik, Z.; Schopka, S.; Debl, K.; Schmid, C.; et al. Abnormal P-Wave Terminal Force in Lead V(1) Is a Marker for Atrial Electrical Dysfunction but Not Structural Remodelling. *ESC Heart Fail.* **2021**, *8*, 4055–4066. [CrossRef]
170. Li, R.; Yang, X.; Jia, M.; Wang, D.; Cui, X.; Bai, L.; Zhao, L.; Zhang, J. Effectiveness of P-Wave ECG Index and Left Atrial Appendage Volume in Predicting Atrial Fibrillation Recurrence after First Radiofrequency Catheter Ablation. *BMC Cardiovasc. Disord.* **2021**, *21*, 164. [CrossRef]
171. Huang, Z.; Zheng, Z.; Wu, B.; Tang, L.; Xie, X.; Dong, R.; Luo, Y.; Li, S.; Zhu, J.; Liu, J. Predictive Value of P Wave Terminal Force in Lead V1 for Atrial Fibrillation: A Meta-Analysis. *Ann. Noninvasive Electrocardiol.* **2020**, *25*, e12739. [CrossRef] [PubMed]
172. Leshem, E.; Zilberman, I.; Tschabrunn, C.M.; Barkagan, M.; Contreras-Valdes, F.M.; Govari, A.; Anter, E. High-Power and Short-Duration Ablation for Pulmonary Vein Isolation: Biophysical Characterization. *JACC Clin. Electrophysiol.* **2018**, *4*, 467–479. [CrossRef]
173. Müller-Edenborn, B.; Minners, J.; Kocher, S.; Chen, J.; Zeh, W.; Lehrmann, H.; Allgeier, J.; Neumann, F.-J.; Arentz, T.; Jadidi, A. Amplified P-Wave Duration Predicts New-Onset Atrial Fibrillation in Patients with Heart Failure with Preserved Ejection Fraction. *Clin. Res. Cardiol.* **2020**, *109*, 978–987. [CrossRef]
174. Alexander, B.; Milden, J.; Hazim, B.; Haseeb, S.; Bayes-Genis, A.; Elosua, R.; Martínez-Sellés, M.; Yeung, C.; Hopman, W.; Bayes de Luna, A.; et al. New Electrocardiographic Score for the Prediction of Atrial Fibrillation: The MVP ECG Risk Score (Morphology-Voltage-P-Wave Duration). *Ann. Noninvasive Electrocardiol.* **2019**, *24*, e12669. [CrossRef]
175. Park, J.-K.; Park, J.; Uhm, J.-S.; Joung, B.; Lee, M.-H.; Pak, H.-N. Low P-Wave Amplitude (<0.1 MV) in Lead I Is Associated with Displaced Inter-Atrial Conduction and Clinical Recurrence of Paroxysmal Atrial Fibrillation after Radiofrequency Catheter Ablation. *Europace* **2016**, *18*, 384–391. [CrossRef] [PubMed]
176. Zeng, C.; Wei, T.; Zhao, R.; Wang, C.; Chen, L.; Wang, L. Electrocardiographic Diagnosis of Left Atrial Enlargement in Patients with Mitral Stenosis: The Value of the P-Wave Area. *Acta Cardiol.* **2003**, *58*, 139–141. [CrossRef] [PubMed]
177. Tse, G.; Lakhani, I.; Zhou, J.; Li, K.H.C.; Lee, S.; Liu, Y.; Leung, K.S.K.; Liu, T.; Baranchuk, A.; Zhang, Q. P-Wave Area Predicts New Onset Atrial Fibrillation in Mitral Stenosis: A Machine Learning Approach. *Front. Bioeng. Biotechnol.* **2020**, *8*, 479. [CrossRef]
178. Dilaveris, P.E.; Gialafos, E.J.; Sideris, S.K.; Theopistou, A.M.; Andrikopoulos, G.K.; Kyriakidis, M.; Gialafos, J.E.; Toutouzas, P.K. Simple Electrocardiographic Markers for the Prediction of Paroxysmal Idiopathic Atrial Fibrillation. *Am. Heart J.* **1998**, *135*, 733–738. [CrossRef]
179. Marks, D.; Ho, R.; Then, R.; Weinstock, J.L.; Teklemariam, E.; Kakadia, B.; Collins, J.; Andriulli, J.; Hunter, K.; Ortman, M.; et al. Real-World Experience with Implantable Loop Recorder Monitoring to Detect Subclinical Atrial Fibrillation in Patients with Cryptogenic Stroke: The Value of p Wave Dispersion in Predicting Arrhythmia Occurrence. *Int. J. Cardiol.* **2021**, *327*, 86–92. [CrossRef]
180. İçen, Y.K.; Koca, H.; Sümbül, H.E.; Yıldırım, A.; Koca, F.; Yıldırım, A.; Ardıc, M.L.; Coşkun, M.; Uğurlu, M.; Koç, M. Relationship between Coarse F Waves and Thromboembolic Events in Patients with Permanent Atrial Fibrillation. *J. Arrhythmia* **2020**, *36*, 1025–1031. [CrossRef]
181. Aysha, M.H.; Hassan, A.S. Diagnostic Importance of Fibrillatory Wave Amplitude: A Clue to Echocardiographic Left Atrial Size and Etiology of Atrial Fibrillation. *J. Electrocardiol.* **1988**, *21*, 247–251. [CrossRef] [PubMed]

182. Park, J.; Lee, C.; Leshem, E.; Blau, I.; Kim, S.; Lee, J.M.; Hwang, J.-A.; Choi, B.-I.; Lee, M.-H.; Hwang, H.J. Early Differentiation of Long-Standing Persistent Atrial Fibrillation Using the Characteristics of Fibrillatory Waves in Surface ECG Multi-Leads. *Sci. Rep.* **2019**, *9*, 2746. [CrossRef]
183. Kawaji, T.; Ogawa, H.; Hamatani, Y.; Kato, M.; Yokomatsu, T.; Miki, S.; Abe, M.; Akao, M. Fine Fibrillatory Wave as a Risk Factor for Heart Failure Events in Patients With Atrial Fibrillation: The Fushimi Atrial Fibrillation (AF) Registry. *J. Am. Heart Assoc.* **2022**, *11*, e024341. [CrossRef] [PubMed]
184. Zarzoso, V.; Latcu, D.G.; Hidalgo-Muñoz, A.R.; Meo, M.; Meste, O.; Popescu, I.; Saoudi, N. Non-Invasive Prediction of Catheter Ablation Outcome in Persistent Atrial Fibrillation by Fibrillatory Wave Amplitude Computation in Multiple Electrocardiogram Leads. *Arch. Cardiovasc. Dis.* **2016**, *109*, 679–688. [CrossRef] [PubMed]
185. Lankveld, T.; Zeemering, S.; Scherr, D.; Kuklik, P.; Hoffmann, B.A.; Willems, S.; Pieske, B.; Haïssaguerre, M.; Jaïs, P.; Crijns, H.J.; et al. Atrial Fibrillation Complexity Parameters Derived From Surface ECGs Predict Procedural Outcome and Long-Term Follow-Up of Stepwise Catheter Ablation for Atrial Fibrillation. *Circ. Arrhythm. Electrophysiol.* **2016**, *9*, e003354. [CrossRef]
186. Dhala, A.; Underwood, D.; Leman, R.; Madu, E.; Baugh, D.; Ozawa, Y.; Kasamaki, Y.; Xue, Q.; Reddy, S. Signal-Averaged P-Wave Analysis of Normal Controls and Patients with Paroxysmal Atrial Fibrillation: A Study in Gender Differences, Age Dependence, and Reproducibility. *Clin. Cardiol.* **2002**, *25*, 525–531. [CrossRef]
187. Okumura, Y.; Watanabe, I.; Ohkubo, K.; Ashino, S.; Kofune, M.; Hashimoto, K.; Shindo, A.; Sugimura, H.; Nakai, T.; Kasamaki, Y.; et al. Prediction of the Efficacy of Pulmonary Vein Isolation for the Treatment of Atrial Fibrillation by the Signal-Averaged P-Wave Duration. *Pacing Clin. Electrophysiol.* **2007**, *30*, 304–313. [CrossRef]
188. Zink, M.D.; Laureanti, R.; Hermans, B.J.M.; Pison, L.; Verheule, S.; Philippens, S.; Pluymaekers, N.; Vroomen, M.; Hermans, A.; van Hunnik, A.; et al. Extended ECG Improves Classification of Paroxysmal and Persistent Atrial Fibrillation Based on P- and f-Waves. *Front. Physiol.* **2022**, *13*, 779826. [CrossRef]
189. Py, A.; Schaaf, M.; Duhamel, S.; Si-Mohamed, S.; Daher, J.; Altman, M.; de Breyne, B.; Mechtouff, L.; Placide, J.; Chauveau, S.; et al. Atrial Premature Activity Detected after an Ischaemic Stroke Unveils Atrial Myopathy. *Arch. Cardiovasc. Dis.* **2020**, *113*, 227–236. [CrossRef]
190. Farinha, J.M.; Gupta, D.; Lip, G.Y.H. Frequent Premature Atrial Contractions as a Signalling Marker of Atrial Cardiomyopathy, Incident Atrial Fibrillation, and Stroke. *Cardiovasc. Res.* **2023**, *119*, 429–439. [CrossRef]
191. Karakasis, P.; Theofilis, P.; Sagris, M.; Pamporis, K.; Stachteas, P.; Sidiropoulos, G.; Vlachakis, P.K.; Patoulias, D.; Antoniadis, A.P.; Fragakis, N. Artificial Intelligence in Atrial Fibrillation: From Early Detection to Precision Therapy. *J. Clin. Med.* **2025**, *14*, 2627. [CrossRef] [PubMed]
192. Attia, Z.I.; Noseworthy, P.A.; Lopez-Jimenez, F.; Asirvatham, S.J.; Deshmukh, A.J.; Gersh, B.J.; Carter, R.E.; Yao, X.; Rabinstein, A.A.; Erickson, B.J.; et al. An Artificial Intelligence-Enabled ECG Algorithm for the Identification of Patients with Atrial Fibrillation during Sinus Rhythm: A Retrospective Analysis of Outcome Prediction. *Lancet* **2019**, *394*, 861–867. [CrossRef] [PubMed]
193. Noseworthy, P.A.; Attia, Z.I.; Behnken, E.M.; Giblon, R.E.; Bews, K.A.; Liu, S.; Gosse, T.A.; Linn, Z.D.; Deng, Y.; Yin, J.; et al. Artificial Intelligence-Guided Screening for Atrial Fibrillation Using Electrocardiogram during Sinus Rhythm: A Prospective Non-Randomised Interventional Trial. *Lancet* **2022**, *400*, 1206–1212. [CrossRef] [PubMed]
194. Liao, S.; Ragot, D.; Nayyar, S.; Suszko, A.; Zhang, Z.; Wang, B.; Chauhan, V.S. Deep Learning Classification of Unipolar Electrograms in Human Atrial Fibrillation: Application in Focal Source Mapping. *Front. Physiol.* **2021**, *12*, 704122. [CrossRef]
195. Lebert, J.; Ravi, N.; Fenton, F.H.; Christoph, J. Rotor Localization and Phase Mapping of Cardiac Excitation Waves Using Deep Neural Networks. *Front. Physiol.* **2021**, *12*, 782176. [CrossRef]

diagnostics

MDPI

Review

Acute Heart Failure and Non-Ischemic Cardiomyopathies: A Comprehensive Review and Critical Appraisal

Lina Manzi [1,†], Federica Buongiorno [1,†], Viviana Narciso [1], Domenico Florimonte [1], Imma Forzano [1], Domenico Simone Castiello [1], Luca Sperandeo [1], Roberta Paolillo [1], Nicola Verde [1,2], Alessandra Spinelli [1], Stefano Cristiano [1], Marisa Avvedimento [1], Mario Enrico Canonico [1], Luca Bardi [1], Giuseppe Giugliano [1] and Giuseppe Gargiulo [1,*]

1 Department of Advanced Biomedical Sciences, Federico II University of Naples, 80131 Naples, Italy; lina.manzi93@gmail.com (L.M.); federicabuongiorno2@gmail.com (F.B.); viviana.narciso@gmail.com (V.N.); florimontedomenico@gmail.com (D.F.); imma.forzano@gmail.com (I.F.); ds.castiello@gmail.com (D.S.C.); luca.sperandeo95@gmail.com (L.S.); robe.paolillo@gmail.com (R.P.); nicoverde91@gmail.com (N.V.); alessandra.spinelli@unina.it (A.S.); stefano.cristiano@unina.it (S.C.); m.avvedimento@gmail.com (M.A.); marioenrico.canonico@unina.it (M.E.C.); bardiluca@me.com (L.B.); giuseppe.giugliano@unina.it (G.G.)

2 Department of Cardiology, AORN Cardarelli, 80131 Naples, Italy

* Correspondence: giuseppe.gargiulo1@unina.it; Tel./Fax: +39-0817462253

† These authors contributed equally to this work.

Abstract: Acute heart failure (AHF) is a complex clinical syndrome characterized by the rapid or gradual onset of symptoms and/or signs of heart failure (HF), leading to an unplanned hospital admission or an emergency department visit. AHF is the leading cause of hospitalization in patients over 65 years, thus significantly impacting public health care. However, its prognosis remains poor with high rates of mortality and rehospitalization. Many pre-existing cardiac conditions can lead to AHF, but it can also arise de novo due to acute events. Therefore, understanding AHF etiology could improve patient management and outcomes. Cardiomyopathies (CMPs) are a heterogeneous group of heart muscle diseases, including dilated cardiomyopathy (DCM), hypertrophic cardiomyopathy (HCM), restrictive cardiomyopathy (RCM), non-dilated cardiomyopathy (NDLVC), and arrhythmogenic right ventricular cardiomyopathy (ARVC), that frequently present with HF. Patients with CMPs are under-represented in AHF studies compared to other etiologies, and therefore therapeutic responses and prognoses remain unknown. In DCM, AHF represents the most frequent cause of death despite treatment improvements. Additionally, DCM is the first indication for heart transplant (HT) among young and middle-aged adults. In HCM, the progression to AHF is rare and more frequent in patients with concomitant severe left ventricle (LV) obstruction and hypertrophy or severe LV systolic dysfunction. HF is the natural evolution of patients with RCM and HF is associated with poor outcomes irrespective of RCM etiology. Furthermore, while the occurrence of AHF is rare among patients with ARVC, this condition in NDLVC patients is currently unknown. In this manuscript, we assessed the available evidence on AHF in patients with CMPs. Data on clinical presentation, therapeutic management, and clinical outcomes according to specific CMPs are limited. Future HF studies assessing the clinical presentation, treatment, and prognosis of specific CMPs are warranted.

Keywords: acute heart failure; cardiomyopathies; dilated cardiomyopathy; hypertrophic cardiomyopathy; restrictive cardiomyopathy; non-dilated left ventricular cardiomyopathy; arrhythmogenic right ventricular cardiomyopathy

1. Introduction

Acute heart failure (AHF) is a complex syndrome characterized by the rapid onset or worsening of symptoms and signs of heart failure (HF), necessitating urgent medical intervention. AHF is the leading cause of unplanned hospital admissions among individuals over 65 years. However, the prognosis remains poor, with high rates of mortality and rehospitalization. Patients with pre-existing cardiac diseases could develop AHF, but it can also occur de novo in response to acute events. Given its prevalence and prognosis, understanding AHF etiology is essential to improving patient outcomes and reducing associated healthcare costs (Figure 1) [1,2]. Furthermore, patients with cardiomyopathies (CMPs), a heterogeneous group of heart muscle diseases, including dilated cardiomyopathy (DCM), hypertrophic cardiomyopathy (HCM), restrictive cardiomyopathy (RCM), non-dilated cardiomyopathy (NDLVC), and arrhythmogenic right ventricular cardiomyopathy (ARVC), may develop AHF. However, when compared to other etiologies of HF, the therapeutic response and prognosis of patients affected by CMPs remain unknown. The aim of this article is to review the evidence on AHF in patients with non-ischemic CMPs and to summarize the data from the main observational and randomized trials on AHF, highlighting the differences according to patient characteristics.

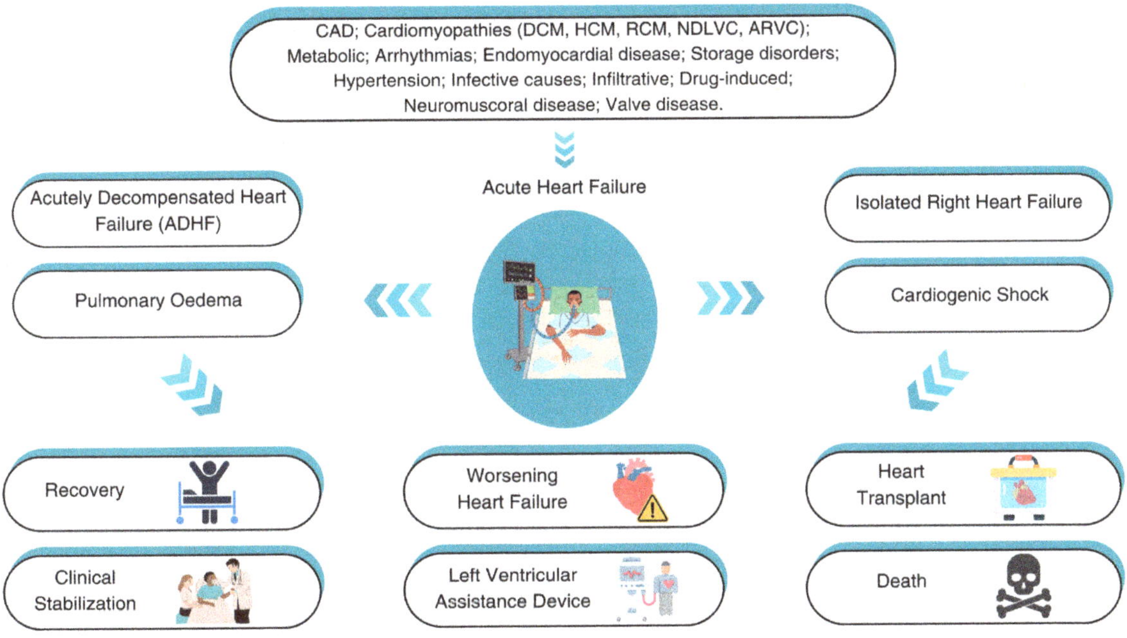

Figure 1. Etiology, clinical presentation, and outcomes of acute heart failure. The figure summarizes the multiple clinical conditions that may determine AHF, as well as the various clinical presentations of a patient with AHF, and this variability is inevitably associated with different prognostic implications. Abbreviations: ARVC = arrhythmogenic right ventricular cardiomyopathy; CAD = coronary artery disease; DCM = dilated cardiomyopathy; HCM = hypertrophic cardiomyopathy; NDLVC = non-dilated left ventricular cardiomyopathy; RCM = restrictive cardiomyopathy.

2. Acute Heart Failure

AHF refers to the rapid or gradual onset of symptoms and/or signs of HF severe enough for the patient to seek urgent medical attention, leading to an unplanned hospital admission or an emergency department visit. Patients with AHF require urgent evaluation with subsequent initiation and/or intensification of treatment, including IV therapies or procedures [1].

2.1. Clinical Presentation

Four clinical presentations of AHF can be described, with possible overlap among them. According to the European Society of Cardiology (ESC) Heart Failure Long-Term Registry, 61% of AHF patients present with acutely decompensated heart failure (ADHF), 13% with acute pulmonary edema, 3.5% with isolated right ventricular (RV) failure, and 3% with cardiogenic shock (CS) [3].

Generally, patients with ADHF have a worsening of chronic HF and have moderately to severely reduced left ventricular (LV) systolic function [4]. These patients often present with systemic congestion with or without hypotension [5]. Acute pulmonary edema is due to fluid redistribution to the lungs, leading to acute respiratory failure. It has a faster onset compared to ADHF and is a life-threatening condition if not promptly treated with oxygen support, diuretics, vasodilators, or inotropes/vasopressors [6]. Isolated RV failure is characterized by increased RV and atrial pressure, RV dysfunction, and systemic congestion, finally leading to a reduction in LV cardiac output (CO) through ventricular interdependence. RV failure is mainly caused by a pulmonary embolism (PE), right myocardial infarction (MI), or myocarditis [7]. Finally, CS is the worst clinical presentation, characterized by inadequate CO, leading to a life-threatening state of tissue hypoperfusion that can evolve into multi-organ failure and death [8–10].

2.2. Epidemiology and Prognosis

AHF represents the most common cause of hospitalization in individuals aged 65 years or older in Western countries, accounting for >1 million hospitalizations per year in the United States. AHF causes substantial morbidity and mortality and is a remarkable burden on health economics worldwide [11]. The prevalence of AHF is approximately 1–2% and increases to >10% among older people (>70 years); however, these estimates may vary according to the study population [12]. The prognosis is often poor and AHF is associated with high mortality and rehospitalization rates. In-hospital mortality ranges from 4% to 10% and death or rehospitalization rates may reach 45% [3,5,13]. This poor prognosis is also related to the characteristics of the people admitted for HF, who are often old, with relevant comorbidities and high rates of admissions due to non-cardiac reasons. Therefore, strengthening continuity of care and managing comorbidities to prevent acute conditions are needed.

2.3. Etiology

Most AHF patients (65–75%) have a known history of HF or cardiovascular disease (CVD). Around 50% of patients are male and have arterial hypertension (70%), coronary artery disease (CAD) (50–60%), or atrial fibrillation (AF) (30–40%). Non-cardiovascular comorbidities include diabetes mellitus in approximately 40% of patients, renal dysfunction in 20–30%, chronic obstructive pulmonary disease (COPD) in 20–30%, and anemia in 15–30% [1]. A systematic review of worldwide risk factors for HF showed that CAD is the major underlying contributor to AHF admissions in >50% of patients in high-income regions, as well as eastern and central European regions. In Asia and Latin America, CAD contributes to 30–40% of admissions, while it contributes to <10% in sub-Saharan Africa [14]. For a practical and rapid assessment, when a patient with AHF is admitted to hospital, specific causes can be addressed with the CHAMPIT acronym (acute coronary syndrome (ACS), hypertensive emergency, arrhythmias, acute mechanical cause, PE, infection, and tamponade) [1]. Once these specific etiologies are excluded, the management of AHF should be individualized according to clinical presentation and phenotype.

2.4. Evidence on AHF in Observational Studies

We searched for data regarding the incidence, prevalence, etiology, and outcome of AHF patients, trying to investigate the key differences in terms of outcomes in patients with a non-ischemic AHF etiology. The ADHERE (Acute Decompensated Heart Failure National Registry) registry enrolled 65,180 patients through 263 hospitals in the United States. The median age of patients was 75.2 years, and more than half (52%) were female. Most patients (75%) had a history of HF, 23% had been hospitalized for AHF within the prior 6 months, and 3% had three or more HF-related hospitalizations in that period. Regarding medical history, 72% had a history of arterial hypertension, while other common conditions were CAD (58%), diabetes (44%), AF (31%), and COPD or asthma (31%) [15]. The OPTIMIZE-HF (Organized Program to Initiate Lifesaving Treatment in Hospitalized Patients with Heart Failure) registry included 48,612 patients, of whom 52% were women and 74% were Caucasian. Their underlying diseases were CAD (in 46% of all patients), AF (in 32%), COPD (in 28%), arterial hypertension (in 23%), insulin-independent diabetes (in 25%), and insulin-dependent diabetes (in 17%). Interestingly, multivariable predictors of mortality included age, heart rate, systolic blood pressure (SBP), sodium, creatinine, HF as the primary cause of hospitalization, and the presence of LV systolic dysfunction [16]. The AHEAD (Acute Heart Failure Database) registry comprised patients hospitalized for AHF in seven Czech Republic hospitals. Of 4153 patients, the most frequent etiologies were ACS (36.2%), chronic coronary syndrome (CCS) (19.9%), valvular disease (10.4%), arrhythmias (7.9%), and arterial hypertension (5.7%). Patients with ACS had lower mortality than those without ACS (9.7% versus 18.1%), while patients with CS had the highest mortality (62.7%) [11]. The EHFS II (EuroHeart Failure Survey II) is a European survey that included 3580 patients hospitalized for AHF. The most common underlying diseases were arterial hypertension (62%), CAD (53.6%), AF (38%), valvular disease (34%), and diabetes mellitus (32%). Concomitant precipitant diseases were ACS (30%), arrhythmias (32%), valvular cause (26%), infection (17%), and non-compliance to therapy (22%) [13]. The international ALARM (Acute Heart Failure Global Registry of Standard Treatment) registry described the characteristics and management of AHF among various countries in order to compare patients with de novo AHF to patients with a pre-existing episode of AHF. The most common clinical presentation was ADHF (38.6%), followed by pulmonary edema (36.7%), CS (11.7%), hypertensive HF (7.4%), right HF (4.5%), and high-output HF (1.1%) The most common underlying diseases were hypertension (70%), diabetes mellitus (45%), AF (38%), chronic HF (37%), CAD (30%), COPD (24%), chronic renal disease (21%), and non-ischemic CMPs (12.6%), while concomitant precipitant diseases were ACS (37%), arrhythmias (27%), infection (16%), and non-compliance with therapy (13%) [17]. In a real-world cohort of 47,241 patients with heart failure with reduced ejection fraction (HFrEF) from the Swedish Heart Failure Registry (SwedeHF), hospitalization with first-episode HF and a higher risk of CV mortality were more frequent in patients with a concomitant glomerular filtration rate (eGFR) of 30–60 mL/min/1.73 m^2 and AF. Furthermore, the authors identified nine profiles with the highest event rates, representing only 5% of the study population, characterized by no hyperkaliemia, an even distribution among the

SBP strata, eGFR <30 mL/min/1.73 m^2, and AF [18]. A recent multicenter retrospective study focused on the management of CS with the use of mechanical circulatory support (MCS) in non-ischemic patients. In these settings, the use of such devices was associated with lower 30-day mortality (HR 0.76, 95% CI 0.59–0.97); however, it was also associated with more complications (severe bleeding (16.5% for MCS vs. 6.4% for non-MCS patients) and access-site-related ischemia (6.7% for MCS vs. 0% for non-MCS patients). Surprisingly, although specific data on non-ischemic CS patients are of paramount importance, no data were reported on the characterization of these patients and details of their etiologies [19].

What emerges from these registries is the fact that patients with AHF are very heterogeneous and this reflects the differences in prognosis. CAD is the most common primary cause in patients with AHF, while ACS is a frequent precipitant factor. In the context of de novo HF, a large amount of evidence suggests that it may have a worse prognosis compared to ADHF. Indeed, in ALARM HF, in-hospital mortality was 12%, the majority of which was due to CS (43%) and in patients with de novo AHF (14.2%). In addition, in the EHFS II registry (overall in-hospital mortality of 6.7%), de novo AHF patients had higher in-hospital mortality (8.1%). In AHEAD, the overall hospital mortality of 12.7% was comparable with the mortality of ALARM-HF, while in-hospital mortality rates for ADHERE (4%) and OPTIMIZE-HF (3.8%) were similar [11,13,15,17].

2.5. Evidence on AHF in Randomized Clinical Trials

Studying AHF as a unique entity is challenging and represents a problem when trying to obtain data from trials that should fit all AHF patients. We attempted to summarize the data from the main AHF trials, highlighting the differences in patient characteristics. Table 1 reports details of the main randomized AHF trials. Here, we mainly focused on the data on patients with non-ischemic CMPs. In the ADVOR (acetazolamide in decompensated heart failure with volume overload) trial, the underlying diseases of patients reflected the ones of the registries discussed before (43% CAD, 20% valvular heart disease, 24% DCM, and 9% arterial hypertension) [20]. In the CLOROTIC (combining loop with thiazide diuretics for decompensated heart failure) trial, the patient population had a high burden of comorbidities and high-risk features: diabetes (56.9% and 56.1% in HCTZ and placebo, respectively), AF (62.9% and 74.6%, respectively), anemia (45.7% and 43.9%, respectively), CAD (25.2% and 40.4%, respectively), and COPD (21.6% and 23.7%, respectively). Interestingly, in the subgroup analysis, patients with both ischemic and non-ischemic CMPs benefitted from the addition of hydrochlorothiazide, although non-ischemic patients showed a better response [21]. Most patients in the COACH (in the comparison of outcomes and access to care for heart failure) trial had a history of HF (63.5%) or hypertension (75.2%); however, it is not known whether non-ischemic HF patients had a better prognosis and earlier discharge [22]. In the EMPULSE (EMPagliflozin outcome trial in patients hospitalized for acute heart failure with preserved and reduced ejection fraction) trial, although clinical benefits were observed for both AHF and ADHF irrespective of LV ejection fraction (LVEF), there are no data about the role of gliflozins in AHF patients based on etiology [23]. In the AFFIRM-AHF (a randomized, double-blind placebo-controlled trial comparing the effect of intravenous ferric carboxymaltose on hospitalizations and mortality in iron-deficient subjects admitted for acute heart failure) trial, 71.3% of patients had a previous history of HF and 47.1% had an ischemic CMP. Furthermore, in the subgroup analysis, ischemic HF patients benefitted more from iron administration compared to non-ischemic HF patients. Therefore, HF etiology could be an effect modifier, given that non-ischemic HF etiology comprises a heterogeneous group of patients, in whom the pathophysiological consequences of iron deficiency are not yet well-established [24]. In addition, the ASTRONAUT (the aliskiren trial on acute heart failure outcomes) trial included predominantly ischemic patients (63.6%) rather than non-ischemic patients. Furthermore, in the prespecified subgroup analysis of the primary endpoint, the negative finding was consistent across ischemic and non-ischemic groups [25]. In the STRONG-HF (the safety, tolerability, and efficacy of rapid optimization, helped by NT-proBNP testing, of heart failure therapies) trial, the percentages of patients with ischemic vs. non-ischemic CMPs were similar (48% and 52%, respectively); however, no data on outcomes according to etiology are available [26]. In the PIONEER (comparison of sacubitril–valsartan versus enalapril on effect on NT-proBNP in patients stabilized from an acute heart failure episode) trial, the baseline characteristics of the enrolled patients were: previous MI (7%), hyperlipidemia (37.1%), hypertension (85.5%), diabetes mellitus (19.1%), previous stroke (9.9%), AF (35.4%), chronic renal insufficiency (28.3%), and current smoking (22.5%). However, there is no clear etiology reported and no data on the effectiveness of sacubitril–valsartan according to HF etiologies in patients with AHF [27]. The subgroup analysis of SURVIVE (survival of patients with acute heart failure in need of intravenous inotropic support) trial revealed that a prior history of HF at baseline influenced the between-group difference at 31 days (treatment x prior HF interaction, $p = 0.05$) but not at 180 days, while MI as a primary cause of hospitalization did not influence the between-group difference at 31 days and 180 days (treatment x prior HF interaction, $p = 0.82$ and $p = 0.23$, respectively) [28]. In the CHAMPION (CardioMEMS heart sensor allows monitoring of pressure to improve outcomes in NYHA class III heart failure patients) trial, the comorbidities of the population were hypertension (77% vs. 79%, respectively in the treatment vs. placebo groups), CAD (67% vs. 72%), diabetes mellitus (48% vs. 50%), atrial flutter or fibrillation (44% vs. 48%), COPD (28% vs. 30%), and chronic kidney disease (20% vs. 19%), but again, the outcomes of the different etiologies of HF are unknown [29]. The subgroup analysis of CV death in the TRUE-AHF (ularitide efficacy and safety in acute heart failure) trial explored the drug's effect according to the presence or absence of CAD history, with a negative interaction suggesting that the lack of a significant benefit of ularitide was consistent regardless of AHF etiology [30]. The RELAX-AHF 2 (effects of serelaxin in patients with acute heart failure) trial also reported the percentage of ischemic patients with AHF (54.5% vs. 53.1, respectively, in serelaxin vs. placebo groups), although outcomes stratified by etiology are not reported [31]. In a post hoc subgroup analysis of the SOLOIST-WHF (sotagliflozin in patients with diabetes and recent worsening heart failure) trial, the primary endpoint was assessed according to the etiology of HF, comparing the drug effect in patients with ischemic versus non-ischemic HF, suggesting that the greatest benefit was observed in ischemic HF ($n = 712$ vs. 503; HR (95% CI) for sotagliflozin vs. placebo 0.55; 0.41, 0.74 vs. 0.88; 0.59, 1.31) [32]. Surprisingly, in the REVIVE I and II (effect of levosimendan on the short-term clinical course of patients with acutely decompensated heart failure), GALACTIC (effect of a strategy of comprehensive vasodilation vs. usual care on mortality and heart failure rehospitalization among patients with acute heart failure), and VICTORIA (vericiguat in patients with heart failure and reduced ejection fraction) trials, there are no data on differences in outcomes between ischemic vs. non-ischemic HF patients [33–35]. On the contrary, the LeoDOR (repetitive levosimendan infusions for patients with advanced chronic heart failure in the weakening post-discharge period) trial reports the percentages of patients with non-ischemic HF treated with levosimendan compared with placebo (ischemic HF 38.7% vs. 50%, $p = 0.127$; DCM 69.9% vs. 53.8%, $p = 0.04$; restrictive cardiomyopathy (RCM) 2.2% vs. 0% $p = 0.410$) [36]. In the subgroup analyses of the PUSH-AHF (natriuresis-guided diuretic therapy in acute heart failure: a pragmatic randomized trial) trial, the main findings (better natriuresis without significant benefit in clinical endpoints) were consistent across subgroups based on ischemic or non-ischemic etiology with negative interaction p-values for both dual primary endpoints (24 h urinary sodium excretion and a combined endpoint of time to all-cause mortality or adjudicated heart failure rehospitalization at 180 days) [37]. In the subgroup analysis of the PREMIER (in-hospital initiation of angiotensin receptor–neprilysin inhibition in acute heart failure) trial, the benefit of sacubitril/valsartan on NT-proBNP levels at week 8 was consistent in patients with ischemic and non-ischemic HF (p for interaction = 0.781) [38]. In a subgroup analysis of the UNLOAD (ultrafiltration versus intravenous diuretics for patients hospitalized for acute decompensated heart failure) trial, compared with intravenous diuretics, ultrafiltration resulted in lower HF rehospitalization rates, and this benefit seemed to be greater in non-ischemic HF patients than in ischemic HF patients (HR 0.21, CI: 0.05–1.00 vs. 0.72, CI: 0.33–1.56) [39]. In the CUORE (continuous ultrafiltration for congestive heart failure) trial, only the percentages of ischemic HF vs. non-ischemic HF patients in the control vs. ultrafiltration groups were reported (55% vs. 59%, p = 0.55; 17% vs. 15%, p = 0.81) [40].

Studying AHF as a unique entity is challenging and represents a problem when trying to obtain data from trials that should fit all AHF patients. We attempted to summarize the data from the main AHF trials, highlighting the differences in patient characteristics. Table 1 reports details of the main randomized AHF trials. Here, we mainly focused on the data on patients with non-ischemic CMPs. In the ADVOR (acetazolamide in decompensated heart failure with volume overload) trial, the underlying diseases of patients reflected the ones of the registries discussed before (43% CAD, 20% valvular heart disease, 24% DCM, and 9% arterial hypertension) [20]. In the CLOROTIC (combining loop with thiazide diuretics for decompensated heart failure) trial, the patient population had a high burden of comorbidities and high-risk features: diabetes (56.9% and 56.1% in HCTZ and placebo, respectively), AF (62.9% and 74.6%, respectively), anemia (45.7% and 43.9%, respectively), CAD (25.2% and 40.4%, respectively), and COPD (21.6% and 23.7%, respectively). Interestingly, in the subgroup analysis, patients with both ischemic and non-ischemic CMPs benefitted from the addition of hydrochlorothiazide, although non-ischemic patients showed a better response [21]. Most patients in the COACH (in the comparison of outcomes and access to care for heart failure) trial had a history of HF (63.5%) or hypertension (75.2%); however, it is not known whether non-ischemic HF patients had a better prognosis and earlier discharge [22]. In the EMPULSE (EMPagliflozin outcome trial in patients hospitalized for acute heart failure with preserved and reduced ejection fraction) trial, although clinical benefits were observed for both AHF and ADHF irrespective of LV ejection fraction (LVEF), there are no data about the role of gliflozins in AHF patients based on etiology [23]. In the AFFIRM-AHF (a randomized, double-blind placebo-controlled trial comparing the effect of intravenous ferric carboxymaltose on hospitalizations and mortality in iron-deficient subjects admitted for acute heart failure) trial, 71.3% of patients had a previous history of HF and 47.1% had an ischemic CMP. Furthermore, in the subgroup analysis, ischemic HF patients benefitted more from iron administration compared to non-ischemic HF patients. Therefore, HF etiology could be an effect modifier, given that non-ischemic HF etiology comprises a heterogeneous group of patients, in whom the pathophysiological consequences of iron deficiency are not yet well-established [24]. In addition, the ASTRONAUT (the aliskiren trial on acute heart failure outcomes) trial included predominantly ischemic patients (63.6%) rather than non-ischemic patients. Furthermore, in the prespecified subgroup analysis of the primary endpoint, the negative finding was consistent across ischemic and non-ischemic groups [25]. In the STRONG-HF (the safety, tolerability, and efficacy of rapid optimization, helped by NT-proBNP testing, of heart failure therapies) trial, the percentages of patients with ischemic vs. non-ischemic CMPs were similar (48% and 52%, respectively); however, no data on outcomes according to etiology are available [26]. In the PIONEER (comparison of sacubitril–valsartan versus enalapril on effect on NT-proBNP in patients stabilized from an acute heart failure episode) trial, the baseline characteristics of the enrolled patients were: previous MI (7%), hyperlipidemia (37.1%), hypertension (85.5%), diabetes mellitus (19.1%), previous stroke (9.9%), AF (35.4%), chronic renal insufficiency (28.%), and current smoking (22.5%). However, there is no clear etiology reported and no data on the effectiveness of sacubitril–valsartan according to HF etiologies in patients with AHF [27]. The subgroup analysis of SURVIVE (survival of patients with acute heart failure in need of intravenous inotropic support) trial revealed that a prior history of HF at baseline influenced the between-group difference at 31 days (treatment x prior HF interaction, $p = 0.05$) but not at 180 days, while MI as a primary cause of hospitalization did not influence the between-group difference at 31 days and 180 days (treatment x prior HF interaction, $p = 0.82$ and $p = 0.23$, respectively) [28]. In the CHAMPION (CardioMEMS heart sensor allows monitoring of pressure to improve outcomes in NYHA class III heart failure patients) trial, the comorbidities of the population were hypertension (77% vs. 79%, respectively in the treatment vs. placebo groups), CAD (67% vs. 72%), diabetes mellitus (48% vs. 50%), atrial flutter or fibrillation (44% vs. 48%), COPD (28% vs. 30%), and chronic kidney disease (20% vs. 19%), but again, the outcomes of the different etiologies of HF are unknown [29]. The subgroup analysis of CV death in the TRUE-AHF (ularitide efficacy and safety in acute heart failure) trial explored the drug's effect according to the presence or absence of CAD history, with a negative interaction suggesting that the lack of a significant benefit of ularitide was consistent regardless of AHF etiology [30]. The RELAX-AHF 2 (effects of serelaxin in patients with acute heart failure) trial also reported the percentage of ischemic patients with AHF (54.5% vs. 53.1, respectively, in serelaxin vs. placebo groups), although outcomes stratified by etiology are not reported [31]. In a post hoc subgroup analysis of the SOLOIST-WHF (sotagliflozin in patients with diabetes and recent worsening heart failure) trial, the primary endpoint was assessed according to the etiology of HF, comparing the drug effect in patients with ischemic versus non-ischemic HF, suggesting that the greatest benefit was observed in ischemic HF ($n = 712$ vs. 503; HR (95% CI) for sotagliflozin vs. placebo 0.55; 0.41, 0.74 vs. 0.88; 0.59, 1.31) [32]. Surprisingly, in the REVIVE I and II (effect of levosimendan on the short-term clinical course of patients with acutely decompensated heart failure), GALACTIC (effect of a strategy of comprehensive vasodilation vs. usual care on mortality and heart failure rehospitalization among patients with acute heart failure), and VICTORIA (vericiguat in patients with heart failure and reduced ejection fraction) trials, there are no data on differences in outcomes between ischemic vs. non-ischemic HF patients [33–35]. On the contrary, the LeoDOR (repetitive levosimendan infusions for patients with advanced chronic heart failure in the weakening post-discharge period) trial reports the percentages of patients with non-ischemic HF treated with levosimendan compared with placebo (ischemic HF 38.7% vs. 50%, $p = 0.127$; DCM 69.9% vs. 53.8%, $p = 0.04$; restrictive cardiomyopathy (RCM) 2.2% vs. 0% $p = 0.410$) [36]. In the subgroup analyses of the PUSH-AHF (natriuresis-guided diuretic therapy in acute heart failure: a pragmatic randomized trial) trial, the main findings (better natriuresis without significant benefit in clinical endpoints) were consistent across subgroups based on ischemic or non-ischemic etiology with negative interaction p-values for both dual primary endpoints (24 h urinary sodium excretion and a combined endpoint of time to all-cause mortality or adjudicated heart failure rehospitalization at 180 days) [37]. In the subgroup analysis of the PREMIER (in-hospital initiation of angiotensin receptor–neprilysin inhibition in acute heart failure) trial, the benefit of sacubitril/valsartan on NT-proBNP levels at week 8 was consistent in patients with ischemic and non-ischemic HF (p for interaction = 0.781) [38]. In a subgroup analysis of the UNLOAD (ultrafiltration versus intravenous diuretics for patients hospitalized for acute decompensated heart failure) trial, compared with intravenous diuretics, ultrafiltration resulted in lower HF rehospitalization rates, and this benefit seemed to be greater in non-ischemic HF patients than in ischemic HF patients (HR 0.21, CI: 0.05–1.00 vs. 0.72, CI: 0.33–1.56) [39]. In the CUORE (continuous ultrafiltration for congestive heart failure) trial, only the percentages of ischemic HF vs. non-ischemic HF patients in the control vs. ultrafiltration groups were reported (55% vs. 59%, p = 0.55; 17% vs. 15%, p = 0.81) [40].

Finererone, a non-steroidal mineralcorticoid receptor antagonist, is currently under investigation in the MOONRAKER program, including four randomized phase III clinical trials in HF settings, including AHF. In particular, the REDEFINE-HF (NCT06008197) trial is comparing finerenone vs. placebo in patients hospitalized for acute HF with HFmrEF and HFpEF, focusing on in-hospital initiation to improve CV outcomes. Moreover, CONFIRMATION-HF (NCT06024746) is assessing the combination of finerenone and an SGLT2 inhibitor, compared with standard care, in patients with acute HF, regardless of ejection fraction, to investigate the benefits of early, intensive combination therapy. Patients with known CMPs will be excluded from both trials.

Table 2 reports the details of the main randomized AHF trials related to CS. This is a peculiar setting characterized by critical patients with high rates of mortality. Most of the patients included in these trials present with CS related to MI and there are limited data specifically related to non-ischemic CS and to possible differences among the characteristics and outcomes according to different etiologies. Only the ECMO-CS (extracorporeal membrane oxygenation in the therapy of cardiogenic shock) trial reported the percentage of patients with DCM (10.9% in the VA-ECMO group vs. 15.5% in the conservative group, p = 0.471), but again, clinical outcomes according to CS etiology remain unknown [41].

Table 1. Main randomized trials of acute heart failure.

	Patients (*n*)	Year of Publication	Population Characteristics	Randomized Arms	Primary Endpoint	Main Results
SURVIVE trial [28]	1327	2007	Patients hospitalized with ADHF who required inotropic support	To receive intravenous levosimendan vs. dobutamine	All-cause mortality at 180 days	26% vs. 28% (HR 0.91; 95% CI, 0.74–1.13; $p = 0.40$)
UNLOAD trial [39]	200	2007	Patients hospitalized for HF with ≥2 signs of hypervolemia	To ultrafiltration or intravenous diuretics	(1) Weight loss at 48 h (2) Dyspnea assessment at 48 h	(1) 5.0 ± 3.1 kg vs. 3.1 ± 3.5 kg; $p = 0.001$) (2) Dyspnea scores (5.4 ± 1.1 [range 3 to 7] vs. 5.2 ± 1.2 [range 1 to 7]; $p = 0.588$)
CHAMPION trial [29]	550	2011	Patients with NYHA class III symptoms, irrespective of the LVEF and with a previous hospital admission for HF	To management with a W-IHM system (treatment group) or without (control group)	The rate of HF-related hospitalizations at 6 months	84 vs. 120 (rate 0.32 vs. 0.44, HR 0.72, 95% CI 0.60–0.85, $p = 0.0002$)
DOSE trial [42]	308	2011	ADHF patients	To receive furosemide administered intravenously by means of either a bolus every 12 h or continuous infusion and at either a low dose (equivalent to the patient's previous oral dose) or a high dose (2.5 times the previous oral dose)	(1) Patients' global assessment of symptoms, quantified as the AUC of the score on a visual analog scale over the course of 72 h (2) The change in the serum creatinine level from baseline to 72 h	(1) In the comparison of bolus with continuous infusion, patients' global assessment of symptoms (mean AUC, 4236 ± 1440 vs. 4373 ± 1404, respectively; $p = 0.47$) or in the mean change in the creatinine level (0.05 ± 0.3 mg per deciliter [4.4 ± 26.5 µmol per liter] vs. 0.07 ± 0.3 mg per deciliter [6.2 ± 26.5 µmol per liter], respectively; $p = 0.45$). (2) In the comparison of the high-dose strategy with the low-dose strategy, patients' global assessment of symptoms in the high-dose group (mean AUC, 4430 ± 1401 vs. 4171 ± 1436; $p = 0.06$). In the mean change in the creatinine level (0.08 ± 0.3 mg per deciliter [7.1 ± 26.5 µmol per liter] vs. 0.04 ± 0.3 mg per deciliter [3.5 ± 26.5 µmol per liter], $p = 0.21$)
CARRESS-HF trial [43]	188	2012	Patients with ADHF, worsened renal function (increase in serum creatitine >0.3 mg/dL from baseline), and signs and symptoms of persistent congestion	To a strategy of stepped pharmacologic therapy or ultrafiltration	The bivariate change from baseline in serum creatinine levels and body weight, as assessed 96 h after randomization	(1) At 96 h, the mean change in the creatinine level was −0.04 ± 0.53 mg per deciliter (−3.5 ± 46.9 µmol per liter) in the pharmacologic therapy group vs. +0.23 ± 0.70 mg per deciliter (20.3 ± 61.9 µmol per liter) in the ultrafiltration group ($p = 0.003$). (2) Difference in weight loss at 96 h in the pharmacologic therapy group vs. in the ultrafiltration group (a loss of 5.5 ± 5.1 kg [12.1 ± 11.3 lb] and 5.7 ± 3.9 kg [12.6 ± 8.5 lb], respectively; $p = 0.58$)
REVIVE I and II trials [30]	100 (I) + 600 (II)	2013	ADHF patients with LVEF < 35%	To receive intravenous levosimendan or placebo for 24 h in addition to standard treatment	A composite that evaluated changes in clinical status during the first 5 days after randomization (time points 6 h, 24 h, and 5 days)	58 patients in the levosimendan group vs. 82 in the placebo group experienced clinical worsening ($p = 0.015$ for the difference between the groups)
ASTRONAUT trial [25]	1615	2013	Patients with EF and LVEF of 40% or less, BNP ≥ 400 pg/mL or NT-proBNP ≥ 1600 pg/mL, and signs and symptoms of fluid overload	To receive 150 mg (increased to 300 mg as tolerated) of aliskiren or placebo daily, in addition to standard therapy	CV death or HF rehospitalization at 6 months and 12 months	24.9% of patients receiving aliskiren vs. 26.5% of patients receiving placebo at 6 months (HR, 0.92; 95% CI, 0.76–1.12; $p = 0.41$). At 12 months, the event rates were 35.0% for the aliskiren group vs. 37.3% for the placebo group (HR, 0.93; 95% CI, 0.79–1.09; $p = 0.36$)

Table 1. *Cont.*

	Patients (*n*)	Year of Publication	Population Characteristics	Randomized Arms	Primary Endpoint	Main Results
ROSE trial [44]	360	2013	Hospitalized patients with AHF and renal dysfunction (eGFR of 15–60 mL/min/1.73 m^2)	1:1 allocation ratio to the dopamine or nesiritide strategy. Within each strategy, participants were randomized in a double-blind, 2:1 ratio to active treatment or placebo	(1) 72 h cumulative urine volume (decongestion endpoint) (2) the change in serum cystatin C from enrollment to 72 h (renal function endpoint)	(1) Low-dose dopamine had no significant effect on 72 h cumulative urine volume vs. placebo (dopamine, 8524 mL; 95% CI, 7917–9131 vs. placebo, 8296 mL; 95% CI, 7762–8830; difference, 229 mL; 95% CI, −714 to 1171 mL; $p = 0.59$) or on the change in cystatin C level (dopamine, 0.12 mg/L; 95% CI, 0.06–0.18 vs. placebo, 0.11 mg/L; 95% CI, 0.06–0.16; difference, 0.01; 95% CI, −0.08 to 0.10; $p = 0.72$). (2) Low-dose nesiritide had no significant effect on 72 h cumulative urine volume vs. placebo (nesiritide, 8574 mL; 95% CI, 8014–9134 vs. placebo, 8296 mL; 95% CI, 7762–8830; difference, 279 mL; 95% CI, −613 to 1176 mL; $p = 0.49$) or on the change in cystatin C level (nesiritide, 0.07 mg/L; 95% CI, 0.01–0.13 vs. placebo, 0.11 mg/L; 95% CI, 0.06–0.16; difference, −0.04; 95% CI, −0.13 to 0.05; $p = 0.36$)
CUORE Trial [40]	56	2014	Patients with severe congestive HF	To receive standard medical therapy or ultrafiltration as first-line treatment	Rehospitalizations for congestive HF during 1 year of FU	A lower incidence of rehospitalizations for HF in the ultrafiltration-treated patients (HR:0.14, 95% CI 0.04–0.48; $p = 0.002$)
AVOID-HF trial [45]	224	2015	Patients hospitalized for congested HF	To AUF or ALD	First HF event within 90 days after hospital discharge	Estimated days to first HF event for the AUF and ALD group are, respectively, 62 vs. 34 ($p = 0.106$)
BLAST-AHF trial [46]	621	2017	Patients hospitalized with AHF, as evidenced by elevated natriuretic peptides and at least two physical HF signs including congestion on chest radiograph, rales, edema, and/or elevated jugular venous pressure	To randomly assigned to one of the following four (1:1:1:1: placebo, 1, 5, or 25 mg/h of TRV027 for at least 48 h and up to 96 h)	(1) Time from baseline to death through day 30 (2) Time from baseline to HF rehospitalization through day 30 (3) The first assessment time point following worsening of HF through day 5 (4) Change in dyspnea VAS score calculated as AUC representing the change from baseline over time through day 5 (5) The length of initial hospital stays (in days) from baseline	No significant differences were observed between any of the dose groups compared with placebo with regard to the primary endpoint
TRUE-AHF trial [30]	2157	2017	AHF patients defined as: an unplanned emergency department visit or hospitalization for AHF, dyspnea at rest that had worsened during the previous week, evidence of HF on chest radiography, a blood BNP level of more than 500 pg per milliliter, or an NT-proBNP level of more than 2000 pg per milliliter. All patients that continued to have dyspnea at rest for at least 2 h after IV furosemide at a dose of 40 mg (or equivalent) with sBP between 116–180 mmHg	To receive a continuous intravenous infusion of either Ularitide at a dose of 15 ng per kilogram of body weight per minute or matching placebo for 48 h, in addition to accepted therapy	Death from CV causes during a median follow-up of 15 months and clinical composite endpoint during the first 48 h	236 vs. 225 (21.7% vs. 21.0%; HR: 1.03; 96% CI, 0.85 to 1.25; $p = 0.75$). No change in clinical course.

Table 1. *Cont.*

	Patients (*n*)	Year of Publication	Population Characteristics	Randomized Arms	Primary Endpoint	Main Results
PIONEER trial [27]	881	2019	Patients with HF with reduced EF who were hospitalized for ADHF (NT-proBNP concentration of 1600 pg per milliliter or more or a BNP) concentration of 400 pg per milliliter or more)	To receive sacubitril–valsartan vs. enalapril	The time-averaged proportional change in NT-pro-BNP concentration from baseline through weeks 4 and 8	0.53 in the sacubitril–valsartan group vs. 0.75 in the enalapril group (percent change, −46.7% vs. −25.3%; ratio of change with sacubitril–valsartan vs. enalapril, 0.71; 95% CI, 0.63 to 0.81; $p < 0.001$)
RELAX AHF 2 trial [31]	6545	2019	Patients with AHF and dyspnea, vascular congestion on chest radiography, increased BNP, mild-to-moderate renal insufficiency, and a systolic blood pressure of at least 125 mmHg	To receive either a 48 h intravenous infusion of serelaxin or placebo, in addition to standard care	(1) Death from CV causes at 180 days (2) Worsening HF at 5 days	(1) 8.7% vs. 8.9% (HR: 0.98; 95% CI: 0.83 to 1.15; $p = 0.77$) (2) 6.9% vs. 7.7% (HR: 0.89; 95% CI, 0.75 to 1.07; $p = 0.19$)
GALACTIC trial [34]	781	2019	AHF patients with dyspnea, increased plasma concentrations of natriuretic peptides, and systolic blood pressure of at least 100 mm Hg	To a strategy of early intensive and sustained vasodilation (individualized doses of sublingual and transdermal nitrates, low-dose oral hydralazine for 48 h, and rapid up-titration of angiotensin-converting enzyme inhibitors, angiotensin receptor blockers, or sacubitril-valsartan) vs. usual care	Composite of all-cause mortality or rehospitalization for AHF at 180 days	30.6 % vs. 27.8% (absolute difference for the primary endpoint, 2.8% [95% CI, −3.7% to 9.3%]; adjusted HR: 1.07 [95% CI, 0.83–1.39]; $p = 0.59$)
AFFIRM-AHF trial [24]	1108	2020	AHF with concomitant iron deficiency (defined as ferritin < 100 μg/L, or 100–299 μg/L with transferrin saturation < 20%) and LVEF < 50%	To receive intravenous ferric carboxymaltose or placebo for up to 24 weeks, dosed according to the extent of iron deficiency	A composite of total hospitalizations for HF and CV death up to 52 weeks after randomization	32% vs. 38% of patients in the placebo group (HR 0.80, 95% CI 0.66–0.98, $p = 0.030$)
VICTORIA trial [35]	5050	2020	Patients with HF and LVEF > 45% who had recently been hospitalized or had received intravenous diuretic therapy	To receive vericiguat (target dose, 10 mg once daily) or placebo, in addition to guideline-based medical therapy	A composite of death from CV causes or first hospitalization for HF	37.9% vs. 40.9% in the placebo group (HR: 0.90; 95% CI, 0.83 to 0.98; $p = 0.02$)
Sean P. Collins et al. trial [47]	479	2020	AHF patients with a history of HF and a discharge within 23 h or less	To usual care vs. a tailored self-care intervention	A global rank of CV death, HF-related events (unscheduled clinic visit due to HF, ED revisit, or hospitalization), and changes in the KCCQ-12 summary score at 90 days	No significant difference in the primary outcome between patients in the intervention vs. usual care arms ([HR], 0.89; 95% CI, 0.73–1.10; $p = 0.028$)
SOLOIST-WHF trial [32]	1222	2021	Patients with type 2 diabetes mellitus who were recently hospitalized for worsening HF	To receive sotagliflozin or placebo	The total number of deaths from CV causes and hospitalizations and urgent visits for HF	51.0 vs. 76.3 (HR: 0.67; 95% CI: 0.52 to 0.85; $p < 0.001$)
ADVOR trial [20]	519	2022	ADHF patients with clinically evident volume overload, elevated NT-pro-BNP, and loop diuretic therapy for at least 1 month before randomization	To receive either intravenous acetazolamide vs. placebo added to standardized intravenous loop diuretics	Successful decongestion assessed by a dedicated score indicating no more than trace edema within 3 days after randomization	42.2% vs. 30.5% ($p < 0.001$)
STRONG-HF trial [26]	1078	2022	Patients with AHF not treated with full doses of guideline-directed drug treatment	Rapid up-titration of treatments before discharge vs. standard approach	180-day readmission to hospital due to HF or all-cause death	15.2% vs. 23.3% (adjusted risk difference 8.1% [95% CI 2.9–13.2]; $p = 0.0021$; risk ratio 0.66 [95% CI 0.50–0.86])

Table 1. *Cont.*

	Patients (*n*)	Year of Publication	Population Characteristics	Randomized Arms	Primary Endpoint	Main Results
EMPULSE trial [23]	530	2022	Primary diagnosis of AHF or ADHF regardless of LVEF, randomized when clinically stable (after 3 days)	To receive empagliflozin 10 mg once daily or placebo for 90 days	Clinical benefit, defined as a hierarchical composite of death from any cause, number of heart failure events and time to first heart failure event, or a 5 point or greater difference in change from baseline in the Kansas City Cardiomyopathy Questionnaire Total Symptom Score at 90 days	More patients treated with empagliflozin displayed a clinical benefit compared with placebo (stratified win ratio, 1.36; 95% confidence interval, 1.09–1.68; $p = 0.0054$), meeting the primary endpoint.
CLOROTIC trial [21]	230	2023	Patients with CHF treated with oral furosemide (80–240 mg/day) for 1 month before admission for AHF	To receive HCTZ vs. placebo in addition to an intravenous furosemide regimen	Changes in: (1) Body weight (2) Patient-reported dyspnea 72 h after randomization	(1) HCTZ group vs. placebo group: −2.3 vs. −1.5 kg; (95% CI: −1.14 (−1.84 to −0.42); $p = 0.002$) (2) HCTZ group vs. placebo group: AUC: 960 vs. 720; $p = 0.497$
COACH trial [22]	5452	2023	Patients with AHF who were seeking emergency care	In the intervention phase, low-risk patients (according to a point-of-care algorithm) were discharged early (in ≤3 days) vs. high-risk patients, who were admitted to the hospital	A composite of death from any cause or hospitalization for CV causes within: (1) 30 days after presentation (2) 20 months	(1) Within 30 days, 12.1% patients enrolled during the intervention phase vs. 14.5% in the control phase (adjusted HR: 0.88; 95% CI, 0.78 to 0.99; $p = 0.04$) (2) Within 20 months, 54.4% in the intervention phase vs. 56.2% in the control phase (adjusted HR: 0.95; 95% CI, 0.92 to 0.99)
LeoDOR trial [36]	148	2023	AHF patients requiring IV diuretics, IV inotropic therapy, IV vasodilators, or any combination of these. LVEF < 30%. At least one hospitalization or clinic visit for HF within 12 months before admission. NTproBNP elevated or NYHA III-IV symptoms after stabilization.	Patients were randomly assigned (2:1) to receive either levosimendan or placebo	A global rank endpoint in which all participants, regardless of treatment assignment, were ranked across three hierarchical tiers: • Tier 1 = time to death or urgent heart transplantation or implantation of a ventricular assist device (VAD). • Tier 2 = time to non-fatal HF requiring IV vasoactive therapy. • Tier 3 = time-averaged proportional change in NT-proBNP from baseline to week 14, each multiplied by the number of elapsed times in weeks, divided by the total number of 14 weeks of observation from baseline to 14 weeks.	The mean rank score was 72.55 for the levosimendan group vs. 73.81 for the placebo group ($p = 0.863$)
PUSH-AHF trial [37]	310	2023	AHF patients requiring treatment with intravenous loop diuretics	To receive natriuresis-guided therapy or standard of care	The dual primary endpoints are: (1) 24 h urinary sodium excretion and (2) A combined endpoint of time to all-cause mortality or adjudicated HF rehospitalization at 180 days	(1) 24 h urinary sodium excretion is 409 ± 178 mmol in the natriuresis-guided therapy group versus 345 ± 202 mmol in the control group, respectively ($p = 0.0061$) (2) Combined endpoint of time to all-cause mortality or first HF rehospitalization occurred in 31% and 31% of patients in the natriuresis-guided and control arms, respectively (HR: 0.92 (3) [95% CI 0.62–1.38], $p = 0.6980$)

Table 1. *Cont.*

	Patients (n)	Year of Publication	Population Characteristics	Randomized Arms	Primary Endpoint	Main Results
SSU-AHF trial [48]	193	2024	Lower-risk patients who presented to the ED with signs and symptoms of AHF	To SSU or hospital admission from the ED	(1) DAOOH at 30-day follow-up (2) Quality of life measured as KCCQ-12	(1) 1.6 more DAOOH at 30-day follow-up in the SSU arm vs. those in the hospitalization arm (median [IQR], 26.9 [24.4–28.8] vs. 25.4 [22.0–27.7] days; $p = 0.02$) (2) The mean (SD) KCCQ-12 summary score between the SSU and hospitalization arms is 51.3 (25.7) vs. 45.8 (23.8) points, respectively ($p = 0.19$)
COLICA trial [49]	278	2024	AHF patients with evidence of congestion requiring at least 40 mg of intravenous (i.v.) furosemide and elevated concentrations of NT-proBNP > 900 pg/mL	To receive either colchicine (loading dose 2 mg, followed by 0.5 mg every 12 h for 8 weeks) or placebo within the first 24 h of presentation	The time-averaged proportional change in NT-pro-BNP concentration from baseline through weeks 4 and 8	Colchicine group [−62.2%, 95% CI −68.9% to −54.2%] vs. the placebo group (−62.1%, 95% CI −68.6% to −54.3%)
PREMIER trial [38]	400	2024	In patients stabilized after hospitalization for AHF irrespective of LVEF	To continue angiotensin-converting enzyme inhibitor or angiotensin receptor blocker (control group) or to switch to Sac/Val (Sac/Val group)	8-week proportional change in geometric means of NT-pro-BNP levels	The percent changes in NT-pro-BNP level geometric means at weeks 4/8 are −35%/−45% (Sac/Val group) and −18%/−32% (control group), and their group ratio (Sac/Val vs. control) is 0.80 (95% CI 0.68–0.94; $p = 0.008$) at week 4 and 0.81 (95% CI 0.68–0.95; $p = 0.012$) at week 8, respectively
DICTATE-AHF trial [50]	240	2024	Hypervolemic hospitalized AHF	To dapagliflozin 10 mg once daily or structured usual care with protocolized diuretic titration until day 5 or hospital discharge	Diuretic efficiency expressed as cumulative weight change per cumulative loop diuretic dose	No difference between dapagliflozin vs. usual care in diuretic efficiency (OR: 0.65; 95% CI: 0.41–1.02; $p = 0.06$)

ADHF: acutely decompensated heart failure, AHF: acute heart failure; ALD: adjustable intravenous loop diuretics; UC: area under the curve; AUF: adjustable ultrafiltration; BNP: B-type natriuretic peptide; CHF: chronic heart failure; CI: confidence interval; CV: cardiovascular; DAOOH: days alive and out of hospital; ED: emergency department; eGFR: estimated glomerular filtration rate; HCTZ: hydrochlorothiazide; HF: heart failure; HR: hazard ratio; LVEF: left ventricle ejection fraction; KCCQ-12: 12-item Kansas City Cardiomyopathy Questionnaire; NYHA: New York Heart Association; NT-pro-BNP: N-terminal pro-B-type natriuretic peptide; W-IHM: wireless implantable hemodynamic monitoring; OR: odds ratio; SSU: short-stay units; VAS: visual analog scale; TRV027: a biased ligand of the angiotensin II type 1 receptor (AT1R).

Table 2. Main recent randomized trials on cardiogenic shock.

	Patients (n)	Year of Publication	Population Characteristics	Randomized Arms	Primary Endpoint	Main Results
ISAR-SHOCK trial [51]	25	2008	AMI patients with CS	To receive Impella LP 2.5 (Abiomed Europe GmbH, Aachen, Germany) or IABP	Change in the CI from baseline to 30 min after implantation	Impella: ΔCI: 0.49 ± 0.46 l/min/m^2; IABP: ΔCI: 0.11± 0.31 l/min/m^2; $p = 0.02$
Joerg T. Fuhrmann et al. trial [52]	32	2008	Persistent refractory CS (systolic blood pressure < 90 mmHg or requirement of inotropic amines and vasopressors to maintain an unaugmented systolic blood pressure of at least 90 mm Hg, a CI below 2.5 L/min/m^2, a PCOP above 18 mm Hg, and clinical signs of hypoperfusion) within 2 h after PCI	Infusion of either levosimendan or enoximone after initiation of current therapy, always including revascularization, intra-aortic balloon pump counterpulsation, and inotropes	All-cause mortality at 30 days	Survival rate at 30 days in the levosimendan-treated group (69%) vs. enoximone group (37%, $p = 0.023$)
IABP SHOCK I trial [53]	45	2010	Patients undergoing PCI for AMI with CS	To receive IABP or not	APACHE II score after 4 days	In the IABP group, the APACHE II score was 18.2 ± 3.7 vs. 20.0 ± 2.4 in the standard treatment group. Hospital mortality was 36.8% in the IABP group vs. 28.6% in the standard treatment group.

Table 2. *Cont.*

	Patients (*n*)	Year of Publication	Population Characteristics	Randomized Arms	Primary Endpoint	Main Results
SOAP II trial [54]	1679	2010	Patients with CS, defined as mean arterial pressure less than 70 mmHg or systolic blood pressure less than 100 mm Hg, despite an adequate amount of fluids, and with signs of tissue hypoperfusion	To receive either dopamine or norepinephrine as first-line vasopressor therapy to restore and maintain blood pressure	The rate of death at 28 days after randomization	Rate of death at 28 days (52.5% in the dopamine group and 48.5% in the norepinephrine group; odds ratio with dopamine: 1.17; 95% CI: 0.97 to 1.42; $p = 0.10$)
IABP-SHOCK II trial [55]	600	2012	Patients undergoing revascularization for AMI with CS	To IABP or no IABP	30-day all-cause mortality	39.7% in the IABP group vs. 41.3% in the control group (RR with IABP, 0.96; 95% CI, 0.79 to 1.17; $p = 0.69$)
IMPRESS shock trial [56]	48	2017	AMI patients with severe CS undergoing immediate revascularization	To receive Impella CP or IABP	30-day all-cause mortality	Mortality in patients treated with either IABP or Impella CP was similar (50% and 46%, respectively; HR with Impella CP: 0.96; 95% CI: 0.42 to 2.18; $p = 0.92$)
Bruno Levy et al. trial [57]	57	2018	CS due to successfully revascularized AMI	To receive epinephrine or norepinephrine	(1) The change in cardiac index (2) The incidence of refractory CS	(1) Cardiac index evolution was similar between the 2 groups (p: 0.43) (2) Higher incidence of refractory shock in the epinephrine group (37% vs. 7% in the norepinephrine group; $p = 0.008$)
IMPELLA-STIC trial [58]	12	2019	Patients with AMI complicated by CS	To receive IABP vs. Impella LP 5.0 + IABP	Change in CPI from baseline to 12 h after implantation, measured with a Swan–Ganz catheter	IABP group: CPI = 0.08 ± 0.08W/m^2 Impella LP5.0 + IABP group: CPI = −0.02 ± 0.25 W/m^2; $p = 0.4$
ECMO-CS trial [41]	117	2022	Patients with either rapidly deteriorating or severe CS	To immediate VA-ECMO or non-immediate VA-ECMO	The composite of death from any cause, resuscitated circulatory arrest, and implementation of another mechanical circulatory support device at 30 days	63.8% of patients in the immediate VA-ECMO group vs. 71.2% of patients in the non-immediate VA-ECMO group, respectively (HR: 0.72 [95% CI, 0.46–1.12]; $p = 0.21$)
ECLS-SHOCK trial [59]	420	2023	Patients with AMI complicated by CS	To receive early ECLS plus usual medical treatment or usual medical treatment alone	Death from any cause at 30 days	47.8% in the ECLS group vs. 49.0% in the control group (RR: 0.98; 95% CI: 0.80 to 1.19; $p = 0.81$)
EURO SHOCK trial [60]	35	2023	AMI patients with persistent CS 30 min after PPCI of the culprit lesion	To receive either VA-ECMO or continue with standard therapy	30-day all-cause mortality	43.8% of patients in the VA-ECMO group vs. 61.1% of patients in standard therapy (HR: 0.56, 95% CI: 0.21–1.45; $p = 0.22$)
DanGer Shock trial [61]	355	2024	Patients with AMI complicated by CS	To receive Impella CP plus standard care or standard care alone	Death from any cause at 180 days	45.8% in the Impella group vs. 58.5% in the standard care group (HR: 0.74; 95% CI: 0.55 to 0.99; $p = 0.04$)

AMI: acute myocardial infarction; APACHE: Acute Physiology and Chronic Health Evaluation (APACHE); PI: cardiac power index; CI: confidence interval; CS: cardiogenic shock; HR: hazard ratio; IABP: intra-aortic balloon pump; PCOP: post-Capillary Pulmonary Hypertension; PPCI: primary percutaneous coronary intervention; RR: relative risk; VA-ECMO: venoarterial extracorporeal membrane oxygenation.

2.6. Evidence on ADHF Rates in Randomized Trials of Chronic HF

Table 3 reports the main randomized trials on chronic HF, with a focus on the rate of ADHF during follow-up. However, again, the different etiologies of patients included in the trials are often unreported. Among DAPA-HF (dapagliflozin and prevention of adverse outcomes in heart failure) trial patients (n = 4744), 1687 had non-ischemic CMP, but the rates of idiopathic DCM or other types of CMPs, especially the differential impact of etiology on the hospitalization rate, are unknown [62]. In the subgroup analysis of the EMPEROR reduced (effect of empagliflozin on the clinical stability of patients with heart failure and a reduced ejection fraction) trial, the benefit from the addition of empagliflozin was consistent in patients with both ischemic and non-ischemic HF [63]. In the IRONMAN (intravenous ferric derisomaltose in patients with heart failure and iron deficiency) trial, the percentages of ischemic and non-ischemic HF patients are reported (58% vs. 56% in the ferric derisomaltose group; 31% vs. 35% in the usual care group), but without specific details on CMP type or outcomes [64]. The TRANSFORM-HF (effect of torsemide vs. furosemide after discharge on all-cause mortality in patients hospitalized with heart failure) trial only reported the percentages of ischemic HF patients (29.8% in the torsemide group vs. 26.7% in the furosemide group) [65]. Similarly, in the DIAMOND (patiromer for the management of hyperkalemia in heart failure with reduced ejection fraction) trial, ischemic HF patients made up 72.2% of the patiromer group vs. 70.6% of the placebo group [66]. In the PARADIGM-HF (angiotensin–neprilysin inhibition versus enalapril in heart failure) trial, ischemic HF patients made up 59.9% of the angiotensin receptor–neprilysin inhibitor group vs. 60.1% of the enalapril group [67]. Conversely, the subgroup analysis of the primary efficacy endpoint of the SOLOIST-WHF (sotagliflozin in patients with diabetes and recent worsening heart failure) trial showed a consistent treatment effect across the subgroups stratified according to ischemic vs. non-ischemic HF etiology [32]. Finally, the recent RESHAPE-HF2 (randomized investigation of the MitraClip device in heart failure: second trial in patients with clinically significant functional mitral regurgitation) and MATTERHORN (transcatheter repair versus mitral valve surgery for secondary mitral regurgitation) trials, although providing important evidence on the effectiveness of mitral transcatheter edge-to-edge repair (M-TEER) in symptomatic HF with functional mitral regurgitation, did not specifically analyze the outcomes based on ischemic or non-ischemic etiology [68,69].

Table 3. Main recent randomized chronic heart failure trials with rates of ADHF.

	Patients (*n*)	Year of Publication	Population Characteristics	Randomized Arms	Primary Endpoint	Rates of ADHF and Main Results
SENIORS trial [70]	2128	2005	Patients aged >70 years and with a history of HF (hospital admission for HF within the previous year or known EF < 35%)	To receive nebivolol or placebo	A composite of all-cause mortality or CV hospital admission	31.1% of patients in the nebivolol group vs. 35.3% in the placebo group (HR 0.86, 95% CI 0.74–0.99; $p = 0.039$)
PARADIGM-HF trial [67]	8442	2014	Patients who had HF with a reduced EF of 40% or less	To receive either angiotensin receptor–neprilysin inhibitor (at a dose of 200 mg twice daily) or enalapril (at a dose of 10 mg twice daily)	A composite of death from CV causes or hospitalization for HF	(1) 21.8% in the angiotensin receptor–neprilysin inhibitor group and 26.5% in the enalapril group (HR, 0.80; 95% CI, 0.73 to 0.87, $p < 0.001$) (2) Angiotensin receptor–neprilysin inhibitor reduced the risk of hospitalization for HF by 21% ($p < 0.001$) and decreased the symptoms and physical limitations of HF ($p = 0.001$)
COMMANDER HF trial [71]	5022	2018	Patients with worsening chronic HF, a left ventricular EF of 40% or less, and underlying CAD	To receive rivaroxaban at a dose of 2.5 mg twice daily or placebo in addition to standard care after treatment for an episode of worsening HF	A composite of death from any cause, MI, or stroke	(1) 25.0% of patients in the rivaroxaban group and 26.2% of patients in the placebo group (HR, 0.94; 95% CI, 0.84 to 1.05; $p = 0.27$) (2) The composite outcome of CV death or rehospitalization for HF occurred in 37.2% in the rivaroxaban group and in 36.9% of the placebo group (HR, 0.99; 95% CI, 0.91 to 1.09). (3) 27.5% of patients in rivaroxaban group vs. 27.5% in the placebo group (HR, 95% CI, 0.98 (0.89–1.09) had a rehospitalization for worsening HF
MITRA-FR trial [72]	152	2018	Patients with chronic HF with reduced left ventricular EF (15–40%) and severe secondary mitral valve regurgitation	To undergo percutaneous mitral valve repair in addition to receiving medical therapy (intervention group) or to receive medical therapy alone (control group)	Composite of death from any cause or unplanned hospitalization for HF at 12 months	(1) 54.6% in the intervention group and 51.3% in the control group (odds ratio, 1.16; 95% CI, 0.73 to 1.84; $p = 0.53$) (2) The rate of unplanned hospitalization for HF was 48.7% in the intervention group and 47.4% in the control group (HR, 1.13; 95% CI, 0.81 to 1.56)
COAPT trial [73]	614	2018	Patients with HF with reduced EF (20–50%) and moderate-to-severe or severe secondary mitral regurgitation who remained symptomatic despite the use of maximal doses of guideline-directed medical therapy	To transcatheter mitral valve repair plus medical therapy (device group) or medical therapy alone (control group)	All hospitalizations for HF within 24 months of follow-up	35.8% per patient-year in the device group vs. 67.9% per patient-year in the control group (HR, 0.53; 95% CI, 0.40 to 0.70; $p < 0.001$)
DAPA-HF trial [62]	4744	2019	Patients with NYHA class II, III, or IV HF and an EF of 40% or less	To receive either dapagliflozin (at a dose of 10 mg once daily) or placebo, in addition to recommended therapy	A composite of worsening HF (hospitalization or an urgent visit resulting in intravenous therapy for HF) or CV death over a median of 18.2 months	(1) 16.3% of patients in the dapagliflozin group and 21.2% of patients in the placebo group (HR, 0.74; 95% CI, 0.65 to 0.85; $p < 0.001$) (2) A first worsening HF event occurred in 10.0% of patients in the dapagliflozin group and in 13.7% of patients in the placebo group (HR, 0.70; 95% CI, 0.59 to 0.83)
EMPEROR-Reduced trial [63]	3730	2020	Patients with class II to IV HF with an EF of ≤40%	To empagliflozin (10 mg once daily) or placebo in addition to recommended treatments for HF	The composite of CV death or hospitalization for HF during a median follow-up of 16 months	(1) 415 empagliflozin patients vs. 519 placebo patients; (HR, 0.76; 95% CI, 0.67–0.87; $p < 0.0001$) (2) Less HF hospitalizations that required intensive care (HR, 0.67; 95% CI, 0.50–0.90; $p = 0.008$) and that required a vasopressor or positive inotropic drug or mechanical or surgical intervention in the empagliflozin group (HR, 0.64; 95% CI, 0.47–0.87; $p = 0.005$)

Table 3. *Cont.*

	Patients (*n*)	Year of Publication	Population Characteristics	Randomized Arms	Primary Endpoint	Rates of ADHF and Main Results
SOLOIST-WHF trial [32]	1222	2021	Patients with type 2 diabetes mellitus who were recently hospitalized for worsening HF	To receive sotagliflozin or placebo	The total number of deaths from CV causes and hospitalizations and urgent visits for HF within a median of 9 months	The rate (the number of events per 100 patient-years) of primary endpoint events was lower in the sotagliflozin group (51.0 in the sotagliflozin group vs. 76.3 in the placebo group (HR: 0.67; 95% CI: 0.52 to 0.85; $p < 0.001$))
IRONMAN trial [64]	1137	2022	Patients with HF (EF $\leq$ 45%) and transferrin saturation less than 20% or serum ferritin less than 100 μg/L	To intravenous ferric derisomaltose or usual care	Recurrent hospital admissions for HF and CV death within a median follow-up of 2.7 years	22.4% of patients in the ferric derisomaltose group and 27.5% of patients in the usual care group (RR: 0.82 [95% CI 0.66 to 1.02]; $p = 0.070$)
DIAMOND trial [66]	1642	2022	Patients with HF and reduced EF and current or a history of RAASi-related hyperkalemia	To receive Patiromer or placebo	Mean change in serum potassium from baseline over a median follow-up of 27 weeks	(1) +0.03 mmol/l in the patiromer group and +0.13 mmol/l in the placebo group [−0.10 mmol/l (95% CI −0.13, 0.07); $p < 0.001$] (2) Total HF hospitalizations were 17 out of 439 in the patiromer group vs. 20 out of 439 in the placebo group (HR, 95% CI 0.79 (0.36, 1.71) $p = 0.544$)
TRANSFORM-HF trial [65]	2859	2023	Patients hospitalized with HF (regardless of EF)	To receive torsemide or furosemide	All-cause mortality over a median follow-up of 17.4 months	(1) 26.1% of patients in the torsemide group and 26.2% of patients in the furosemide group (HR, 1.02 [95% CI, 0.89–1.18]) (2) All-cause mortality or all-cause hospitalization occurred in 47.3% of patients in the torsemide group and in 49.3% in the furosemide group (HR, 0.92 [95% CI, 0.83–1.02]) (3) 940 total hospitalizations among 536 participants in the torsemide group and 987 total hospitalizations among 577 participants in the furosemide group (RR, 0.94 [95% CI, 0.84–1.07])
TRILUMINATE trial [74]	350	2023	Patients with severe tricuspid regurgitation, symptomatic, in NYHA class II, III, or IV, a pulmonary artery systolic pressure of less than 70 mm Hg, and in stable ($\geq$30 days) guideline-directed medical therapy for HF	To receive either TEER or medical therapy (control)	Hierarchical composite of death from any cause or tricuspid valve surgery, hospitalization for HF, and an improvement in quality of life as measured with KCCQ at the 1-year follow-up	11,348 wins for the TEER group, 7643 wins for the control group, and 11,634 ties between the groups (win ratio: 1.48 (95% CI, 1.06 to 2.13; $p = 0.02$). The annualized rate of hospitalization for HF was 0.21 events per patient-year in the TEER group vs. 0.17 events per patient-year in the control group
RESHAPE HF2 trial [68]	505	2024	Patients with HF and moderate-to-severe functional mitral regurgitation	To either transcatheter mitral valve repair and guideline-recommended medical therapy (device group) or medical therapy alone (control group)	(1) The rate of the composite of first or recurrent hospitalizations for HF or CV death during 24 months (2) The rate of first or recurrent hospitalizations for HF during 24 months (3) The change from baseline to 12 months in the score on the KCCQ-OS	(1) 37 events per 100 patient-years in the device group and 58.9 events per 100 patient-years in the control group (RR, 0.64; 95% CI, 0.48 to 0.85; $p = 0.002$) (2) 26.9 events per 100 patient-years in the device group and 46.6 events per 100 patient-years in the control group (RR, 0.59; 95% CI, 0.42 to 0.82; $p = 0.002$) (3) The KCCQ-OS score increased by a mean (±SD) of 21.6 ± 26.9 points in the device group and 8.0 ± 24.5 points in the control group (mean difference, 10.9 points; 95% CI, 6.8 to 15.0; $p < 0.001$)

Table 3. *Cont.*

	Patients (*n*)	Year of Publication	Population Characteristics	Randomized Arms	Primary Endpoint	Rates of ADHF and Main Results
MATTERHORN trial [69]	210	2024	Patients with HF and secondary mitral regurgitation who have symptoms despite guideline-directed medical therapy	To undergo either transcatheter edge-to-edge repair (intervention group) or surgical mitral valve repair or replacement (surgery group)	A composite of death, hospitalization for HF, mitral valve reintervention, implantation of an assist device, or stroke within 1 year after the procedure	(1) 16.7% of patients in the intervention group vs. 22.5% in the surgery group (estimated mean difference, −6 percentage points; 95% CI, −17 to 6; $p < 0.001$ for noninferiority) (2) Rehospitalization because of congestive HF: 1.0% in the interventional group vs. 3.5% in the surgery group (difference (95% CI) −3 (−9 to 3)

ADHF: acutely decompensated heart failure; CAD: coronary artery disease; CI: confidence interval; CV: cardiovascular; EF: ejection fraction; HF: heart failure; HR: hazard ratio; MI: myocardial infarction; KCCQ-OS: Kansas City Cardiomyopathy Questionnaire–Overall Summary; NYHA: New York Heart Association functional class; RAASi: renin–angiotensin–aldosterone system inhibitor; RR: rate ratio; TEER: tricuspid transcatheter edge-to-edge repair.

3. AHF and Non-Ischemic Cardiomyopathies

CMPs are a heterogeneous group of myocardial disorders in which the heart muscle is structurally and functionally abnormal in the absence of CAD, hypertension, valvular disease, and congenital heart disease (CHD), leading to the observed myocardial disease [75]. The current European guidelines on CMPs identify five main clinical phenotypes: DCM, HCM, RCM, NDLVC, and ARVC [75]. CMPs may be inherited (genetic/familial) or acquired and represent one of the main causes of HF in young populations [76–78]. However, specific data on patients with CMPs that evolve into AHF are limited. As mentioned above, HF studies often omit the characterization of HF etiology, and when reported, a generic distinction between ischemic and non-ischemic groups is made. A study based on data from the ESC Heart Failure registries was designed to evaluate the prevalence, clinical characteristics, management, and outcomes of patients with two main etiologies of HF (ischemic and non-ischemic CMPs) and showed that ischemic CMP patients were older and had more comorbidities, while non-ischemic patients had worse systolic EFs. There were no other differences in terms of using HF guideline-recommended medications, implantable cardioverter defibrillators (ICDs), or cardiac resynchronization therapy. Furthermore, all-cause death, as well as all-cause death and readmissions for HF at 12 months, occurred more frequently in the ischemic group compared with the non-ischemic group (15.9% vs. 10%, $p = 0.016$; and 40.9% vs. 28.6%, $p = 0.00089$, respectively) [79]. Cardiac magnetic resonance (CMR) is crucial in these patients. A study of 168 patients assessed the role of late gadolinium enhancement (LGE) between ischemic and non-ischemic CMP patients and demonstrated that in ischemic patients, LGE extent (≥median) and inferior wall LGE independently predicted the primary endpoint (ventricular tachycardia (VT)). In non-ischemic patients, LGE extent (≥median, number of LGE segments, LGE stratified per 5% increase) and mid-wall LGE were independent predictors of the primary endpoint. Therefore, LGE extent, location, and pattern characteristics were greater predictors of worse outcomes in patients with non-ischemic rather than ischemic CMP [80].

3.1. Dilated Cardiomyopathy

DCM is defined by the presence of LV dilatation and systolic dysfunction unexplained solely by abnormal loading conditions or CAD [75]. The estimated prevalence is 0.036–0.4% [81,82]. DCM may be inherited in 30–50% of cases. In another 40% of cases, it is caused by genetic mutations in sarcomere proteins, the cytoskeleton, the nuclear envelope, sarcolemma, ion channels, and/or intercellular junction molecules or acquired due to infection, exposure to toxic agents, endocrine or metabolic pathologies, or peripartum conditions [75,83,84]. As regards the genetic form, most of the mutations have an autosomal dominant inheritance transmission. A truncating titin mutation is the most frequent mutation (13% and 25% of non-familial and familial cases of DCM, respectively [85]) and desmoplakin (DSP), filamin C (FLCN), lamin A/C (LMNA), phospholamban (PLN), and RNA binding motif protein 20 (RMB20) mutations are strong predictors of sudden cardiac death (SCD) [75].

3.1.1. AHF in DCM

In patients with DCM, advanced HF represents the most frequent cause of death, while SCD affects only 30% of patients [86,87]. Determining the incidence and prevalence of AHF in DCM is challenging because HF etiology is under-reported in clinical trials and observational studies, as mentioned above. When the etiology of HF is reported, it is mainly described as ischemic or non-ischemic, but what the latter group includes often remains unknown. In the ESC-HF-LT registry, a prospective observational study that collected hospitalization and 1-year follow-up data from 6629 AHF patients, 13.6%

had DCM, and among these, the most common presentation of AHF was CS or ADHF (18.4% vs. 18.5%, respectively) [3]. In a retrospective observational study on patients admitted for AHF to intensive care units (ICUs), in which the aim was to investigate the prevalence of acute kidney injury (AKI), many patients were affected by idiopathic DCM (47%) [88]. In a multicenter randomized trial that compared the use of dopamine and norepinephrine in the treatment of CS, out of a total of 1679 patients, only 44 patients had DCM [54], while in an open-label randomized trial of 85 patients admitted to the ICU for CS, eight had DCM [89]. However, none of these studies report the outcome of patients affected by AHF due to DCM. A meta-analysis of three trials (DOSE [diuretic strategies in patients with acute decompensated heart failure] [42], ROSE [acute heart failure randomized trial] [44], and CARRESS-HF [ultrafiltration in decompensated heart failure with cardiorenal syndrome [43]) evaluated the progression to AHF in ischemic vs. non-ischemic patients. Among 762 patients, 43.1% had a history of non-ischemic CMP and these patients had a similar course of decongestion, self-assessment of well-being, and dyspnea, with no significant differences in 60-day composite all-cause mortality or hospitalization for HF compared to ischemic patients. However, non-ischemic patients with LVEFs < 40% scored more highly regarding their global sense of well-being (global visual analog scale; +25.5 vs. +19.1, p = 0.023) and improvement in serum creatinine (0.031 mg/100 mL vs. +0.042 mg/100 mL, p = 0.009) at 72 h (Table 1) [90]. Different outcomes were found by Cherbi et al. in the prospective FRENSHOCK registry, which includes 772 CS patients and analyzes the 1-year outcomes (rehospitalizations, mortality, heart transplants (HTs), left ventricular assist devices (LVADs)) of idiopathic DCM patients (10.1%). Among patients with CS, idiopathic DCM was a very common scenario, with higher rates of 1-year death or CV rehospitalizations (adjusted odds ratio of 4.77 [95% CI 1.13 to 20.1], p = 0.03) and higher rates of HT or LVAD for patients aged < 65 years (adjusted odds ratio of 2.68 [1.21 to 5.91], p = 0.02) [91]. Furthermore, in the same registry, the authors evaluated the prevalence of potential ventricular arrhythmias (VA) as a trigger of CS, demonstrating that in the VA-triggered CS group (12.2%), there was no 1-year mortality difference between ischemic and non-ischemic DCMs (42.5 vs. 42.6%, HR 0.97 (0.52–1.81), p = 0.92). However, idiopathic DCMs led to higher rates of HT and LVAD implantation (25.9 vs. 5%, p = 0.02) [92]. Meanwhile, Miric et al. demonstrated that, among patients hospitalized for AHF, patients with CMD had more severe RV dysfunction, measured by RV free wall strain, compared to ischemic patients with the same functional class and outcome [93]. DCM patients with advanced HF may have three different clinical evolutions: (1) structural and functional recovery following an incident of HF, (2) remission of HF symptoms and stabilization of the LVEF, or (3) progression to HT (Figure 1) [94]. In recent years, outcomes have been improved by advances in HF treatment. In a cohort of 373 DCM patients, after guideline-directed medical therapy (GDMT), transplant-free survival at 1, 2, and 4 years of follow-up was 94%, 92%, and 88%, respectively, and survival free from HF hospitalization was 88%, 82%, and 78%, respectively [95]. Importantly, the continuation of HF treatment is crucial in patients who have recovered from DCM; the TRED-HF (withdrawal of pharmacological treatment for heart failure in patients recovered from dilated cardiomyopathy) trial showed that the withdrawal of GDMT for HF was associated with a 40% relapse of LV dysfunction within 6 months [96]. Furthermore, non-ischemic HF etiology has been reported as an independent predictor of LV functional recovery in a cohort of 3994 HF patients [97]. On the contrary, male sex, age > 60 years, Black race, lower baseline LVEF, higher New York Heart Association functional classification (NYHA) class, significant mitral regurgitation, and higher natriuretic peptide levels have been associated with adverse outcomes in non-ischemic DCM patients [86,95,98]. Furthermore, extended mid-wall myocardial LGE on CMR in DCM patients is associated with higher HF mortality and hospitalizations [87]. CMR is also important for recognizing "hot phases" of CMPs (also in HCM, ARCV, and NDLVC), characterized by worsening clinical status, including evolution into AHF or life-threatening arrhythmias. Indeed, in these patients, the presence of specific abnormalities on CMR, such as myocardial edema and an LGE with a ring-like pattern (involving mostly LV myocardial segments), has been associated with a worse prognosis [99].

3.1.2. Diagnosis of DCM

The essential elements of the diagnostic work-up for DCM patients include clinical and family history, laboratory tests, ECGs, Holter monitoring, cardiac imaging, and genetic testing. Echocardiography is the cornerstone for DCM diagnosis, as it provides an overall evaluation of LV anatomy and function, associated valvular disease, atrial and RV function, and pulmonary pressure [100]. Nowadays, CMR has even greater value for two main reasons: (1) the distribution and extent of LGE has prognostic value for predicting arrhythmic risk and the severity of HF [101,102], and (2) it provides additional information on tissue characterization, generating etiological hypotheses (e.g., the presence of myocardial edema may suggest myocarditis or an inflammatory cause, while LGE distribution may exclude MI or generate etiological hypotheses such as subepicardial distribution in post-myocarditis forms, patchy in sarcoidosis, inferolateral in dystrophinopathies, septal mid-wall in LMNA carriers, ring-like in DSP and FLNC-truncating variant carriers, and iron deposition in hemochromatosis) [100,103].

3.1.3. Treatment of DCM

In patients with HF and DCM, treatment should be guided by HF guidelines, while in AHF, in-hospital treatment with intravenous diuretics, vasodilators, or inotropes may be required, although there is no evidence of improved outcomes with these drugs [1]. Primary prevention ICD implantation is currently recommended in DCM patients with HF (NYHA class II–III) and an LVEF $\leq$ 35% on GDMT [1,75]. However, the DANISH (Danish study to assess the efficacy of ICDs in patients with non-ischemic systolic heart failure on mortality) trial failed to demonstrate a mortality benefit of primary prevention ICD implantation in patients with non-ischemic DCM and an EF < 35%, suggesting that more detailed and personalized stratification (accounting for etiology, LGE distribution, or high-risk genes such as DSP, FLCN, LMNA, PLN, and RMB) may be necessary in order to identify a subgroup of patients at a high risk of SCD who may benefit from primary prevention ICD implantation regardless of their LVEF [104–106]. Current knowledge facilitates therapies that can ameliorate the clinical consequences of causative gene mutations in some patients with DCM. For example, LMNA mutations are known to predispose individuals to VAs and SCD, regardless of LV function [107], while patients with PLN and RBM20 mutations should allow for more intensive follow-up and the timely provision of advanced HF therapies [75]. Accordingly, Paldino et al. demonstrated that genotypic-based classification showed higher precision in predicting the outcome of patients with CMP than phenotype-based classification. In particular, genotype-based classification, but not phenotype-based classification, was predictive of SCD/VAs, and LMNA showed the worst trends in terms of death/HT or death for HF/HT/LVAD [108]. In the future, genetic mutations may be targets for innovative therapeutic interventions.

3.1.4. MCS and HT in DCM

Recently, the proportion of DCM patients with AHF requiring HT increased compared with other HF etiologies, and DCM is the third most common indication for heart and lung transplantation in adults [109–112]. In young and middle-aged adults, 64% and 51% of HT cases, respectively, are attributed to DCM [111], while, among older adults,

DCM remains the second most common indication for HT after CAD. Similarly, >40% of DCM patients receive MCS, either as a bridge to HT or as a destination therapy [113,114]. Regarding the prognosis, favorable outcomes (with a median survival of 12.2 years) have been demonstrated after HT in patients with DCM [109–111]. Few data are available for DCM patients with AHF who are receiving LVAD. In a contemporary analysis of 24,809 adults listed for HT, 2.7% of genetic DCM patients were less frequently delisted for clinical deterioration or death and more likely to be transplanted compared with those with non-ischemic DCM [HR: 0.617, 95% CI: 0.47–0.81; HR: 1.25, 95%CI: 1.14–1.37, respectively], perhaps because they were younger and without CAD and related organ dysfunction [115,116].

3.1.5. Prognosis of DCM

Despite the advances made in pharmacological and interventional management, AHF remains the most frequent cause of death in DCM, overtaking mortality in other non-ischemic HF etiologies [86,87]. Furthermore, non-cardiac mortality should not be neglected in these patients, as approximately one-third of patients die of cancer, infections, pulmonary disease, or hemorrhage [86].

3.2. *Hypertrophic Cardiomyopathy*

HCM is defined by the presence of increased LV wall thickness ($\geq$15 mm in adults and $\geq$13 mm in adults with first-degree relatives with HCM) or mass that is not solely explained by abnormal loading conditions. Approximately 40–70% of patients have obstructive HCM, diagnosed by an LV outflow tract (LVOT) gradient $\geq$30 mmHg at rest or during exercise [75,117–119]. HCM is the most common inherited heart disease among young adults, with a prevalence of 0.2% and 0.029% in adults and children, respectively, and is caused by genetic mutations affecting sarcomere genes in more than half of cases. These mutations are inherited in an autosomal dominant pattern [75]. Between 5 and 10% of cases of HCM may be caused by other etiologies: hereditary syndromes, neuromuscular disorders, glycogen and lysosomal storage diseases (e.g., Fabry, Pompe, or Danon disease), or mitochondrial disorders (the so-called phenocopies). In the remaining 25–30% of cases, the etiology is unknown.

3.2.1. AHF in HCM

Patients with advanced HCM may present with two clinical phenotypes:

(1) The hyperkinetic-restrictive form is characterized by a small, stiff, and hypertrophic LV with severe diastolic dysfunction and a normal ejection fraction. In this case, HF presents as heart failure with a preserved ejection fraction (HfpEF) phenotype [120].
(2) The hypokinetic-dilated form is characterized by an increase in volume and a reduction in LV wall thickness. This clinical presentation may be similar to DCM and the differential diagnosis may be challenging in the absence of prior documentation of asymmetrical LV hypertrophy or a family history. Patients with this phenotype, characterized by fibrotic replacement, cardiomyocyte necrosis, and microvascular ischemia, develop AHF more frequently [121]. Furthermore, patients with mutations in thin-filament genes (TNNT2, TNNI3, TPM1, and ACTC) or multiple gene mutations in sarcomeric proteins are more susceptible to progression towards AHF compared with other mutations [122].

The prevalence of HF In patients with HCM is around 67%, and it is more prevalent in patients with obstructive HCM compared with non-obstructive HCM (only 10%) [123]. This is because pressure overload caused by dynamic LVOT obstruction during systole exacerbates the severity of HF [124]. However, the progression to AHF is infrequent and occurs only in 3% to 15% of HCM patients with an incidence ranging from 0.5% to 1.5% per year due to severe LV obstruction and hypertrophy or severe LV systolic dysfunction [121,125]. However, these data may be underestimated since patients with HCM are under-represented in AHF studies.

3.2.2. Diagnosis of HCM

Multimodality imaging provides a comprehensive characterization for HCM diagnosis. The evaluation of LV wall thickness, LVEF estimation, the definition of LV apical aneurysms, LVOT obstruction, and the extent of LGE is essential for defining the diagnosis and the arrhythmic risk of HCM, as well as for predicting the progression to AHF. Echocardiography is essential for diagnosing the hypokinetic-dilated phenotype, reported in patients with end-stage HCM, characterized by regression of hypertrophy, LV spherical remodeling, and volume increase. The diagnostic work-up continues with a systematic approach that includes echocardiography during a Valsalva maneuver in the sitting, semi-supine, or standing positions to identify LVOT obstruction. Indeed, a gradient $\geq$30 mmHg at rest independently predicts HF progression and mortality [126], with progression to advanced HF in 20% of patients with severe LVOT obstruction [127]. Furthermore, exercise stress echocardiography is recommended in symptomatic patients without LVOT obstruction at rest [128], since induced exercise LVOT obstruction is associated with an increased risk of symptom onset at follow-up (3.2% vs. 1.6% per year for non-obstructive patients) [129]. CMR is recommended in all patients with HCM. The evaluation of LGE and the identification of edema help to define both arrhythmic risk and progression to AHF. Indeed, the extension of LGE is associated with a higher mortality for refractory HF [130]. In a histological study of explanted hearts with end-stage HCM, fibrosis involved more than one-third of the LV, particularly the LV apex and the mid-wall [131].

3.2.3. Treatment of HCM

The management of LVOT obstruction is based on drug therapy with beta blockers, calcium channel blockers, disopyramide, and a cardiac myosin ATPase inhibitor (mavacamten) or, if refractory, on invasive surgical or percutaneous septal reduction therapies. Regarding end-stage HCM resulting in severe HF, there are no strong therapeutic recommendations. Current guidelines recommend the same management for HF regardless of the etiology [1]. However, these patients may experience numerous adverse effects of drugs: vasodilators can cause hypotension, while diuretics and negative inotropic drugs used against LVOT obstruction could be poorly tolerated. Furthermore, the current guidelines confirm the indication of primary prevention ICD implantation in patients with a 5-year risk of SCD $\geq$6% (Class IIa Level B) and with a 5-year risk of SCD $\geq$4% and <6% (Class IIb Level B). It also advises the consideration of primary prevention ICD implantation in the presence of LGE $\geq$ 15% or with LVEF <50%, regardless of the risk score (Class IIb Level B) [75].

3.2.4. MCS and HT in HCM

HCM patients who progress to AHF are a difficult population to treat with LVAD therapy, since the small LV cavity and severely impaired diastolic filling pressure may cause inflow cannula obstruction or a risk of ventricular and atrial suction events and lead to low pump flows, VA, and pump thrombosis more frequently. However, a recent analysis evaluated outcomes of RCM and HCM after LVAD, demonstrating that in selected patients with non-obstructive HCM and larger LV cavities, overall survival was similar to DCM patients, with a 1-year survival of 50% on pumps [132]. Since the progression of the disease is rapid from the onset of symptoms to the development of AHF, early inclusion on an HT list should be considered, especially in young patients and patients with a family history, an EF <50%, a severe restrictive diastolic pattern, and a high distribution of LGE [125].

3.2.5. Prognosis of HCM

Patients with end-stage HCM have a very poor prognosis with greater mortality (11% per year vs. 1% per year) compared with the overall HCM population [125]. However, as recently reported, data on survival after HT are reassuring, with 1-, 5-, and 10-year survival rates after HF of 89%, 79%, and 66%, respectively [133].

3.3. Restrictive Cardiomyopathy

RCMs are a rare and heterogeneous group of diseases characterized by restrictive left and/or right ventricle pathophysiology in the presence of normal or reduced diastolic and systolic volumes (of one or both ventricles) and normal ventricular wall thickness [75]. The spectrum of RCM includes idiopathic or genetic forms, endomyocardial disorders such as endomyocardial fibrosis and Loffler endocarditis, and myocardial extracellular matrix disorders such as hyperoxaluria and radiation exposition. In addition, myocardial diseases, often in the context of LV hypertrophy, such as amyloidosis, sarcoidosis, or hemochromatosis, can progress to occasional restrictive physiology [75]. Around 30% of patients with RCM have mutations in sarcomeric, cytoskeletal, nuclear envelope, filamin, titin, or desmin genes, which have autosomal dominant patterns of transmission and incomplete penetrance [134]. Rapezzi et al. proposed a classification for RCM according to myocardial histology, genetic basis, and the transient or permanent nature of restriction. They only identify apolipoprotein A (APOA), variant amyloid transthyretin (ATTRv) and AL amyloidosis, genetic RCM, radiation-mediated RCM, tropical and non-tropical endomyocardial fibrosis, and hypereosinophilic syndrome (HES) as "true RCM" [135].

3.3.1. AHF in RCM

RCM is associated with the worst prognosis of all the CMPs. The prognosis of RCM depends on the restrictive physiology, regardless of etiology [136,137]. Indeed, the increased stiffness of the myocardium and the rising filling pressure, without a compensatory increase in volume, may lead to pulmonary and systemic congestion and, in the end, systolic LV dysfunction. The prevalence of HF in RCM patients is high. In a large European registry of CMPs, HF is reported in 83% of RCM patients (present in 41%, 40%, and 1.6% of NYHA class II, III, and IV patients, respectively) [123]. In a cohort of 97 patients with genetic RCM, 81% developed HF, 53% demonstrated NYHA class II symptoms, and 28% demonstrated NYHA class III–IV symptoms [137]. Among children with RCM, more than 75% demonstrate advanced HF and poor prognoses (death or HT within a few years after diagnosis) [136,138]. The clinical features associated with an increased risk of death or transplantation include HF symptoms, a reduced EF, increased left atrial size, and impaired LV diastolic function on echocardiography [138–140]. However, available data on AHF in patients with RCM are limited and not fully defined. Only the LeoDOR trial reports a percentage of 2.2% RCM patients in the levosimendan group, but no specific data are available for their outcomes [36]. Considering recent trials on amyloid CMP, tafamidis and, more recently, vutrisiran reduced mortality and HF rehospitalizations. In a multicenter randomized trial of 441 patients with transthyretin amyloid CMP (either variant or wild-type ATTR amyloidosis, of which 76% were wild-type), tafamidis, which binds to transthyretin and prevents tetramer dissociation and amyloidogenesis, was associated with lower all-cause mortality than the placebo (78 of 264 [29.5%] vs. 76 of 177 [42.9%]; HR: 0.70; 95% CI 0.51 to 0.96), as well as a lower rate of CV-related hospitalizations, with a relative risk ratio of 0.68 (0.48 per year vs. 0.70 per year; 95% CI 0.56 to 0.81). Interestingly, heart and liver transplantation and LVAD use were treated as death for the purposes of this analysis. Rehospitalization rates due to CV-related causes were lower with tafamidis but were frequent in both groups (52.3% vs. 60.5%) [141]. Recently, in 665 patients with ATTR-CM (either variant or wild-type ATTR amyloidosis, 89% were wild-type), vutrisiran, a subcutaneously administered RNA interference therapeutic agent that inhibits hepatic synthesis of both wild-type and variant TTR messenger RNA, led to a lower risk of all-cause death and recurrent CV events (defined as hospitalizations for CV causes or urgent visits for HF) compared with the placebo (HR in the overall population, 0.72; 95% CI 0.56 to 0.93; $p = 0.01$). Yet, recurrent hospitalizations for CV-related causes or urgent visits for HF were 34% vs. 41% (HR in the overall population, 0.73; 95% CI 0.61 to 0.88; $p = 0.001$) [142].

3.3.2. Diagnosis of RCM

The diagnosis algorithm should include clinical examinations, an ECG, and cardiac imaging [143]. The most frequent ECG abnormality is AF, particularly in end-stage RCM (>50%) [137]. Echocardiography is essential in advanced stages of RCM demonstrating increased left atrial volume and restrictive LV filling with elevated ventricular filling pressures. CMR is essential for the tissue characterization of the disease, while an endomyocardial biopsy may be indicated in complex cases such as endomyocardial disorders. Right heart catheterization (RHC) may help in differentiating the alternative diagnosis of constrictive pericarditis [75].

3.3.3. Treatment of RCM

The prognosis remains poor despite medical therapy. Due to the restrictive physiology in RCM, standard heart failure treatment requires modification. In patients with pronounced restrictive filling, heart-rate-lowering medications such as beta blockers should be avoided or used with extreme caution due to the risk of hemodynamic deterioration. Similarly, since RCM patients often have low-to-normal blood pressure, drugs acting on the renin–angiotensin–aldosterone system are frequently poorly tolerated due to hypotension. While loop diuretics remain the primary therapy for relieving congestion, careful monitoring is essential to avoid overdiuresis, since even mild hypovolemia can provoke hypotension and hypoperfusion [144].

3.3.4. MCS and HT in RCM

RCM patients are frequently excluded from LVAD implantation due to small cavities and restrictive physiologies that impair pump function [144]. However, in a small cohort of RCM patients treated with LVAD, improved survival was reported, especially among patients with larger LV dimensions [145]. In another cohort of patients, 60% of RCM patients survived at least 1 year after LVAD implantation [132]. HT may be considered for symptomatic RCM patients without pulmonary hypertension or in patients with end-stage HF [144]. During the planning of an HT, an etiological diagnosis is essential in order to identify forms that benefit from specific therapies [75]. In a cohort of 544 RCM patients, the prevalence of HT was 1.4%, with reduced survival after the HT (1-, 5-, and 10-year overall survival after HT of 84%, 66%, and 45%, respectively) [146].

3.3.5. Prognosis of RCM

RCM is associated with the worst prognosis among CMPs. A total of 50% of patients die within two years from diagnosis and the prognosis is worse with increasing NYHA class, irrespective of other characteristics [134].

3.4. Non-Dilated Left Ventricular Cardiomyopathy

The 2023 ESC Guidelines for the Management of CMPs have introduced a new CMP subtype called NDLVC. This phenotype is characterized by the presence of non-ischemic

LV scarring or fatty replacement in the absence of LV dilatation, with or without global or regional wall motion abnormalities, or isolated global LV hypokinesia without scarring (as assessed by the presence of LGE on CMR) that is unexplained solely by abnormal loading conditions (hypertension, valve disease, or CAD) [75].

3.4.1. AHF in NDLVC

There are no data evaluating the progression to advanced HF in patients with NDLVC. NDLVC patients are a heterogeneous group in which it is necessary to provide thorough characterization and assess clinical outcomes. Indeed, in these patients, who are not dilated and have a normal or less reduced EF, the evolution into AHF is not as concerning as the arrhythmic risk. In a survival analysis of patients with or without dysfunction and with or without LGE, patients with dysfunction (EF < 50%) and LGE showed a lower survival rate than patients without dysfunction and with or without LGE [147]. Castrichini et al. first performed a clinical characterization of patients with NDLVC, demonstrating that septal LGE, together with LV dilatation, age, advanced disease, and frequent and repetitive VAs, was a predictor of major arrhythmic events [148].

3.4.2. Diagnosis of NDLVC

Diagnosis is based on a multimodality imaging approach and includes echocardiography to provide all relevant information on global and regional LV anatomy, function, and hemodynamics, as well as CMR, which is the foremost imaging modality in NDLVC patients as it provides confirmation of the presence of LGE and its localization. Numerous studies have demonstrated the association between LGE and a worse prognosis, emphasizing that a greater extension of the LGE pattern or its distribution at the level of the anterior septum is associated with a worse prognosis regardless of the LVEF [106,149,150]. In line with previous studies, Castrichini et al. demonstrated that the presence of LGE is associated with a high risk of major arrhythmic events and that septal LGE is an independent predictor of outcome, irrespective of the underlying genetic substrate [148]. A recent observational, longitudinal cohort study of 42 patients meeting the criteria for NDLVC demonstrated a heterogeneous etiology. Over one-third of cases had an underlying genetic etiology, nearly one-fifth had an inflammatory or neuromuscular condition, and the remaining cases lacked an identified underlying etiology after a comprehensive diagnostic work-up. A genetic cause was more frequent in patients with LGE at CMR compared with patients with isolated LV systolic dysfunction without fibrosis. Myocarditis-like onset was a common clinical presentation, with nearly one-third of cases uncovering a pathogenic (P)/likely pathogenic (LP) variant in DSP, confirming the emerging notion that myocarditis can uncover a genetic CMP. Finally, while most patients with genetic NDLVC fulfilled diagnostic criteria for arrhythmogenic left ventricular cardiomyopathy (ALVC) according to the Padua criteria, the diagnosis of ALVC was rarely observed in patients with an inflammatory or neuromuscular etiology [151]. Furthermore, the authors of the DERIVATE study, selecting patients with NDLVC criteria, demonstrated that patients with "hypokinetic" NDLVC (without LGE) had significantly lower rates of major adverse arrhythmic cardiac events (MAACE) than those with non-ischemic DCM ($p = 0.001$). However, patients with "fibrotic" NDLVC (with LGE) had similar rates of all-cause mortality ($p = 0.48$) and MAACE ($p = 0.616$) compared with non-ischemic DCM patients [152]. The ESC CMP guidelines' core message is that phenotype must guide specific disease definitions. These data support the concept that NDLVC might be considered an umbrella term encompassing patients with various etiologies and differing prognoses. The ESC guidelines emphasize that this phenotype is a crucial first step in the diagnostic pathway, not a diagnosis itself, and its identification should trigger an investigation into the underlying etiology [75].

3.4.3. Treatment, MCS, and HT in NDLVC

Validated criteria for the stratification of arrhythmic risk are lacking. However, as for DCM patients, clinical and family history, the extent and localization of LGE, and genetic mutations should be evaluated to consider primary prevention ICD implantation [75]. The clinical management of HF should be based on HF guidelines [1]. HT is the treatment of choice in patients with end-stage HF, whereas LVADs are valid alternatives in patients ineligible for HT or with AHF.

3.4.4. Prognosis of NDLVC

As reported previously, factors associated with poor outcomes are septal LGE, LV dilatation, age, advanced disease, and frequent and repetitive VAs [148]. The evolution into AHF in NDLVC patients is currently unknown.

3.5. *Arrhythmogenic Right Ventricular Cardiomyopathy*

ARVC is a heart muscle disease structurally characterized by progressive myocardial atrophy with fibro-fatty replacement of RV myocardium, causing RV dilatation and/or dysfunction and predisposing to fatal VAs [75]. Over time, the focus of ARVC has moved from a severe RV disease with malignant VAs to a broader concept that includes biventricular or even left-dominant disease [153]. ARVC has a prevalence of 0.078% and 60% of patients inherit gene mutations (mainly in desmosomal genes such as DSP, PKP2, DSG2, DSC2, and JUP) with an autosomal dominant pattern. Patients with multiple mutations have a worse prognosis [75,154].

3.5.1. AHF in ARVC

Three stages of the disease can be described:

(1) First stage: asymptomatic, but still at risk of SCD;
(2) Electrical phase: characterized by symptomatic monomorphic ventricular VT;
(3) Advanced phase: characterized by advanced HF. In this phase, VT and ventricular fibrillation due to fibrosis and reactive inflammation are possible and LV involvement is frequent. Notably, LV involvement in the final stage of the disease is an old concept, as emerging data suggest that it can be found in the early stages of the disease [155]. Only 3.9% of ARVC patients manifest advanced HF, while specific data regarding AHF are scarce and often derived from studies not exclusively focused on this condition [123].

3.5.2. Diagnosis of ARVC

The diagnostic criteria used for recognizing ARVC patients were published by Marcus et al. in 2010 and are still used [156]. The Padua criteria, instead, have offered an integration to include LV involvement [153]. The diagnostic work-up includes ECG (more frequently T-wave inversion in V1–V3, VT, and ventricular ectopic beats with typical left-bundle branch morphology), Holter monitoring, cardiac imaging, genetic testing, and, rarely, endomyocardial biopsy [75]. Echocardiography is essential for recognizing RV dilatation, akinetic–dyskinetic wall motion abnormalities, and myocardial aneurysms and biventricular dysfunction that characterize the evolution to AHF. However, CMR is considered the first-line test for the assessment of the RV's functional and structural abnormalities [156]. Moreover, subepicardial LGE is typical of ARVC, particularly in cases of DSP variants [156,157].

3.5.3. Treatment, MCS, and HT in ARVC

In ARVC patients who are developing AHF, the recommendations should not deviate from AHF guidelines [1]. ICD implantation is recommended in patients with a history of cardiac arrest or sustained VT and should be considered in patients with high-risk features (arrhythmic syncope, non-sustained VT, RVEF < 40%, LVEF < 45%, or sustained monomorphic VT) [75]. Only a small portion of patients undergo HT, while LVAD could be used as a bridge to HT. However, the involvement of the RV contraindicates the exclusive use of left-side MCS [158].

3.5.4. Prognosis of ARVC

A young age is the only independent predictor of worse outcomes. The main indication for HT is HF and the survival at 5 years is 91% [158].

4. Conclusions

AHF is a complex and life-threatening syndrome requiring urgent hospitalization and medical intervention. However, AHF is an extremely heterogeneous umbrella term. Indeed, the rapid onset of symptoms and signs of HF can be due to several causes and can present with different clinical scenarios, thus significantly influencing clinical management and prognosis (Figure 1). For example, the de novo presentation of AHF with pulmonary edema in a patient with a hypertensive emergency or tachi-bradyarrhythmia can often be easily managed and can potentially lead to a better prognosis compared to a patient with CS. Yet, even in the latter context, ACS as a leading cause can lead to a potentially better prognosis compared with a patient with end-stage/advanced HF, as the potential treatment options can solve the root cause behind CS. Within the AHF scenario, patients with specific CMPs are also included; however, CMPs are often under-represented compared to ischemic etiology and are mixed with other etiologies in the so-called "non-ischemic HF" group in many studies on the management of and therapies for chronic or acute HF. Therefore, future studies should be more detailed in exploring the etiology, clinical presentation, treatment, and outcomes of patients presenting with AHF, and specific data on the characteristics of AHF presentation, treatment, and outcomes of individual CMPs are warranted.

Author Contributions: Conceptualization, L.M., F.B. and G.G. (Giuseppe Gargiulo); methodology, L.M., F.B. and G.G. (Giuseppe Gargiulo); writing—original draft preparation, L.M., F.B. and G.G. (Giuseppe Gargiulo); writing—review and editing, L.M., F.B., V.N., I.F., D.F., D.S.C., L.S., R.P., A.S., S.C., M.E.C., M.A., N.V., G.G. (Giuseppe Giugliano), L.B. and G.G. (Giuseppe Gargiulo); visualization, L.M., V.N. and G.G. (Giuseppe Gargiulo); supervision, G.G. (Giuseppe Gargiulo); project administration, G.G. (Giuseppe Gargiulo). All authors have read and agreed to the published version of the manuscript.

Funding: This study was internally supported by the University of Naples Federico II and Azienda Ospedaliera Universitaria Federico II. No direct or indirect external funding was received.

Institutional Review Board Statement: Not applicable.

Informed Consent Statement: Not applicable.

Data Availability Statement: Not applicable.

Acknowledgments: Nicola Verde and Domenico Simone castiello are supported by a research grant provided by the Cardiopath PhD program. Lina Manzi is support by a research grant provided by the Medicina Clinica e Sperimentale PhD program.

Conflicts of Interest: The authors declare no conflicts of interest.

Abbreviations

ACS	acute coronary syndrome
ADHF	acutely decompensated heart failure
AF	atrial fibrillation
AHF	acute heart failure
ALVC	arrhythmogenic left ventricular cardiomyopathy
AKI	acute kidney injury
APOA	apolipoprotein A (APOA)
ARVC	arrhythmogenic right ventricular cardiomyopathy
ATTRv	variant amyloid transthyretin
CAD	coronary artery disease
CCS	chronic coronary syndrome
CHD	congenital heart disease
CMP	Cardiomyopathy
CMR	cardiac magnetic resonance
CO	cardiac output
COPD	chronic obstructive pulmonary disease
CS	cardiogenic shock
CVD	cardiovascular disease
DCM	dilated cardiomyopathy
DSP	desmoplakin
ESC	European Society of Cardiology
FLCN	filamin C
GDMT	guideline-directed medical therapy
eGFR	Glomerular Filtration Rate
HCM	hypertrophic cardiomyopathy
HES	hypereosinophilic syndrome
HFpEF	heart failure with preserved ejection fraction
HFrEF	Heart Failure with Reduced Ejection Fraction
HF	heart failure
HT	heart transplant
ICD	Implantable Cardioverter Defibrillator
ICU	intensive care unit
LGE	late gadolinium enhancement
LMNA	lamin A/C
LP	likely pathogenic
LV	left ventricular
LVAD	Left Ventricular Assist Device
LVEF	left ventricle ejection fraction
MAACE	Major adverse arrhythmic cardiac events
MCS	Mechanical circulatory support
MI	myocardial infarction
M-TEER	mitral transcatheter edge-to-edge repair
NDLVC	non-dilated left ventricular cardiomyopathy
NYHA	New York Heart Association Functional Classification
P	pathogenic
PE	pulmonary embolism
PLN	phospholamban
RCM	restrictive cardiomyopathy
RHC	Right heart catheterization
RMB	RNA binding motif protein
RV	right ventricular

SBP	systolic blood pressure
SCD	sudden cardiac death
VA	ventricular arrhythmia
VT	ventricular tachycardia

References

1. McDonagh, T.A.; Metra, M.; Adamo, M.; Baumbach, A.; Böhm, M.; Burri, H.; Čelutkiene, J.; Chioncel, O.; Cleland, J.G.F.; Coats, A.J.S.; et al. 2021 ESC Guidelines for the Diagnosis and Treatment of Acute and Chronic Heart Failure. *Eur. Heart J.* **2021**, *42*, 3599–3726. [CrossRef] [PubMed]
2. Beghini, A.; Sammartino, A.M.; Papp, Z.; von Haehling, S.; Biegus, J.; Ponikowski, P.; Adamo, M.; Falco, L.; Lombardi, C.M.; Pagnesi, M.; et al. 2024 Update in Heart Failure. *ESC Heart Fail.* **2025**, *12*, 8–42. [CrossRef] [PubMed]
3. Chioncel, O.; Mebazaa, A.; Harjola, V.P.; Coats, A.J.; Piepoli, M.F.; Crespo-Leiro, M.G.; Laroche, C.; Seferovic, P.M.; Anker, S.D.; Ferrari, R.; et al. Clinical Phenotypes and Outcome of Patients Hospitalized for Acute Heart Failure: The ESC Heart Failure Long-Term Registry. *Eur. J. Heart Fail.* **2017**, *19*, 1242–1254. [CrossRef] [PubMed]
4. O'Connor, C.M.; Stough, W.G.; Gallup, D.S.; Hasselblad, V.; Gheorghiade, M. Demographics, Clinical Characteristics, and Outcomes of Patients Hospitalized for Decompensated Heart Failure: Observations from the IMPACT-HF Registry. *J. Card. Fail.* **2005**, *11*, 200–205. [CrossRef]
5. Chioncel, O.; Mebazaa, A.; Maggioni, A.P.; Harjola, V.P.; Rosano, G.; Laroche, C.; Piepoli, M.F.; Crespo-Leiro, M.G.; Lainscak, M.; Ponikowski, P.; et al. Acute Heart Failure Congestion and Perfusion Status—Impact of the Clinical Classification on in-Hospital and Long-Term Outcomes; Insights from the ESC-EORP-HFA Heart Failure Long-Term Registry. *Eur. J. Heart Fail.* **2019**, *21*, 1338–1352. [CrossRef]
6. Marini, M.; Manfredi, R.; Battistoni, I.; Francioni, M.; Vittoria Matassini, M.; Pongetti, G.; Angelini, L.; Shkoza, M.; Bontempo, A.; Belfioretti, L.; et al. Acute Heart Failure: Differential Diagnosis and Treatment. *Eur. Heart J. Suppl.* **2023**, *25*, C276–C282. [CrossRef]
7. Harjola, V.P.; Mebazaa, A.; Čelutkiene, J.; Bettex, D.; Bueno, H.; Chioncel, O.; Crespo-Leiro, M.G.; Falk, V.; Filippatos, G.; Gibbs, S.; et al. Contemporary Management of Acute Right Ventricular Failure: A Statement from the Heart Failure Association and the Working Group on Pulmonary Circulation and Right Ventricular Function of the European Society of Cardiology. *Eur. J. Heart Fail.* **2016**, *18*, 226–241. [CrossRef]
8. Chioncel, O.; Parissis, J.; Mebazaa, A.; Thiele, H.; Desch, S.; Bauersachs, J.; Harjola, V.P.; Antohi, E.L.; Arrigo, M.; Gal, T.B.; et al. Epidemiology, Pathophysiology and Contemporary Management of Cardiogenic Shock—A Position Statement from the Heart Failure Association of the European Society of Cardiology. *Eur. J. Heart Fail.* **2020**, *22*, 1315–1341. [CrossRef]
9. Thiele, H.; Ohman, E.M.; De Waha-Thiele, S.; Zeymer, U.; Desch, S. Management of Cardiogenic Shock Complicating Myocardial Infarction: An Update 2019. *Eur. Heart. J.* **2019**, *40*, 2671–2683. [CrossRef]
10. Manzi, L.; Sperandeo, L.; Forzano, I.; Castiello, D.S.; Florimonte, D.; Paolillo, R.; Santoro, C.; Mancusi, C.; Di Serafino, L.; Esposito, G.; et al. Contemporary Evidence and Practice on Right Heart Catheterization in Patients with Acute or Chronic Heart Failure. *Diagnostics* **2024**, *14*, 136. [CrossRef]
11. Spinar, J.; Parenica, J.; Vitovec, J.; Widimsky, P.; Linhart, A.; Fedorco, M.; Malek, F.; Cihalik, C.; Spinarová, L.; Miklik, R.; et al. Baseline Characteristics and Hospital Mortality in the Acute Heart Failure Database (AHEAD) Main Registry. *Crit. Care* **2011**, *15*, R291. [CrossRef] [PubMed]
12. Kurmani, S.; Squire, I. Acute Heart Failure: Definition, Classification and Epidemiology. *Curr. Heart Fail. Rep.* **2017**, *14*, 385–392. [CrossRef] [PubMed]
13. Nieminen, M.S.; Brutsaert, D.; Dickstein, K.; Drexler, H.; Follath, F.; Harjola, V.P.; Hochadel, M.; Komajda, M.; Lassus, J.; Lopez-Sendon, J.L.; et al. EuroHeart Failure Survey II (EHFS II): A Survey on Hospitalized Acute Heart Failure Patients: Description of Population. *Eur. Heart J.* **2006**, *27*, 2725–2736. [CrossRef] [PubMed]
14. Agarwal, M.A.; Fonarow, G.C.; Ziaeian, B. National Trends in Heart Failure Hospitalizations and Readmissions From 2010 to 2017. *JAMA Cardiol.* **2021**, *6*, 952–956. [CrossRef]
15. Fonarow, G.C. The Acute Decompensated Failure National Registry(ADHERETM): Opportunities Improve Care of Patients With Acute Heart Failure. *Rev. Cardiovasc. Med.* **2003**, *4*, 21–30.
16. Abraham, W.T.; Fonarow, G.C.; Albert, N.M.; Stough, W.G.; Gheorghiade, M.; Greenberg, B.H.; O'Connor, C.M.; Sun, J.L.; Yancy, C.W.; Young, J.B. Predictors of In-Hospital Mortality in Patients Hospitalized for Heart Failure: Insights from the Organized Program to Initiate Lifesaving Treatment in Hospitalized Patients with Heart Failure (OPTIMIZE-HF). *J. Am. Coll. Cardiol.* **2008**, *52*, 347–356. [CrossRef]
17. Follath, F.; Yilmaz, M.B.; Delgado, J.F.; Parissis, J.T.; Porcher, R.; Gayat, E.; Burrows, N.; Mclean, A.; Vilas-Boas, F.; Mebazaa, A. Clinical Presentation, Management and Outcomes in the Acute Heart Failure Global Survey of Standard Treatment (ALARM-HF). *Intensive Care Med.* **2011**, *37*, 619–626. [CrossRef]

18. Musella, F.; Rosano, G.M.C.; Hage, C.; Benson, L.; Guidetti, F.; Moura, B.; Sibilio, G.; Boccalatte, M.; Dahlström, U.; Coats, A.J.S.; et al. Patient Profiles in Heart Failure with Reduced Ejection Fraction: Prevalence, Characteristics, Treatments and Outcomes in a Real-World Heart Failure Population. *Eur. J. Heart Fail.* **2023**, *25*, 1246–1253. [CrossRef]
19. Schrage, B.; Sundermeyer, J.; Beer, B.N.; Bertoldi, L.; Bernhardt, A.; Blankenberg, S.; Dauw, J.; Dindane, Z.; Eckner, D.; Eitel, I.; et al. Use of Mechanical Circulatory Support in Patients with Non-Ischaemic Cardiogenic Shock. *Eur. J. Heart Fail.* **2023**, *25*, 562–572. [CrossRef]
20. Mullens, W.; Dauw, J.; Martens, P.; Meekers, E.; Nijst, P.; Verbrugge, F.H.; Chenot, F.; Moubayed, S.; Dierckx, R.; Blouard, P.; et al. Acetazolamide in Decompensated Heart Failure with Volume Overload Trial (ADVOR): Baseline Characteristics. *Eur. J. Heart Fail.* **2022**, *24*, 1601–1610. [CrossRef]
21. Trullàs, J.C.; Morales-Rull, J.L.; Casado, J.; Carrera-Izquierdo, M.; Sánchez-Marteles, M.; Conde-Martel, A.; Dávila-Ramos, M.F.; Llácer, P.; Salamanca-Bautista, P.; Pérez-Silvestre, J.; et al. Combining Loop with Thiazide Diuretics for Decompensated Heart Failure: The CLOROTIC Trial. *Eur. Heart J.* **2023**, *44*, 411–421. [CrossRef] [PubMed]
22. Lee, D.S.; Straus, S.E.; Farkouh, M.E.; Austin, P.C.; Taljaard, M.; Chong, A.; Fahim, C.; Poon, S.; Cram, P.; Smith, S.; et al. Trial of an Intervention to Improve Acute Heart Failure Outcomes. *N. Engl. J. Med.* **2023**, *388*, 22–32. [CrossRef] [PubMed]
23. Voors, A.A.; Angermann, C.E.; Teerlink, J.R.; Collins, S.P.; Kosiborod, M.; Biegus, J.; Ferreira, J.P.; Nassif, M.E.; Psotka, M.A.; Tromp, J.; et al. The SGLT2 Inhibitor Empagliflozin in Patients Hospitalized for Acute Heart Failure: A Multinational Randomized Trial. *Nat. Med.* **2022**, *28*, 568–574. [CrossRef] [PubMed]
24. Ponikowski, P.; Kirwan, B.A.; Anker, S.D.; McDonagh, T.; Dorobantu, M.; Drozdz, J.; Fabien, V.; Filippatos, G.; Göhring, U.M.; Keren, A.; et al. Ferric Carboxymaltose for Iron Deficiency at Discharge after Acute Heart Failure: A Multicentre, Double-Blind, Randomised, Controlled Trial. *Lancet* **2020**, *396*, 1895–1904. [CrossRef]
25. Gheorghiade, M.; Böhm, M.; Greene, S.J.; Fonarow, G.C.; Lewis, E.F.; Zannad, F.; Solomon, S.D.; Baschiera, F.; Botha, J.; Hua, T.A.; et al. Effect of Aliskiren on Postdischarge Mortality and Heart Failure Readmissions among Patients Hospitalized for Heart Failure: The ASTRONAUT Randomized Trial. *JAMA* **2013**, *309*, 1125–1135. [CrossRef]
26. Mebazaa, A.; Davison, B.; Chioncel, O.; Cohen-Solal, A.; Diaz, R.; Filippatos, G.; Metra, M.; Ponikowski, P.; Sliwa, K.; Voors, A.A.; et al. Safety, Tolerability and Efficacy of up-Titration of Guideline-Directed Medical Therapies for Acute Heart Failure (STRONG-HF): A Multinational, Open-Label, Randomised, Trial. *Lancet* **2022**, *400*, 1938–1952. [CrossRef]
27. Velazquez, E.J.; Morrow, D.A.; DeVore, A.D.; Duffy, C.I.; Ambrosy, A.P.; McCague, K.; Rocha, R.; Braunwald, E. Angiotensin-Neprilysin Inhibition in Acute Decompensated Heart Failure. *N. Engl. J. Med.* **2019**, *380*, 539–548. [CrossRef]
28. Mebazaa, A.; Nieminen, M.S.; Packer, M.; Cohen-Solal, A.; Kleber, F.X.; Pocock, S.J.; Thakkar, R.; Padley, R.J.; Põder, P.; Kivikko, M. Levosimendan vs Dobutamine for Patients with Acute Decompensated Heart Failure: The SURVIVE Randomized Trial. *JAMA* **2007**, *297*, 1883–1891. [CrossRef]
29. Adamson, P.B.; Abraham, W.T.; Aaron, M.; Aranda, J.M.; Bourge, R.C.; Smith, A.; Stevenson, L.W.; Bauman, J.G.; Yadav, J.S. CHAMPION Trial Rationale and Design: The Long-Term Safety and Clinical Efficacy of a Wireless Pulmonary Artery Pressure Monitoring System. *J. Card. Fail.* **2011**, *17*, 3–10. [CrossRef]
30. Packer, M.; O'Connor, C.; McMurray, J.J.V.; Wittes, J.; Abraham, W.T.; Anker, S.D.; Dickstein, K.; Filippatos, G.; Holcomb, R.; Krum, H.; et al. Effect of Ularitide on Cardiovascular Mortality in Acute Heart Failure. *N. Engl. J. Med.* **2017**, *376*, 1956–1964. [CrossRef]
31. Sato, N.; Lam, C.S.P.; Teerlink, J.R.; Greenberg, B.H.; Tsutsui, H.; Oh, B.H.; Zhang, J.; Lefkowitz, M.; Hua, T.A.; Holbro, T.; et al. Evaluating the Efficacy, Safety, and Tolerability of Serelaxin When Added to Standard Therapy in Asian Patients With Acute Heart Failure: Design and Rationale of RELAX-AHF-ASIA Trial. *J. Card. Fail.* **2017**, *23*, 63–71. [CrossRef] [PubMed]
32. Bhatt, D.L.; Szarek, M.; Steg, P.G.; Cannon, C.P.; Leiter, L.A.; McGuire, D.K.; Lewis, J.B.; Riddle, M.C.; Voors, A.A.; Metra, M.; et al. Sotagliflozin in Patients with Diabetes and Recent Worsening Heart Failure. *N. Engl. J. Med.* **2021**, *384*, 117–128. [CrossRef] [PubMed]
33. Packer, M.; Colucci, W.; Fisher, L.; Massie, B.M.; Teerlink, J.R.; Young, J.; Padley, R.J.; Thakkar, R.; Delgado-Herrera, L.; Salon, J.; et al. Effect of Levosimendan on the Short-Term Clinical Course of Patients with Acutely Decompensated Heart Failure. *JACC Heart Fail.* **2013**, *1*, 103–111. [CrossRef] [PubMed]
34. Kozhuharov, N.; Goudev, A.; Flores, D.; Maeder, M.T.; Walter, J.; Shrestha, S.; Gualandro, D.M.; Junior, M.T.d.O.; Sabti, Z.; Müller, B.; et al. Effect of a Strategy of Comprehensive Vasodilation vs Usual Care on Mortality and Heart Failure Rehospitalization Among Patients With Acute Heart Failure: The GALACTIC Randomized Clinical Trial. *JAMA* **2019**, *322*, 2292–2302. [CrossRef]
35. Armstrong, P.W.; Pieske, B.; Anstrom, K.J.; Ezekowitz, J.; Hernandez, A.F.; Butler, J.; Lam, C.S.; Ponikowski, P.; Voors, A.A.; Jia, G.; et al. Vericiguat in Patients with Heart Failure and Reduced Ejection Fraction. *N. Engl. J. Med.* **2020**, *382*, 1883–1893. [CrossRef]

36. Pölzl, G.; Altenberger, J.; Comín-Colet, J.; Delgado, J.F.; Fedele, F.; García-González, M.J.; Gustafsson, F.; Masip, J.; Papp, Z.; Störk, S.; et al. Repetitive Levosimendan Infusions for Patients with Advanced Chronic Heart Failure in the Vulnerable Post-Discharge Period: The Multinational Randomized LeoDOR Trial. *Eur. J. Heart Fail.* **2023**, *25*, 2007–2017. [CrossRef]
37. ter Maaten, J.M.; Beldhuis, I.E.; van der Meer, P.; Krikken, J.A.; Postmus, D.; Coster, J.E.; Nieuwland, W.; van Veldhuisen, D.J.; Voors, A.A.; Damman, K. Natriuresis-Guided Diuretic Therapy in Acute Heart Failure: A Pragmatic Randomized Trial. *Nat. Med.* **2023**, *29*, 2625–2632. [CrossRef]
38. Tanaka, A.; Kida, K.; Matsue, Y.; Imai, T.; Suwa, S.; Taguchi, I.; Hisauchi, I.; Teragawa, H.; Yazaki, Y.; Moroi, M.; et al. In-Hospital Initiation of Angiotensin Receptor-Neprilysin Inhibition in Acute Heart Failure: The PREMIER Trial. *Eur. Heart J.* **2024**, *45*, 4482–4493. [CrossRef]
39. Costanzo, M.R.; Guglin, M.E.; Saltzberg, M.T.; Jessup, M.L.; Bart, B.A.; Teerlink, J.R.; Jaski, B.E.; Fang, J.C.; Feller, E.D.; Haas, G.J.; et al. Ultrafiltration versus Intravenous Diuretics for Patients Hospitalized for Acute Decompensated Heart Failure. *J. Am. Coll. Cardiol.* **2007**, *49*, 675–683. [CrossRef]
40. Marenzi, G.; Muratori, M.; Cosentino, E.R.; Rinaldi, E.R.; Donghi, V.; Milazzo, V.; Ferramosca, E.; Borghi, C.; Santoro, A.; Agostoni, P. Continuous Ultrafiltration for Congestive Heart Failure: The CUORE Trial. *J. Card. Fail.* **2014**, *20*, 9–17. [CrossRef]
41. Ostadal, P.; Rokyta, R.; Karasek, J.; Kruger, A.; Vondrakova, D.; Janotka, M.; Naar, J.; Smalcova, J.; Hubatova, M.; Hromadka, M.; et al. Extracorporeal Membrane Oxygenation in the Therapy of Cardiogenic Shock: Results of the ECMO-CS Randomized Clinical Trial. *Circulation* **2023**, *147*, 454–464. [CrossRef] [PubMed]
42. Felker, G.M.; Lee, K.L.; Bull, D.A.; Redfield, M.M.; Stevenson, L.W.; Goldsmith, S.R.; LeWinter, M.M.; Deswal, A.; Rouleau, J.L.; Ofili, E.O.; et al. Diuretic Strategies in Patients with Acute Decompensated Heart Failure. *N. Engl. J. Med.* **2011**, *364*, E1. [CrossRef] [PubMed]
43. Bart, B.A.; Goldsmith, S.R.; Lee, K.L.; Givertz, M.M.; O'Connor, C.M.; Bull, D.A.; Redfield, M.M.; Deswal, A.; Rouleau, J.L.; LeWinter, M.M.; et al. Ultrafiltration in Decompensated Heart Failure with Cardiorenal Syndrome. *N. Engl. J. Med.* **2012**, *367*, 2296–2304. [CrossRef] [PubMed]
44. Chen, H.H.; Anstrom, K.J.; Givertz, M.M.; Stevenson, L.W.; Semigran, M.J.; Goldsmith, S.R.; Bart, B.A.; Bull, D.A.; Stehlik, J.; LeWinter, M.M.; et al. Low-Dose Dopamine or Low-Dose Nesiritide in Acute Heart Failure with Renal Dysfunction: The ROSE Acute Heart Failure Randomized Trial. *JAMA* **2013**, *310*, 2533–2543. [CrossRef]
45. Costanzo, M.R.; Negoianu, D.; Jaski, B.E.; Bart, B.A.; Heywood, J.T.; Anand, I.S.; Smelser, J.M.; Kaneshige, A.M.; Chomsky, D.B.; Adler, E.D.; et al. Aquapheresis Versus Intravenous Diuretics and Hospitalizations for Heart Failure. *JACC Heart Fail.* **2016**, *4*, 95–105. [CrossRef]
46. Pang, P.S.; Butler, J.; Collins, S.P.; Cotter, G.; Davison, B.A.; Ezekowitz, J.A.; Filippatos, G.; Levy, P.D.; Metra, M.; Ponikowski, P.; et al. Biased Ligand of the Angiotensin II Type 1 Receptor in Patients with Acute Heart Failure: A Randomized, Double-Blind, Placebo-Controlled, Phase IIB, Dose Ranging Trial (BLAST-AHF). *Eur. Heart J.* **2017**, *38*, 2364–2373. [CrossRef]
47. Collins, S.P.; Liu, D.; Jenkins, C.A.; Storrow, A.B.; Levy, P.D.; Pang, P.S.; Chang, A.M.; Char, D.; Diercks, D.J.; Fermann, G.J.; et al. Effect of a Self-Care Intervention on 90-Day Outcomes in Patients With Acute Heart Failure Discharged From the Emergency Department: A Randomized Clinical Trial. *JAMA Cardiol.* **2021**, *6*, 200–208. [CrossRef]
48. Pang, P.S.; Berger, D.A.; Mahler, S.A.; Li, X.; Pressler, S.J.; Lane, K.A.; Bischof, J.J.; Char, D.; Diercks, D.; Jones, A.E.; et al. Short-Stay Units vs Routine Admission From the Emergency Department in Patients With Acute Heart Failure: The SSU-AHF Randomized Clinical Trial. *JAMA Netw. Open* **2024**, *7*, e2350511. [CrossRef]
49. Pascual-Figal, D.; Núñez Villota, J.; Pérez-Martínez, M.T.; González-Juanatey, J.R.; Taibo-Urquía, M.; Llàcer Iborra, P.; González-Martín, J.; Villar, S.; Soler, M.; Mirabet, S.; et al. Colchicine in Acute Heart Failure: Rationale and Design of a Randomized Double-Blind Placebo-Controlled Trial (COLICA). *Eur. J. Heart Fail.* **2024**, *26*, 1999–2007. [CrossRef]
50. Cox, Z.L.; Collins, S.P.; Hernandez, G.A.; McRae, A.T.; Davidson, B.T.; Adams, K.; Aaron, M.; Cunningham, L.; Jenkins, C.A.; Lindsell, C.J.; et al. Efficacy and Safety of Dapagliflozin in Patients With Acute Heart Failure. *J. Am. Coll. Cardiol.* **2024**, *83*, 1295–1306. [CrossRef]
51. Seyfarth, M.; Sibbing, D.; Bauer, I.; Fröhlich, G.; Bott-Flügel, L.; Byrne, R.; Dirschinger, J.; Kastrati, A.; Schömig, A. A Randomized Clinical Trial to Evaluate the Safety and Efficacy of a Percutaneous Left Ventricular Assist Device versus Intra-Aortic Balloon Pumping for Treatment of Cardiogenic Shock Caused by Myocardial Infarction. *J. Am. Coll. Cardiol.* **2008**, *52*, 1584–1588. [CrossRef] [PubMed]
52. Fuhrmann, J.T.; Schmeisser, A.; Schulze, M.R.; Wunderlich, C.; Schoen, S.P.; Rauwolf, T.; Weinbrenner, C.; Strasser, R.H. Levosimendan Is Superior to Enoximone in Refractory Cardiogenic Shock Complicating Acute Myocardial Infarction. *Crit. Care Med.* **2008**, *36*, 2257–2266. [CrossRef] [PubMed]
53. Prondzinsky, R.; Lemm, H.; Swyter, M.; Wegener, N.; Unverzagt, S.; Carter, J.M.; Russ, M.; Schlitt, A.; Buerke, U.; Christoph, A.; et al. Intra-Aortic Balloon Counterpulsation in Patients with Acute Myocardial Infarction Complicated by Cardiogenic Shock: The Prospective, Randomized IABP SHOCK Trial for Attenuation of Multiorgan Dysfunction Syndrome. *Crit. Care Med.* **2010**, *38*, 152–160. [CrossRef] [PubMed]

54. De Backer, D.; Biston, P.; Devriendt, J.; Madl, C.; Chochrad, D.; Aldecoa, C.; Brasseur, A.; Defrance, P.; Gottignies, P.; Vincent, J.-L. Comparison of Dopamine and Norepinephrine in the Treatment of Shock. *N. Engl. J. Med.* **2010**, *362*, 779–789. [CrossRef]
55. Thiele, H.; Zeymer, U.; Neumann, F.-J.; Ferenc, M.; Olbrich, H.-G.; Hausleiter, J.; Richardt, G.; Hennersdorf, M.; Empen, K.; Fuernau, G.; et al. Intraaortic Balloon Support for Myocardial Infarction with Cardiogenic Shock. *N. Engl. J. Med.* **2012**, *367*, 1287–1296. [CrossRef]
56. Ouweneel, D.M.; Eriksen, E.; Sjauw, K.D.; van Dongen, I.M.; Hirsch, A.; Packer, E.J.S.; Vis, M.M.; Wykrzykowska, J.J.; Koch, K.T.; Baan, J.; et al. Percutaneous Mechanical Circulatory Support Versus Intra-Aortic Balloon Pump in Cardiogenic Shock After Acute Myocardial Infarction. *J. Am. Coll. Cardiol.* **2017**, *69*, 278–287. [CrossRef]
57. Levy, B.; Clere-Jehl, R.; Legras, A.; Morichau-Beauchant, T.; Leone, M.; Frederique, G.; Quenot, J.P.; Kimmoun, A.; Cariou, A.; Lassus, J.; et al. Epinephrine Versus Norepinephrine for Cardiogenic Shock After Acute Myocardial Infarction. *J. Am. Coll. Cardiol.* **2018**, *72*, 173–182. [CrossRef]
58. Bochaton, T.; Huot, L.; Elbaz, M.; Delmas, C.; Aissaoui, N.; Farhat, F.; Mewton, N.; Bonnefoy, E. Mechanical Circulatory Support with the Impella® LP5.0 Pump and an Intra-Aortic Balloon Pump for Cardiogenic Shock in Acute Myocardial Infarction: The IMPELLA-STIC Randomized Study. *Arch. Cardiovasc. Dis.* **2020**, *113*, 237–243. [CrossRef]
59. Thiele, H.; Zeymer, U.; Akin, I.; Behnes, M.; Rassaf, T.; Mahabadi, A.A.; Lehmann, R.; Eitel, I.; Graf, T.; Seidler, T.; et al. Extracorporeal Life Support in Infarct-Related Cardiogenic Shock. *N. Engl. J. Med.* **2023**, *389*, 1286–1297. [CrossRef]
60. Banning, A.S.; Sabaté, M.; Orban, M.; Gracey, J.; López-Sobrino, T.; Massberg, S.; Kastrati, A.; Bogaerts, K.; Adriaenssens, T.; Berry, C.; et al. Venoarterial Extracorporeal Membrane Oxygenation or Standard Care in Patients with Cardiogenic Shock Complicating Acute Myocardial Infarction: The Multicentre, Randomised EURO SHOCK Trial. *EuroIntervention* **2023**, *19*, 482–492. [CrossRef]
61. Møller, J.E.; Engstrøm, T.; Jensen, L.O.; Eiskjær, H.; Mangner, N.; Polzin, A.; Schulze, P.C.; Skurk, C.; Nordbeck, P.; Clemmensen, P.; et al. Microaxial Flow Pump or Standard Care in Infarct-Related Cardiogenic Shock. *N. Engl. J. Med.* **2024**, *390*, 1382–1393. [CrossRef] [PubMed]
62. McMurray, J.J.V.; Solomon, S.D.; Inzucchi, S.E.; Køber, L.; Kosiborod, M.N.; Martinez, F.A.; Ponikowski, P.; Sabatine, M.S.; Anand, I.S.; Bělohlávek, J.; et al. Dapagliflozin in Patients with Heart Failure and Reduced Ejection Fraction. *N. Engl. J. Med.* **2019**, *381*, 1995–2008. [CrossRef] [PubMed]
63. Packer, M.; Anker, S.D.; Butler, J.; Filippatos, G.; Ferreira, J.P.; Pocock, S.J.; Carson, P.; Anand, I.; Doehner, W.; Haass, M.; et al. Effect of Empagliflozin on the Clinical Stability of Patients With Heart Failure and a Reduced Ejection Fraction: The EMPEROR-Reduced Trial. *Circulation* **2021**, *143*, 326–336. [CrossRef] [PubMed]
64. Kalra, P.R.; Cleland, J.G.F.; Petrie, M.C.; Thomson, E.A.; Kalra, P.A.; Squire, I.B.; Ahmed, F.Z.; Al-Mohammad, A.; Cowburn, P.J.; Foley, P.W.X.; et al. Intravenous Ferric Derisomaltose in Patients with Heart Failure and Iron Deficiency in the UK (IRONMAN): An Investigator-Initiated, Prospective, Randomised, Open-Label, Blinded-Endpoint Trial. *Lancet* **2022**, *400*, 2199–2209. [CrossRef] [PubMed]
65. Mentz, R.J.; Anstrom, K.J.; Eisenstein, E.L.; Sapp, S.; Greene, S.J.; Morgan, S.; Testani, J.M.; Harrington, A.H.; Sachdev, V.; Ketema, F.; et al. Effect of Torsemide vs Furosemide After Discharge on All-Cause Mortality in Patients Hospitalized With Heart Failure: The TRANSFORM-HF Randomized Clinical Trial. *JAMA* **2023**, *329*, 214–223. [CrossRef]
66. Butler, J.; Anker, S.D.; Lund, L.H.; Coats, A.J.S.; Filippatos, G.; Siddiqi, T.J.; Friede, T.; Fabien, V.; Kosiborod, M.; Metra, M.; et al. Patiromer for the Management of Hyperkalemia in Heart Failure with Reduced Ejection Fraction: The DIAMOND Trial. *Eur. Heart J.* **2022**, *43*, 4362–4373. [CrossRef]
67. Mcmurray, J.J.V.; Packer, M.; Desai, A.S.; Gong, J.; Lefkowitz, M.P.; Rizkala, A.R.; Rouleau, J.L.; Shi, V.C.; Solomon, S.D.; Swedberg, K.; et al. Angiotensin-Neprilysin Inhibition versus Enalapril in Heart Failure. *N. Engl. J. Med.* **2014**, *371*, 132–133. [CrossRef]
68. Anker, S.D.; Friede, T.; von Bardeleben, R.-S.; Butler, J.; Khan, M.-S.; Diek, M.; Heinrich, J.; Geyer, M.; Placzek, M.; Ferrari, R.; et al. Transcatheter Valve Repair in Heart Failure with Moderate to Severe Mitral Regurgitation. *N. Engl. J. Med.* **2024**, *391*, 1799–1809. [CrossRef]
69. Baldus, S.; Doenst, T.; Pfister, R.; Gummert, J.; Kessler, M.; Boekstegers, P.; Lubos, E.; Schröder, J.; Thiele, H.; Walther, T.; et al. Transcatheter Repair versus Mitral-Valve Surgery for Secondary Mitral Regurgitation. *N. Engl. J. Med.* **2024**, *391*, 1787–1798. [CrossRef]
70. Flather, M.D.; Shibata, M.C.; Coats, A.J.S.; Van Veldhuisen, D.J.; Parkhomenko, A.; Borbola, J.; Cohen-Solal, A.; Dumitrascu, D.; Ferrari, R.; Lechat, P.; et al. Randomized Trial to Determine the Effect of Nebivolol on Mortality and Cardiovascular Hospital Admission in Elderly Patients with Heart Failure (SENIORS). *Eur. Heart J.* **2005**, *26*, 215–225. [CrossRef]
71. Pfeffer, M.A.; Tardif, J.-C. COMMANDER HF—A Trial and an Answer. *N. Engl. J. Med.* **2018**, *379*, 1372–1374. [CrossRef] [PubMed]
72. Obadia, J.-F.; Messika-Zeitoun, D.; Leurent, G.; Iung, B.; Bonnet, G.; Piriou, N.; Lefèvre, T.; Piot, C.; Rouleau, F.; Carrié, D.; et al. Percutaneous Repair or Medical Treatment for Secondary Mitral Regurgitation. *N. Engl. J. Med.* **2018**, *379*, 2297–2306. [CrossRef] [PubMed]

73. Stone, G.W.; Lindenfeld, J.; Abraham, W.T.; Kar, S.; Lim, D.S.; Mishell, J.M.; Whisenant, B.; Grayburn, P.A.; Rinaldi, M.; Kapadia, S.R.; et al. Transcatheter Mitral-Valve Repair in Patients with Heart Failure. *N. Engl. J. Med.* **2018**, *379*, 2307–2318. [CrossRef] [PubMed]
74. Sorajja, P.; Whisenant, B.; Hamid, N.; Naik, H.; Makkar, R.; Tadros, P.; Price, M.J.; Singh, G.; Fam, N.; Kar, S.; et al. Transcatheter Repair for Patients with Tricuspid Regurgitation. *N. Engl. J. Med.* **2023**, *388*, 1833–1842. [CrossRef]
75. Arbelo, E.; Protonotarios, A.; Gimeno, J.R.; Arbustini, E.; Barriales-Villa, R.; Basso, C.; Bezzina, C.R.; Biagini, E.; Blom, N.A.; De Boer, R.A.; et al. 2023 ESC Guidelines for the Management of Cardiomyopathies. *Eur. Heart. J.* **2023**, *44*, 3503–3626. [CrossRef]
76. Arbustini, E.; Narula, N.; William Dec, G.; Srinath Reddy, K.; Greenberg, B.; Kushwaha, S.; Marwick, T.; Pinney, S.; Bellazzi, R.; Favalli, V.; et al. Reply: The MOGE(S) Classification for a Phenotype-Genotype Nomenclature of Cardiomyopathy: More Questions than Answers? *J. Am. Coll. Cardiol.* **2014**, *63*, 2584–2586. [CrossRef]
77. Pinto, Y.M.; Elliott, P.M.; Arbustini, E.; Adler, Y.; Anastasakis, A.; Böhm, M.; Duboc, D.; Gimeno, J.; De Groote, P.; Imazio, M.; et al. Proposal for a Revised Definition of Dilated Cardiomyopathy, Hypokinetic Non-Dilated Cardiomyopathy, and Its Implications for Clinical Practice: A Position Statement of the ESC Working Group on Myocardial and Pericardial Diseases. *Eur. Heart J.* **2016**, *37*, 1850–1858. [CrossRef]
78. Seferović, P.M.; Polovina, M.; Bauersachs, J.; Arad, M.; Gal, T.B.; Lund, L.H.; Felix, S.B.; Arbustini, E.; Caforio, A.L.P.; Farmakis, D.; et al. Heart Failure in Cardiomyopathies: A Position Paper from the Heart Failure Association of the European Society of Cardiology. *Eur. J. Heart Fail.* **2019**, *21*, 553–576. [CrossRef]
79. Tymińska, A.; Ozierański, K.; Balsam, P.; Maciejewski, C.; Wancerz, A.; Brociek, E.; Marchel, M.; Crespo-Leiro, M.G.; Maggioni, A.P.; Drożdż, J.; et al. Ischemic Cardiomyopathy versus Non-Ischemic Dilated Cardiomyopathy in Patients with Reduced Ejection Fraction-Clinical Characteristics and Prognosis Depending on Heart Failure Etiology (Data from European Society of Cardiology Heart Failure Registries). *Biology* **2022**, *11*, 341. [CrossRef]
80. Tat, E.; Ball, C.; Camren, G.P.; Wroblewski, I.; Dajani, K.A.; Goldberg, A.; Kinno, M.; Sanagala, T.; Syed, M.A.; Wilber, D.J.; et al. Impact of Late Gadolinium Enhancement Extent, Location, and Pattern on Ventricular Tachycardia and Major Adverse Cardiac Events in Patients with Ischemic vs. Non-Ischemic Cardiomyopathy. *Front. Cardiovasc. Med.* **2022**, *9*, 1026215. [CrossRef]
81. Hershberger, R.E.; Hedges, D.J.; Morales, A. Dilated Cardiomyopathy: The Complexity of a Diverse Genetic Architecture. *Nat. Rev. Cardiol.* **2013**, *10*, 531–547. [CrossRef] [PubMed]
82. Codd, M.B.; Sugrue, D.D.; Gersh, B.J.; Melton, L.J. Epidemiology of Idiopathic Dilated and Hypertrophic Cardiomyopathy. A Population-Based Study in Olmsted County, Minnesota, 1975-1984. *Circulation* **1989**, *80*, 564–572. [CrossRef] [PubMed]
83. Jefferies, J.L.; Towbin, J.A. Dilated Cardiomyopathy. *Lancet* **2010**, *375*, 752–762. [CrossRef] [PubMed]
84. Japp, A.G.; Gulati, A.; Cook, S.A.; Cowie, M.R.; Prasad, S.K. The Diagnosis and Evaluation of Dilated Cardiomyopathy. *J. Am. Coll. Cardiol.* **2016**, *67*, 2996–3010. [CrossRef]
85. Herman, D.S.; Lam, L.; Taylor, M.R.G.; Wang, L.; Teekakirikul, P.; Christodoulou, D.; Conner, L.; DePalma, S.R.; McDonough, B.; Sparks, E.; et al. Truncations of Titin Causing Dilated Cardiomyopathy. *N. Engl. J. Med.* **2012**, *366*, 619–628. [CrossRef]
86. Halliday, B.P.; Gulati, A.; Ali, A.; Newsome, S.; Lota, A.; Tayal, U.; Vassiliou, V.S.; Arzanauskaite, M.; Izgi, C.; Krishnathasan, K.; et al. Sex- and Age-Based Differences in the Natural History and Outcome of Dilated Cardiomyopathy. *Eur. J. Heart Fail.* **2018**, *20*, 1392–1400. [CrossRef]
87. Gulati, A.; Jabbour, A.; Ismail, T.F.; Guha, K.; Khwaja, J.; Raza, S.; Morarji, K.; Brown, T.D.H.; Ismail, N.A.; Dweck, M.R.; et al. Association of Fibrosis with Mortality and Sudden Cardiac Death in Patients with Nonischemic Dilated Cardiomyopathy. *JAMA* **2013**, *309*, 896–908. [CrossRef]
88. Bottiroli, M.; Calini, A.; Morici, N.; Tavazzi, G.; Galimberti, L.; Facciorusso, C.; Ammirati, E.; Russo, C.; Montoli, A.; Mondino, M. Acute Kidney Injury in Patients with Acute Decompensated Heart Failure-Cardiogenic Shock: Prevalence, Risk Factors and Outcome. *Int. J. Cardiol.* **2023**, *383*, 42–49. [CrossRef]
89. Levy, B.; Perez, P.; Perny, J.; Thivilier, C.; Gerard, A. Comparison of Norepinephrine-Dobutamine to Epinephrine for Hemodynamics, Lactate Metabolism, and Organ Function Variables in Cardiogenic Shock. A Prospective, Randomized Pilot Study. *Crit. Care Med.* **2011**, *39*, 450–455. [CrossRef]
90. Akhtar, K.H.; Maqsood, M.H.; Ansari, S.A.; Siddiqi, T.J.; Arshad, M.S.; Greene, S.J.; Butler, J.; Khan, M.S. An Individual Patient-Level Meta-Analysis of Ischemic Versus Nonischemic Cardiomyopathy and Trajectory of Decongestion in Patients With Acute Decompensated Heart Failure. *Am. J. Cardiol.* **2023**, *200*, 32–39. [CrossRef]
91. Cherbi, M.; Gerbaud, E.; Lamblin, N.; Bonnefoy, E.; Bonello, L.; Levy, B.; Ternacle, J.; Schneider, F.; Elbaz, M.; Khachab, H.; et al. Cardiogenic Shock in Idiopathic Dilated Cardiomyopathy Patients: Red Flag for Myocardial Decline. *Am. J. Cardiol.* **2023**, *206*, 89–97. [CrossRef] [PubMed]
92. Cherbi, M.; Roubille, F.; Lamblin, N.; Bonello, L.; Leurent, G.; Levy, B.; Elbaz, M.; Champion, S.; Lim, P.; Schneider, F.; et al. One-Year Outcomes in Cardiogenic Shock Triggered by Ventricular Arrhythmia: An Analysis of the FRENSHOCK Multicenter Prospective Registry. *Front. Cardiovasc. Med.* **2023**, *10*, 1092904. [CrossRef] [PubMed]

93. Miric, D.; Barac, A.; Capkun, V.; Bakovic, D. Right Ventricular Free Wall Strain in Acutely Decompensated Heart Failure Patients with Ischemic and Non-Ischemic Cardiomyopathy. *Echocardiography* **2021**, *38*, 1747–1753. [CrossRef] [PubMed]
94. Wilcox, J.E.; Fonarow, G.C.; Ardehali, H.; Bonow, R.O.; Butler, J.; Sauer, A.J.; Epstein, S.E.; Khan, S.S.; Kim, R.J.; Sabbah, H.N.; et al. "Targeting the Heart" in Heart Failure: Myocardial Recovery in Heart Failure With Reduced Ejection Fraction. *JACC Heart Fail.* **2015**, *3*, 661–669. [CrossRef]
95. McNamara, D.M.; Starling, R.C.; Cooper, L.T.; Boehmer, J.P.; Mather, P.J.; Janosko, K.M.; Gorcsan, J.; Kip, K.E.; Dec, G.W. Clinical and Demographic Predictors of Outcomes in Recent Onset Dilated Cardiomyopathy: Results of the IMAC (Intervention in Myocarditis and Acute Cardiomyopathy)-2 Study. *J. Am. Coll. Cardiol.* **2011**, *58*, 1112–1118. [CrossRef]
96. Halliday, B.P.; Wassall, R.; Lota, A.S.; Khalique, Z.; Gregson, J.; Newsome, S.; Jackson, R.; Rahneva, T.; Wage, R.; Smith, G.; et al. Withdrawal of Pharmacological Treatment for Heart Failure in Patients with Recovered Dilated Cardiomyopathy (TRED-HF): An Open-Label, Pilot, Randomised Trial. *Lancet* **2019**, *393*, 61–73. [CrossRef]
97. Wilcox, J.E.; Fonarow, G.C.; Yancy, C.W.; Albert, N.M.; Curtis, A.B.; Heywood, J.T.; Inge, P.J.; McBride, M.L.; Mehra, M.R.; O'Connor, C.M.; et al. Factors Associated with Improvement in Ejection Fraction in Clinical Practice among Patients with Heart Failure: Findings from IMPROVE HF. *Am. Heart J.* **2012**, *163*, 49–56.e2. [CrossRef]
98. Kubanek, M.; Sramko, M.; Maluskova, J.; Kautznerova, D.; Weichet, J.; Lupinek, P.; Vrbska, J.; Malek, I.; Kautzner, J. Novel Predictors of Left Ventricular Reverse Remodeling in Individuals with Recent-Onset Dilated Cardiomyopathy. *J. Am. Coll. Cardiol.* **2013**, *61*, 54–63. [CrossRef]
99. Todiere, G.; Barison, A.; Baritussio, A.; Cipriani, A.; Guaricci, A.I.; Pica, S.; Indolfi, C.; Pontone, G.; Dellegrottaglie, S. Acute Clinical Presentation of Nonischemic Cardiomyopathies: Early Detection by Cardiovascular Magnetic Resonance. *J. Cardiovasc. Med.* **2023**, *24*, E36–E46. [CrossRef]
100. Donal, E.; Delgado, V.; Bucciarelli-Ducci, C.; Galli, E.; Haugaa, K.H.; Charron, P.; Voigt, J.U.; Cardim, N.; Masci, P.G.; Galderisi, M.; et al. Multimodality Imaging in the Diagnosis, Risk Stratification, and Management of Patients with Dilated Cardiomyopathies: An Expert Consensus Document from the European Association of Cardiovascular Imaging. *Eur. Heart J. Cardiovasc. Imaging* **2019**, *20*, 1075–1093. [CrossRef]
101. Di Marco, A.; Anguera, I.; Schmitt, M.; Klem, I.; Neilan, T.; White, J.A.; Sramko, M.; Masci, P.G.; Barison, A.; Mckenna, P.; et al. Late Gadolinium Enhancement and the Risk for Ventricular Arrhythmias or Sudden Death in Dilated Cardiomyopathy: Systematic Review and Meta-Analysis. *JACC Heart Fail.* **2017**, *5*, 28–38. [CrossRef] [PubMed]
102. Chen, W.; Qian, W.; Zhang, X.; Li, D.; Qian, Z.; Xu, H.; Liao, S.; Chen, X.; Wang, Y.; Hou, X.; et al. Ring-like Late Gadolinium Enhancement for Predicting Ventricular Tachyarrhythmias in Non-Ischaemic Dilated Cardiomyopathy. *Eur. Heart J. Cardiovasc. Imaging* **2021**, *22*, 1130–1138. [CrossRef] [PubMed]
103. Sepehrkhouy, S.; Gho, J.M.I.H.; van Es, R.; Harakalova, M.; de Jonge, N.; Dooijes, D.; van der Smagt, J.J.; Buijsrogge, M.P.; Hauer, R.N.W.; Goldschmeding, R.; et al. Distinct Fibrosis Pattern in Desmosomal and Phospholamban Mutation Carriers in Hereditary Cardiomyopathies. *Heart Rhythm.* **2017**, *14*, 1024–1032. [CrossRef] [PubMed]
104. Cannatà, A.; De Angelis, G.; Boscutti, A.; Normand, C.; Artico, J.; Gentile, P.; Zecchin, M.; Heymans, S.; Merlo, M.; Sinagra, G. Arrhythmic Risk Stratification in Non-Ischaemic Dilated Cardiomyopathy beyond Ejection Fraction. *Heart* **2020**, *106*, 656–664. [CrossRef]
105. Merlo, M.; Grilli, G.; Cappelletto, C.; Masé, M.; Porcari, A.; Ferro, M.D.; Gigli, M.; Stolfo, D.; Zecchin, M.; De Luca, A.; et al. The Arrhythmic Phenotype in Cardiomyopathy. *Heart Fail. Clin.* **2022**, *18*, 101–113. [CrossRef]
106. Mirelis, J.G.; Escobar-Lopez, L.; Ochoa, J.P.; Espinosa, M.Á.; Villacorta, E.; Navarro, M.; Casas, G.; Mora-Ayestarán, N.; Barriales-Villa, R.; Mogollón-Jiménez, M.V.; et al. Combination of Late Gadolinium Enhancement and Genotype Improves Prediction of Prognosis in Non-Ischaemic Dilated Cardiomyopathy. *Eur. J. Heart Fail.* **2022**, *24*, 1183–1196. [CrossRef]
107. Van Rijsingen, I.A.W.; Arbustini, E.; Elliott, P.M.; Mogensen, J.; Hermans-Van Ast, J.F.; Van Der Kooi, A.J.; Van Tintelen, J.P.; Van Den Berg, M.P.; Pilotto, A.; Pasotti, M.; et al. Risk Factors for Malignant Ventricular Arrhythmias in Lamin a/c Mutation Carriers a European Cohort Study. *J. Am. Coll. Cardiol.* **2012**, *59*, 493–500. [CrossRef]
108. Paldino, A.; Dal Ferro, M.; Stolfo, D.; Gandin, I.; Medo, K.; Graw, S.; Gigli, M.; Gagno, G.; Zaffalon, D.; Castrichini, M.; et al. Prognostic Prediction of Genotype vs Phenotype in Genetic Cardiomyopathies. *J. Am. Coll. Cardiol.* **2022**, *80*, 1981–1994. [CrossRef]
109. Rossano, J.W.; Dipchand, A.I.; Edwards, L.B.; Goldfarb, S.; Kucheryavaya, A.Y.; Levvey, R.N.B.J.; Lund, L.H.; Meiser, B.; Yusen, R.D.; Stehlik, J. The Registry of the International Society for Heart and Lung Transplantation: Nineteenth Pediatric Heart Transplantation Report-2016; Focus Theme: Primary Diagnostic Indications for Transplant. *J. Heart Lung Transplant.* **2016**, *35*, 1185–1195. [CrossRef]
110. Yusen, R.D.; Edwards, L.B.; Dipchand, A.I.; Goldfarb, S.B.; Kucheryavaya, A.Y.; Levvey, B.J.; Lund, L.H.; Meiser, B.; Rossano, J.W.; Stehlik, J. The Registry of the International Society for Heart and Lung Transplantation: Thirty-Third Adult Lung and Heart-Lung Transplant Report-2016; Focus Theme: Primary Diagnostic Indications for Transplant. *J. Heart Lung Transplant.* **2016**, *35*, 1170–1184. [CrossRef]

111. Lund, L.H.; Edwards, L.B.; Dipchand, A.I.; Goldfarb, S.; Kucheryavaya, A.Y.; Levvey, B.J.; Meiser, B.; Rossano, J.W.; Yusen, R.D.; Stehlik, J. The Registry of the International Society for Heart and Lung Transplantation: Thirty-Third Adult Heart Transplantation Report-2016; Focus Theme: Primary Diagnostic Indications for Transplant. *J. Heart Lung Transplant.* **2016**, *35*, 1158–1169. [CrossRef] [PubMed]
112. Falco, L.; Valente, F.; De Falco, A.; Barbato, R.; Marotta, L.; Soviero, D.; Cantiello, L.M.; Contaldi, C.; Brescia, B.; Coscioni, E.; et al. Beyond Medical Therapy-An Update on Heart Failure Devices. *J. Cardiovasc. Dev. Dis.* **2024**, *11*, 187. [CrossRef] [PubMed]
113. Rose, E.A.; Gelijns, A.C.; Moskowitz, A.J.; Heitjan, D.F.; Stevenson, L.W.; Dembitsky, W.; Long, J.W.; Ascheim, D.D.; Tierney, A.R.; Levitan, R.G.; et al. Long-Term Use of a Left Ventricular Assist Device for End-Stage Heart Failure. *N. Engl. J. Med.* **2001**, *345*, 1435–1443. [CrossRef] [PubMed]
114. Yoshioka, D.; Li, B.; Takayama, H.; Garan, R.A.; Topkara, V.K.; Han, J.; Kurlansky, P.; Yuzefpolskaya, M.; Colombo, P.C.; Naka, Y.; et al. Outcome of Heart Transplantation after Bridge-to-Transplant Strategy Using Various Mechanical Circulatory Support Devices. *Interact. Cardiovasc. Thorac. Surg.* **2017**, *25*, 918–924. [CrossRef] [PubMed]
115. Khayata, M.; Al-Kindi, S.G.; Oliveira, G.H. Contemporary Characteristics and Outcomes of Adults with Familial Dilated Cardiomyopathy Listed for Heart Transplantation. *World J. Cardiol.* **2019**, *11*, 38–46. [CrossRef]
116. Heymans, S.; Lakdawala, N.K.; Tschöpe, C.; Klingel, K. Dilated Cardiomyopathy: Causes, Mechanisms, and Current and Future Treatment Approaches. *Lancet* **2023**, *402*, 998–1011. [CrossRef]
117. Elliott, P.; Andersson, B.; Arbustini, E.; Bilinska, Z.; Cecchi, F.; Charron, P.; Dubourg, O.; Kühl, U.; Maisch, B.; McKenna, W.J.; et al. Classification of the Cardiomyopathies: A Position Statement from the European Society Of Cardiology Working Group on Myocardial and Pericardial Diseases. *Eur. Heart J.* **2008**, *29*, 270–276. [CrossRef]
118. Olivotto, I.; Girolami, F.; Nistri, S.; Rossi, A.; Rega, L.; Garbini, F.; Grifoni, C.; Cecchi, F.; Yacoub, M.H. The Many Faces of Hypertrophic Cardiomyopathy: From Developmental Biology to Clinical Practice. *J. Cardiovasc. Transl. Res.* **2009**, *2*, 349–367. [CrossRef]
119. Maron, M.S.; Olivotto, I.; Zenovich, A.G.; Link, M.S.; Pandian, N.G.; Kuvin, J.T.; Nistri, S.; Cecchi, F.; Udelson, J.E.; Maron, B.J. Hypertrophic Cardiomyopathy Is Predominantly a Disease of Left Ventricular Outflow Tract Obstruction. *Circulation* **2006**, *114*, 2232–2239. [CrossRef]
120. Mancusi, C.; Basile, C.; Spaccarotella, C.; Gargiulo, G.; Fucile, I.; Paolillo, S.; Gargiulo, P.; Santoro, C.; Manzi, L.; Marzano, F.; et al. Novel Strategies in Diagnosing Heart Failure with Preserved Ejection Fraction: A Comprehensive Literature Review. *High. Blood Press. Cardiovasc. Prev.* **2024**, *31*, 127–140. [CrossRef]
121. Olivotto, I.; Cecchi, F.; Poggesi, C.; Yacoub, M.H. Patterns of Disease Progression in Hypertrophic Cardiomyopathy: An Individualized Approach to Clinical Staging. *Circ. Heart Fail.* **2012**, *5*, 535–546. [CrossRef] [PubMed]
122. Morita, H.; Rehm, H.L.; Menesses, A.; McDonough, B.; Roberts, A.E.; Kucherlapati, R.; Towbin, J.A.; Seidman, J.G.; Seidman, C.E. Shared Genetic Causes of Cardiac Hypertrophy in Children and Adults. *N. Engl. J. Med.* **2008**, *358*, 1899–1908. [CrossRef] [PubMed]
123. Charron, P.; Elliott, P.M.; Gimeno, J.R.; Caforio, A.L.P.; Kaski, J.P.; Tavazzi, L.; Tendera, M.; Maupain, C.; Laroche, C.; Rubis, P.; et al. The Cardiomyopathy Registry of the EURObservational Research Programme of the European Society of Cardiology: Baseline Data and Contemporary Management of Adult Patients with Cardiomyopathies. *Eur. Heart J.* **2018**, *39*, 1784–1793. [CrossRef] [PubMed]
124. Maron, B.J.; Rowin, E.J.; Udelson, J.E.; Maron, M.S. Clinical Spectrum and Management of Heart Failure in Hypertrophic Cardiomyopathy. *JACC Heart Fail.* **2018**, *6*, 353–363. [CrossRef]
125. Harris, K.M.; Spirito, P.; Maron, M.S.; Zenovich, A.G.; Formisano, F.; Lesser, J.R.; Mackey-Bojack, S.; Manning, W.J.; Udelson, J.E.; Maron, B.J. Prevalence, Clinical Profile, and Significance of Left Ventricular Remodeling in the End-Stage Phase of Hypertrophic Cardiomyopathy. *Circulation* **2006**, *114*, 216–225. [CrossRef]
126. Maron, M.S.; Olivotto, I.; Betocchi, S.; Casey, S.A.; Lesser, J.R.; Losi, M.A.; Cecchi, F.; Maron, B.J. Effect of Left Ventricular Outflow Tract Obstruction on Clinical Outcome in Hypertrophic Cardiomyopathy. *N. Engl. J. Med.* **2003**, *348*, 465–466. [CrossRef]
127. Melacini, P.; Basso, C.; Angelini, A.; Calore, C.; Bobbo, F.; Tokajuk, B.; Bellini, N.; Smaniotto, G.; Zucchetto, M.; Iliceto, S.; et al. Clinicopathological Profiles of Progressive Heart Failure in Hypertrophic Cardiomyopathy. *Eur. Heart J.* **2010**, *31*, 2111–2123. [CrossRef]
128. Dimitrow, P.P.; Bober, M.; Michałowska, J.; Sorysz, D. Left Ventricular Outflow Tract Gradient Provoked by Upright Position or Exercise in Treated Patients with Hypertrophic Cardiomyopathy without Obstruction at Rest. *Echocardiography* **2009**, *26*, 513–520. [CrossRef]
129. Magrì, D.; Re, F.; Limongelli, G.; Agostoni, P.; Zachara, E.; Correale, M.; Mastromarino, V.; Santolamazza, C.; Casenghi, M.; Pacileo, G.; et al. Heart Failure Progression in Hypertrophic Cardiomyopathy—Possible Insights From Cardiopulmonary Exercise Testing. *Circ. J.* **2016**, *80*, 2204–2211. [CrossRef]

130. Moon, J.C.C.; Reed, E.; Sheppard, M.N.; Elkington, A.G.; Ho, S.Y.; Burke, M.; Petrou, M.; Pennell, D.J. The Histologic Basis of Late Gadolinium Enhancement Cardiovascular Magnetic Resonance in Hypertrophic Cardiomyopathy. *J. Am. Coll. Cardiol.* **2004**, *43*, 2260–2264. [CrossRef]
131. Galati, G.; Leone, O.; Pasquale, F.; Olivotto, I.; Biagini, E.; Grigioni, F.; Pilato, E.; Lorenzini, M.; Corti, B.; Foà, A.; et al. Histological and Histometric Characterization of Myocardial Fibrosis in End-Stage Hypertrophic Cardiomyopathy: A Clinical-Pathological Study of 30 Explanted Hearts. *Circ. Heart Fail.* **2016**, *9*. [CrossRef] [PubMed]
132. Patel, S.R.; Saeed, O.; Naftel, D.; Myers, S.; Kirklin, J.; Jorde, U.P.; Goldstein, D.J. Outcomes of Restrictive and Hypertrophic Cardiomyopathies After LVAD: An INTERMACS Analysis. *J. Card. Fail.* **2017**, *23*, 859–867. [CrossRef] [PubMed]
133. DePasquale, E.C.; Deng, M.; Ardehali, A.; Jacoby, D. Outcomes of Heart Transplantation in Adults with Hypertrophic Cardiomyopathy (HCM): UNOS Registry Analysis. *J. Heart Lung Transplant.* **2016**, *35*, S64. [CrossRef]
134. Kaski, J.P.; Syrris, P.; Burch, M.; Tomé Esteban, M.T.; Fenton, M.; Christiansen, M.; Andersen, P.S.; Sebire, N.; Ashworth, M.; Deanfield, J.E.; et al. Idiopathic Restrictive Cardiomyopathy in Children Is Caused by Mutations in Cardiac Sarcomere Protein Genes. *Heart* **2008**, *94*, 1478–1484. [CrossRef]
135. Rapezzi, C.; Aimo, A.; Barison, A.; Emdin, M.; Porcari, A.; Linhart, A.; Keren, A.; Merlo, M.; Sinagra, G. Restrictive Cardiomyopathy: Definition and Diagnosis. *Eur. Heart J.* **2022**, *43*, 4679–4693. [CrossRef]
136. Mori, H.; Kogaki, S.; Ishida, H.; Yoshikawa, T.; Shindo, T.; Inuzuka, R.; Furutani, Y.; Ishido, M.; Nakanishi, T. Outcomes of Restrictive Cardiomyopathy in Japanese Children—A Retrospective Cohort Study. *Circ. J.* **2022**, *86*, 1943–1949. [CrossRef]
137. Ammash, N.M.; Seward, J.B.; Bailey, K.R.; Edwards, W.D.; Tajik, A.J. Clinical Profile and Outcome of Idiopathic Restrictive Cardiomyopathy. *Circulation* **2000**, *101*, 2490–2496. [CrossRef]
138. Rivenes, S.M.; Kearney, D.L.; Smith, E.O.B.; Towbin, J.A.; Denfield, S.W. Sudden Death and Cardiovascular Collapse in Children with Restrictive Cardiomyopathy. *Circulation* **2000**, *102*, 876–882. [CrossRef]
139. Webber, S.A.; Lipshultz, S.E.; Sleeper, L.A.; Lu, M.; Wilkinson, J.D.; Addonizio, L.J.; Canter, C.E.; Colan, S.D.; Everitt, M.D.; Jefferies, J.L.; et al. Outcomes of Restrictive Cardiomyopathy in Childhood and the Influence of Phenotype: A Report from the Pediatric Cardiomyopathy Registry. *Circulation* **2012**, *126*, 1237–1244. [CrossRef]
140. Anderson, H.N.; Cetta, F.; Driscoll, D.J.; Olson, T.M.; Ackerman, M.J.; Johnson, J.N. Idiopathic Restrictive Cardiomyopathy in Children and Young Adults. *Am. J. Cardiol.* **2018**, *121*, 1266–1270. [CrossRef]
141. Maurer, M.S.; Schwartz, J.H.; Gundapaneni, B.; Elliott, P.M.; Merlini, G.; Waddington-Cruz, M.; Kristen, A.V.; Grogan, M.; Witteles, R.; Damy, T.; et al. Tafamidis Treatment for Patients with Transthyretin Amyloid Cardiomyopathy. *N. Engl. J. Med.* **2018**, *379*, 1007–1016. [CrossRef] [PubMed]
142. Fontana, M.; Berk, J.L.; Gillmore, J.D.; Witteles, R.M.; Grogan, M.; Drachman, B.; Damy, T.; Garcia-Pavia, P.; Taubel, J.; Solomon, S.D.; et al. Vutrisiran in Patients with Transthyretin Amyloidosis with Cardiomyopathy. *N. Engl. J. Med.* **2025**, *392*, 33–44. [CrossRef] [PubMed]
143. Habib, G.; Bucciarelli-Ducci, C.; Caforio, A.L.P.; Cardim, N.; Charron, P.; Cosyns, B.; Dehaene, A.; Derumeaux, G.; Donal, E.; Dweck, M.R.; et al. Multimodality Imaging in Restrictive Cardiomyopathies: An EACVI Expert Consensus Document In Collaboration with the "Working Group on Myocardial and Pericardial Diseases" of the European Society of Cardiology Endorsed by The Indian Academy of Echocardiography. *Eur. Heart J. Cardiovasc. Imaging* **2017**, *18*, 1090–1091. [CrossRef] [PubMed]
144. Muchtar, E.; Blauwet, L.A.; Gertz, M.A. Restrictive Cardiomyopathy: Genetics, Pathogenesis, Clinical Manifestations, Diagnosis, and Therapy. *Circ. Res.* **2017**, *121*, 819–837. [CrossRef]
145. Grupper, A.; Park, S.J.; Pereira, N.L.; Schettle, S.D.; Gerber, Y.; Topilsky, Y.; Edwards, B.S.; Daly, R.C.; Stulak, J.M.; Joyce, L.D.; et al. Role of Ventricular Assist Therapy for Patients with Heart Failure and Restrictive Physiology: Improving Outcomes for a Lethal Disease. *J. Heart Lung Transplant.* **2015**, *34*, 1042–1049. [CrossRef]
146. Depasquale, E.C.; Nasir, K.; Jacoby, D.L. Outcomes of Adults with Restrictive Cardiomyopathy after Heart Transplantation. *J. Heart Lung Transplant.* **2012**, *31*, 1269–1275. [CrossRef]
147. Cheong, B.Y.C.; Muthupillai, R.; Wilson, J.M.; Sung, A.; Huber, S.; Amin, S.; Elayda, M.A.; Lee, V.V.; Flamm, S.D. Prognostic Significance of Delayed-Enhancement Magnetic Resonance Imaging: Survival of 857 Patients with and without Left Ventricular Dysfunction. *Circulation* **2009**, *120*, 2069–2076. [CrossRef]
148. Castrichini, M.; De Luca, A.; De Angelis, G.; Neves, R.; Paldino, A.; Dal Ferro, M.; Barbati, G.; Medo, K.; Barison, A.; Grigoratos, C.; et al. Magnetic Resonance Imaging Characterization and Clinical Outcomes of Dilated and Arrhythmogenic Left Ventricular Cardiomyopathies. *J. Am. Coll. Cardiol.* **2024**, *83*, 1841–1851. [CrossRef]
149. Aquaro, G.D.; Perfetti, M.; Camastra, G.; Monti, L.; Dellegrottaglie, S.; Moro, C.; Pepe, A.; Todiere, G.; Lanzillo, C.; Scatteia, A.; et al. Cardiac MR With Late Gadolinium Enhancement in Acute Myocarditis With Preserved Systolic Function: ITAMY Study. *J. Am. Coll. Cardiol.* **2017**, *70*, 1977–1987. [CrossRef]

150. Augusto, J.B.; Eiros, R.; Nakou, E.; Moura-Ferreira, S.; Treibel, T.A.; Captur, G.; Akhtar, M.M.; Protonotarios, A.; Gossios, T.D.; Savvatis, K.; et al. Dilated Cardiomyopathy and Arrhythmogenic Left Ventricular Cardiomyopathy: A Comprehensive Genotype-Imaging Phenotype Study. *Eur. Heart J. Cardiovasc. Imaging* **2020**, *21*, 326–336. [CrossRef]
151. Monda, E.; Murredda, A.; Rubino, M.; Diana, G.; Palmiero, G.; Verrillo, F.; Cirillo, C.; Cirillo, A.; Fusco, A.; Frisso, G.; et al. Aetiology and Clinical Manifestations of Patients with Non-Dilated Left Ventricular Cardiomyopathy. *Eur. J. Heart Fail.* **2024**, *26*, 2579–2581. [CrossRef] [PubMed]
152. Leo, I.; Dellegrottaglie, S.; Scatteia, A.; Torella, D.; Abete, R.; Aquaro, G.D.; Baggiano, A.; Barison, A.; Bogaert, J.; Calo', L.; et al. CarDiac MagnEtic Resonance for Prophylactic Implantable-CardioVerter DefibrillAtor ThErapy in Non-Dilated Left Ventricular Cardiomyopathy: A Sub-Study from the DERIVATE Registry. *Eur. Heart J. Cardiovasc. Imaging* **2025**. [CrossRef] [PubMed]
153. Corrado, D.; Perazzolo Marra, M.; Zorzi, A.; Beffagna, G.; Cipriani, A.; De Lazzari, M.; Migliore, F.; Pilichou, K.; Rampazzo, A.; Rigato, I.; et al. Diagnosis of Arrhythmogenic Cardiomyopathy: The Padua Criteria. *Int. J. Cardiol.* **2020**, *319*, 106–114. [CrossRef] [PubMed]
154. Bhonsale, A.; Groeneweg, J.A.; James, C.A.; Dooijes, D.; Tichnell, C.; Jongbloed, J.D.H.; Murray, B.; Te Riele, A.S.J.M.; Van Den Berg, M.P.; Bikker, H.; et al. Impact of Genotype on Clinical Course in Arrhythmogenic Right Ventricular Dysplasia/Cardiomyopathy-Associated Mutation Carriers. *Eur. Heart J.* **2015**, *36*, 847–855. [CrossRef]
155. Bauce, B.; Basso, C.; Rampazzo, A.; Beffagna, G.; Daliento, L.; Frigo, G.; Malacrida, S.; Settimo, L.; Danieli, G.A.; Thiene, G.; et al. Clinical Profile of Four Families with Arrhythmogenic Right Ventricular Cardiomyopathy Caused by Dominant Desmoplakin Mutations. *Eur. Heart J.* **2005**, *26*, 1666–1675. [CrossRef]
156. Marcus, F.I.; McKenna, W.J.; Sherrill, D.; Basso, C.; Bauce, B.; Bluemke, D.A.; Calkins, H.; Corrado, D.; Cox, M.G.P.J.; Daubert, J.P.; et al. Diagnosis of Arrhythmogenic Right Ventricular Cardiomyopathy/Dysplasia: Proposed Modification of the Task Force Criteria. *Eur. Heart J.* **2010**, *31*, 806–814. [CrossRef]
157. Towbin, J.A.; McKenna, W.J.; Abrams, D.J.; Ackerman, M.J.; Calkins, H.; Darrieux, F.C.C.; Daubert, J.P.; de Chillou, C.; DePasquale, E.C.; Desai, M.Y.; et al. 2019 HRS Expert Consensus Statement on Evaluation, Risk Stratification, and Management of Arrhythmogenic Cardiomyopathy. *Heart Rhythm* **2019**, *16*, e301–e372. [CrossRef]
158. Gilljam, T.; Haugaa, K.H.; Jensen, H.K.; Svensson, A.; Bundgaard, H.; Hansen, J.; Dellgren, G.; Gustafsson, F.; Eiskjær, H.; Andreassen, A.K.; et al. Heart Transplantation in Arrhythmogenic Right Ventricular Cardiomyopathy—Experience from the Nordic ARVC Registry. *Int. J. Cardiol.* **2018**, *250*, 201–206. [CrossRef]

diagnostics

MDPI

Review

Identification of Pre-Heart Failure in Early Stages: The Role of Six Stages of Heart Failure

Monika Jankajova [1], Ram B. Singh [2], Krasimira Hristova [3], Galal Elkilany [4], Ghizal Fatima [5], Jaipaul Singh [6] and Jan Fedacko [7,*]

1 First Department of Cardiology, East-Slovak Institute of Cardiovascular Diseases, 04011 Kosice, Slovakia; mjankajova@gmail.com
2 Halberg Hospital and Research Institute, Moradabad 244001, India; rbs@tsimtsoum.net
3 Department of Cardiology, Center of Cardiovascular Diseases, 1309 Sofia, Bulgaria; khristovabg@yahoo.com
4 Gulf Medical College, Ajman 4184, United Arab Emirates; galal.elkilany@gmail.com
5 Chronobiology Laboratory, Era University, Lucknow 226003, India; ghizalfatima8@gmail.com
6 School of Pharmacy and Biomedical Sciences, University of Central Lancashire, Preston PR1 2HE, UK; jaipsingh10@gmail.com
7 Department of Gediatrics, PJ Safaric University, 04001 Kosice, Slovakia
* Correspondence: janfedacko@hotmail.com

Abstract: Despite increased availability of effective drug therapy for treatment of heart failure (HF), the morbidity and mortality in chronic heart failure (CHF) are unacceptably high. Therefore, there is an urgent need to ascertain new imaging techniques to identify early sub-clinical forms of cardiac dysfunctions, to guide early relevant treatment. It seems that all the behavioral risk factors—such as tobacco, alcoholism, Western-type diet, sedentary behavior and obesity, emotional disorders, and sleep disorder are associated with early cardiac dysfunction, which may be identified by speckle-tracking echocardiography (STE). Cardiac remodeling can also occur chronologically in association with biological risk factors of CHF, such as diabetes mellitus (DM), hypertension, cardiomyopathy, valvular heart disease, and coronary artery disease (CAD). In these conditions, twisting and untwisting of the heart, cardiac fibrosis, and hypertrophy can be identified early and accurately with 2-Dimentional (2D) and 3D echocardiography (2D echo and 3D echo) with tissue Doppler imaging (TDI), strain imaging via STE, and cardiac magnetic resonance imaging (CMR). Both 2D and 3D echo with STE are also useful in the identification of myocardial damage during chemotherapy and in the presence of risk factors. It is possible that global longitudinal systolic strain (GLS) obtained by STE may be an accurate marker for early identification of the severity of CAD in patients with non-ST segment elevation MI. Left ventricular ejection fraction (LVEF) is not the constant indicator of HF and it is normal in early cardiac dysfunction. In conclusion, this review suggests that GLS can be a useful early diagnostic marker of early or pre-cardiac dysfunction which may be treated by suitable drug therapy of HF along with the causes of HF and adhere to prevention strategies for recurrence. In addition, STE may be a superior clinical tool in the identification of cardiac dysfunction in its early stages compared to ejection fraction (EF) based on conventional echocardiography. Therefore, it is suggested that the chances of either stalling or reversing HF are far better for patients who are identified at an early stage of the disease.

Keywords: myocardial dysfunction; hypertrophy; heart enlargement; stages of heart failure; speckle-tracking echocardiography

1. Introduction

Chronic heart failure (CHF) continues to be associated with considerable worldwide morbidity and mortality [1]. Early diagnosis in the first 2–3 stages of the six stages of HF may be useful in the reversal of HF [2,3]. The Cleveland Clinic and other experts have classified HF into four stages—A, B, C, and D [3–6]—which are different from the

New York Heart Association (NYHA)'s classes I–IV [1]. There are gaps in the knowledge about when to institute drug therapy despite the increased availability of four pillars of pharmaco-therapy for the treatment of HF [7,8]. Early identification of HF requires the establishment of new imaging techniques to guide treatment of patients presenting with biological risk factors—such as obesity, diabetes mellitus (DM), hypertension (HTN), and coronary artery diseases (CAD)—as well as those with acute coronary syndromes [9]. Apart from biological risk factors, behavioral risk factors may also cause pre-heart failure (PHF), which may be diagnosed via speckle-tracking echocardiography (STE) [9–11]. Advanced cardiac imaging has progressed significantly which serves as a major diagnostic, as well as a prognostic tool [11–14]. It seems that patients with CHF need continuous follow-up along with imaging such as 2D echo or STE and stress imaging to identify HF in early stages. Echocardiography can guide therapy and related procedures concerning intervention for the better prognosis of HF patients [11,12]. This communication was designed to highlight the important role of biomedical, biological, and behavioral risk factors in an early stage of HF or PHF and moreover to emphasize on the significance of such cardiac clinical tools and imaging techniques as STE, echocardiography, and CMR to aid early diagnosis process leading to early treatment of the disease.

2. Risk Factors of Pre-Heart Failure (PHF)

There is much evidence that an increase in the dysfunction of myocardial tissue predisposes to reduced pumping activity, which worsens the clinical and biological functions of the heart, because cardiac muscle pumps less blood to organs, leading to the various stages of HF [3–7]. In the treatment of HF, the goal is to reduce worsening of failure from progression through the stages, or to slow down the progression of failure to advanced stages of HF [4–7]. It is crucial to target HF before its clinical manifestation, at the stage of PHF, during asymptomatic phase. There is an unmet need to identify PHF in the early stage in the population via 2D and 3D STE among subjects predisposed to risky health behaviors and risk factors of PHF. The risk factors of PHF could be behavioral as well as biological. However, the risk of these risk factors can be variable due to presence of protective factors, which may blunt the adverse effects of these risk factors on cardiac function. Figure 1 illustrates the pathways for the development of cardiac dysfunction leading to pre-heart failure, while Table 1 shows the risk factors and protective factors of HF.

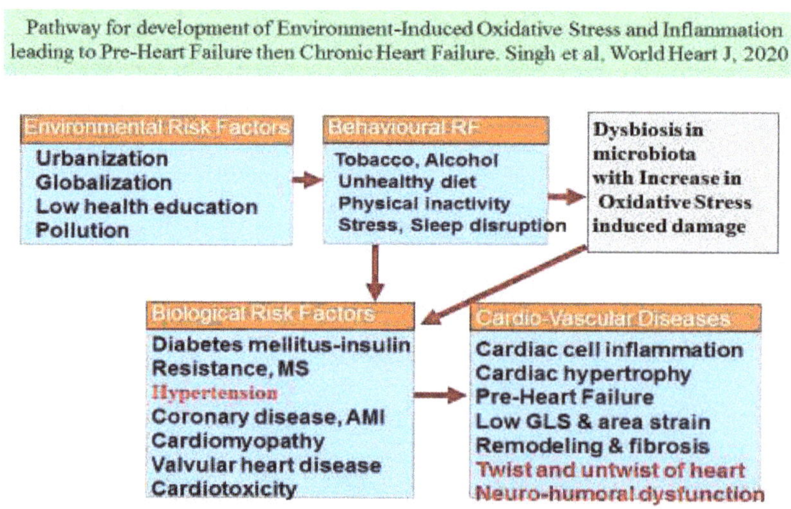

Figure 1. Pathways for the development of cardiac dysfunction leading to pre-heart failure [10].

Table 1. Risk factors and protective factors of heart failure.

Behavioral Risk Factors	Biological Risk Factors	Protective Factors
Western diet	Hypertension	Mediterranean-style diet
Tobacco use	Diabetes mellitus	Exercise training
Alcoholism	Coronary artery disease	Yoga posture
Mental disorders	Metabolic syndrome	Circadian restricted, day eating
Cardiotoxic drugs; cancer drugs	Rheumatic heart disease	Intermittent fasting
Sleep disorders	Cardiomyopathy	Optimal sleep
Sedentary behavior	Obesity	Meditation and prayer

3. Effects of Behavioral Risk Factors on Development of Pre-Heart Failure

Table 2 illustrates the effects of behavioral risk factors on pre-heart failure of the six stages of HF via speckle-tracking echocardiography. The six stages of HF have been proposed to identify earliest changes in cardiac dysfunction because early diagnosis may be useful in the primordial prevention [13,14]. There is much evidence that behavioral risk factors can independently predispose PHF [9,10]. It seems that STE, via 2D and 3D echo, has demonstrated abnormalities in global longitudinal strain (GLS) and area strain, indicating PHF in patients with primary risk factors [1–4]. There is an unmet need to conduct STE studies among patients with these risk factors for early diagnosis of PHF [9,10]. These patients with behavioral risk factors may not reveal any clinical manifestations of HF, and 2D echo may be within normal limits [9,10]. Early diagnosis at the stage of PHF, at first two stages of the six stages of HF, may be useful while treating for the reversal of HF [9,10].

Table 2. Effects of behavioral risk factors on pre-heart failure of the six stages of heart failure via speckle-tracking echocardiography.

Stages of HF	Tobacco Smoking	Alcoholism	Western Diets	Sleep Disorders	Emotional Disorders	Sedentary Behaviors
Stage A	Normal GLS	Normal GLS	Normal GLS	Normal GLS	Normal GLS	Normal GLS
Stage B	Normal GLS	Preserved GLS	Normal GLS	Preserved GLS	Normal GLS	Normal GLS
Stage C	Preserved GLS	Preserved or reduced GLS	Preserved GLS	Preserved or reduced GLS	Preserved GLS	Preserved GLS
Stage D	Slightly Reduced GLS	Reduced GLS	Slightly Reduced GLS	Reduced GLS	Slightly Reduced GLS	Slightly Reduced GLS
Stage E	Moderately reduced GLS	Heavily reduced GLS	Moderately reduced GLS	Heavily reduced GLS	Heavily reduced GLS	Heavily reduced GLS
Stage F	Heavily reduced GLS	Heavily reduced GLS	Heavily reduced GLS	Heavily reduced GLS	Heavily reduced GLS	Heavily reduced GLS

GLS = global longitudinal strain, HF = heart failure.

Previous studies indicated that Western-type diets rich in sweetened products, tobacco and alcoholism, short sleep, stress, and depression are known to be associated with increased risk of cardiovascular diseases (CVDs), including HF [15–27]. All these risk factors are known to cause oxidative dysfunction of the myocardium leading to HF [10,18]. Using 2D-STE myocardial imaging tobacco smoking has been confirmed to have adverse effects on the myocardium [17,28,29]. The global longitudinal strain (GLS), as assessed by 2D speckle-tracking echocardiography, was also significantly reduced in smokers ($p < 0.05$) [28,29]. It is possible that chronic tobacco smoking alters long-axis diastolic and systolic functions in healthy hearts with influence on the right ventricle, which may be detected early with

2D STE [28,29]. In another study of 100 healthy males (50 smokers and 50 non-smokers), all subjects underwent echocardiography to compare diastolic function of the left (LV) and right ventricles (RV) and examined the chronic and acute effects of smoking [28]. The atrial late diastolic mitral inflow velocity (Am) and atrial late diastolic septal mitral annular velocity were high in smokers compared to those of the control group before smoking. After 5 min, isovolumetric relaxation time (IVRT) was prolonged, and septal mitral annular velocity showed an increase as a sign of left ventricular diastolic dysfunction. In the right ventricle, the changes that occurred were similar and in favor of acute diastolic dysfunction. After 30 min, early diastolic tricuspid inflow (Et) decreased and septal mitral annular velocity and late diastolic tricuspid inflow (At) remained high. After 5 min, diastolic blood pressure increased but returned to normal after 30 min. However, pulmonary arterial pressure did not change before and after smoking. Smoking for a longer time causes left ventricular diastolic dysfunction, whereas acute intake causes left and right ventricular diastolic dysfunction [28]. Similar results have been found due to adverse effects of acute smoking in an earlier study [29].

Emotional stress may be a risk factor for CVDs, including HF [19–25]. In a cohort study, among 82,695 men, during a follow-up of mean 7.8 years (646,989 person-years), 3473 males had HF [19]. After controlling for time in sedentary behavior, HTN, DM, dyslipidemia, tobacco smoking, body mass index (BMI), and diet, the hazard ratio (95% confidence interval (CI)) of HF in the category of lower physical activity compared to subjects in the highest category was 1.52 [19]. Those in the category of medium physical activity also had greater risk (HR, 1.17 (95% CI, 1.06–1.29)). Medium sedentary time also conveyed risk (hazard ratio, 1.13) [19]. Sedentary behavior may be a risk factor for CHF. Among patients either with or without CAD, greater mental distress was associated with higher cardiovascular events in the future [22–25].

Increased heart rate can interfere with normal heart function and increase the risk of sudden cardiac arrest in patients with HF. Mental stress may also increase blood pressure in HTN and can lead to coronary disease, as well as weakening of the cardiac muscle and the immune system, leading to HF. There is much evidence that depression may predispose to increased platelet reactivity, decreased heart variability, and increased pro-inflammatory cytokines, which are all risk factors for HF [22–24]. In a cohort study of 136,637 patients with mental disorders—along with patients of post-traumatic stress disorder (PTSD), acute stress reaction, adjustment disorder, and other stress reactions—in which all subjects were followed up for 27 years and including 171,314 unaffected full siblings of these patients and 1,366,370 matched unexposed people from the general population [23]. The crude incidence rates of any CVDs were 10.5, 8.4, and 6.9 per 1000 person years during the follow-up of 27 years. In sibling-based comparisons, the HR for any CVDs was 1.64, with the greater subtype specific HR observed for HF (6.95, 1.88 to 25.68), during the first year after the diagnosis of any stress related disorder [23]. Beyond one year, the HRs became lower (overall 1.29), ranging from 1.12 for arrhythmia to 2.02 for artery thrombosis/embolus. It is possible that stress disorders had strong association with all types of CVDs, including HF, independent of family history of somatic/psychiatric diseases, and psychiatric comorbidity.

In a case–control study, the borderline personality disorder (BPD) group (n = 50) had a higher prevalence of cardiovascular risk factors for CVD compared to the control group (n = 50) [25]. Subjects in the BPD group revealed associations with chronic stress, major risk factors, and abnormalities in the myocardial wall motion as revealed by STE. It seems that chronic stress influences the structures in the cortico-limbic system and the nuclei responsible for the regulation of autonomic function, eliciting an imbalance of sympathetic and vagal activity, leading to adrenergic activation and withdrawal of vagal activity [24]. The adrenergic terminals of the nerves are connected to cardiac cells in a form of quasi-synasis, resembling the neuro-cardiac junction. The release of norepinephrine is increased, during chronic stress, leading to increased stimulation of cardiomyocytes via β_1-adrenergic receptors, causing mainly calcium dynamics and β_2-adrenergic receptors, which control housekeeping functions. There is impairment of the circadian rhythm of cardiomyocytes with elongation of the catabolic compared to the anabolic nocturnal phase, leading to a depletion of cell energy storage, with a reduced turnover of cell constituents [24]. The coupling concerning cardiomyocytes declines while there is an increase in the coupling between cardiomyocytes and fibroblasts, with alteration in the shape and velocity of action potential, fibroblast activation, and deposition of extracellular matrix. These alterations can predispose stress-related arrhythmias and cardiomyopathy.

There is profound adverse impact of disturbed sleep, such as in shift work, on circadian rhythms and of sleep on the cardiovascular system [26,27]. In a clinical study involving 784 consecutive patients, 165 patients had central sleep apnea, 139 had no sleep-disordered breathing; the remainder had obstructive sleep apnea (OSA) [26]. The rate ratio of cardiac readmissions in 6 months was 1.53; $p = 0.03$) in patients with central sleep apnea compared to no sleep-disordered breathing. The results revealed that severe OSA was also an independent predictor of readmissions with an adjusted rate ratio of 1.49 ($p = 0.04$). The impact of breathing due to sleep disorder on cardiac readmissions in HF showed that central sleep apnea was an independent risk factor for 6-month cardiac re-hospitalization. The effect of central sleep apnea on admissions for CVDs including HF exceeded that of all known predictors of readmissions [27].

4. Mechanisms and Pathophysiology of Pre-Heart Failure (PHF)

The mechanisms of adverse effects may be due to micro-inflammation in the mitochondrial signaling pathway [30], redox signaling in the cardiac myocyte [31] and Western diet induced cardiac ceramide [32]. These adverse effects may be inhibited by antioxidants either present in the diet or by antioxidant administration [33]. Oxidative dysfunction and inflammation in the cardiac tissues predisposes intracellular Ca^{2+} overload, via activation of proteases and phospholipases, mitochondrial dysfunction, and. alterations in cardiac gene expression and accompanied with alterations in the molecular structure, biochemical composition and physiology of various subcellular organelles. In turn, all these lead to pathological subcellular remodeling or myocardial deformation.

5. Effect of Biological Risk Factor on Pre-Heart Failure

Acute coronary syndromes (ACS); in-particular myocardial infarction (AMI), is the most common biological risk factor of PHF that may lead to CHF [34]. Other risk factors are obesity [11,35–37], cardiomyopathy [38], valvular heart disease [39], HTN [40–42], and DM [43–45]. On the other hand, AMI is a highly inflammatory state characterized with marked increase in catecholamine, cortisol and aldosterone, insulin, lipoprotein and brain natriuretic peptide. In turn, these induce acute reactants to fight the oxidative stress, which is damaging to cardiac cells, neurons, beta cells, and nephrons [34,35,46,47]. In a clinical study of a cohort of AMI survivors with ventricular dysfunction but no information of HF, plasma aldosterone levels estimated in the initial couple of days after confirmation was an independent indicator of remodeling [34]. Even in the absence of HF, higher plasma aldosterone concentrations—after ST-elevation myocardial infarction (STEMI) and non-STEMI—are associated with greater mortality. There is a consistent relationship between higher aldosterone concentrations and greater adverse remodeling. Remodeling may be due to the activation of both the cortisol and aldosterone limbs of corticosteroid synthesis, which may be responsible for myocardial fibrosis, leading to PHF [34].

6. Speckle-Tracking Echocardiography via 2D and 3D Echocardiography

It seems that 2D echocardiography is the commonest modality for imaging in the identification of cardiac hypertrophy [7,12–14]. However, in sub-clinical states, it is not able to detect subtle changes in cardiac structure. Recently, strain echocardiography, which is a new modality of echocardiography, has been found to identify subclinical dysfunction of the myocardium through measurement of intrinsic deformation of the myocardium. It is known that the co-existing abnormal conventional echocardiographic findings, such as LV systolic and diastolic dysfunction, abnormal LV geometry and LVH, and dilatation of the left atrium, can cause poor prognosis associated with biological risk factors of PHF. Since STE is a non-invasive modality examined using 2D and 3D echocardiography, it can provide information about myocardial deformation. STE may be more accurate in the detection of early cardiac ischemia and dysfunction of the myocardium undetected by the conventional echocardiographic examination before onset of symptoms [7,12–14].

The intrinsic myocardial performance can be measured through assessment of the strain rate and myocardial strain [40]. In addition, 2D-STE can be used to measure myocardial strain, by tracking ultrasonic speckles that are small footprints in the myocardium in 2-D images [48]. The myocardial speckles are tracked in each cardiac cycle, frame-by-frame, and the strain is calculated automatically by measuring the distances in the speckles [48]. It seems that deformation of the myocardium can show directions which include radial strain, longitudinal strain (LS), and circumferential strain (CS), where LS is commonly detected component of deformation assessed from averaging values. The left ventricular LS (LVLS) is under the influence of several factors, such as the geometry of chambers, the characteristics of myocardial tissue hemodynamic factors, and myocardial contraction synchrony. There may be a decline in LVLS due to tachycardia and increase in LV afterload, and the alteration in preload may be related to the change of LVLS [40]. Any increase in the thickness of LV walls or in the dilatation of the cavity of the LV may cause a decline in LS. There may be decline in the regional LS in the areas of infiltration as well as in the myocardial fibrosis after MI [40,48]. In the presence of a left bundle branch block, non-homogeneous myocardial activation can cause differences in the sequence of myocardial contraction of the ventricular septum and the LV lateral wall. In addition, age and gender can also affect LS. It is known that females have better LVLS values compared to males and that older adults have lower LS values than younger adults. Figure 2 shows the structural and functional alterations in various components of the heart due to risk factors of HF.

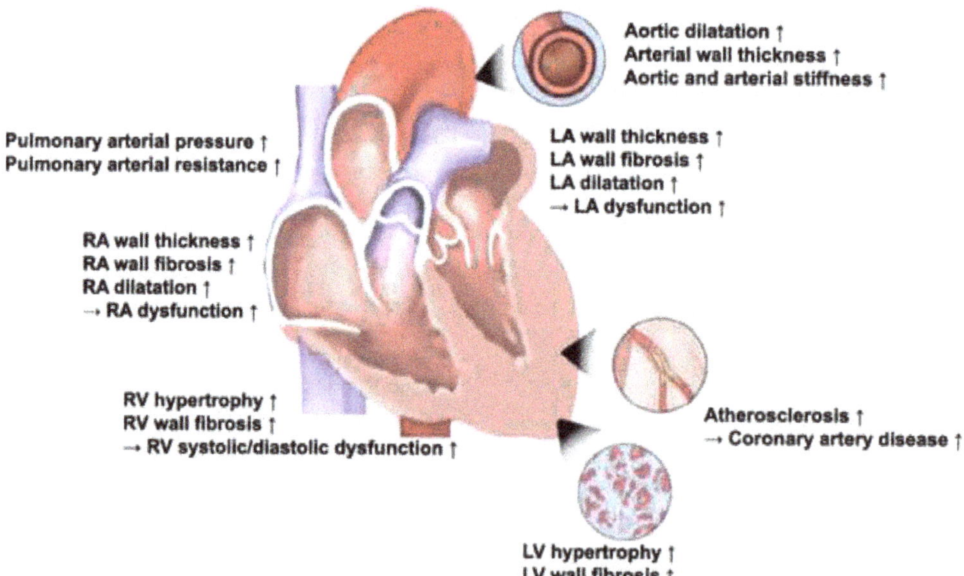

Overview of the structural and functional alterations present in hypertensive heart disease. LA, left atrium; LV, left ventricular; RA, right atrium; RV, right ventricular. Oh JK, Park JH. Role of strain echocardiography in patients with hypertension. Clin Hypertens. 2022 Feb 15;28(1):6. doi: 10.1186/s40885-021-00186-y.

Figure 2. Diagram showing structural and functional alterations in various components of the heart due to risk factors of heart failure. (Adapted from Reference [40], Oh and Park, 2022).

7. Effects of Hypertension (HTN) on Development of Pre-Heart Failure (PHF)

The early identification of cardiac changes due to HTN may allow reversal of the PHF [40–42]. HTN could be associated with complex and diverse alterations in the structure and function of the heart due to direct pressure on the cardiac cells. LV longitudinal strain during systole was impaired in patients with HTN and also in those with no LVH [41]. In LVH, radial strain may be decreased, and circumferential strain and twist may be increased. In hypertensive patients with normal EF, LS was significantly reduced, and basal-to-apical torsion became higher [42]. Higher serum concentrations of the tissue inhibitor matrix metalloproteinase-1, which is known to control myocardial collagen turnover, may be found among patients with impairment in the LS and increased torsion in the LV in patients with hypertension and normal LVEF. Excessive collection of collagens in fibroblasts may progress to myocardial fibrosis, leading to early LV contractile dysfunction. Since high blood pressure can impact all myocardial layers—including epicardial, mid-cardiac, and endocardial—LS may be lower than normal. It seems that epicardial LS could be an important predictor for CV events, including mortality and re-hospitalization in patients with hypertension.

8. Effects of Obesity, Diabetes, and Metabolic Syndrome on Pre-Heart Failure

In individuals with excess body weight, several pathophysiological processes can directly predispose the risk of PHF. These mechanisms are neuro-hormonal activation, hemodynamic alterations, hormonal effects of dysfunctional adipose tissue, ectopic fat deposition with resulting lip toxicity, and microvascular dysfunction. Overweight and obesity may indirectly cause PHF via causal increases in high BP, dyslipidemia, metabolic syndrome, and—most importantly—type 2 diabetes mellitus (T2DM) via producing insulin resistance [11]. Lower physical activity and fitness can further predispose risk of PHF in relation to obesity and insulin resistance (IR). These pathophysiological mechanisms may lead to myocardial-injury-induced cardiac remodeling, which is first physiological but later becomes pathological. The pathological remodeling is reflected via abnormal cardiac biomarkers and cardiovascular (CV) function on STE myocardial imaging.

Recently, Galal and coworkers have reported the prediction of preclinical dysfunction of the myocardium among patients with obesity and diabetes with preserved ejection fraction, by use of STE and tissue Doppler imaging [11]. In their study of 240 subjects with obesity, the results indicate that patients with obesity with T2DM should be advised to undertake early tissue Doppler imaging and STE for early diagnosis of decreased cardiac diastolic and systolic dysfunctions and cardiomyopathy [11]. These conditions are likely to be missed by conventional echocardiography. Significant differences in regional and global strains were also identified between the severely obese diabetic (BMI $\geq$ 35) (global longitudinal pre-systolic strain, -13) patients compared to less obese subjects.

The CARDIOBESE study included 100 obesity patients (body mass index (BMI) $\geq$ 35 kg/m^2) without known CVD and 50 age-matched and gender-matched non-obese controls (BMI $\leq$ 30 kg/m^2) [36]; 59 showed subclinical cardiac dysfunction. The data reveal that 57 patients had decreased global GLS and that two patients with normal GLS had either diastolic dysfunction or increased brain natriuretic peptide (BNP). In patients with obesity, autonomic dysfunction was observed as a significant independent risk factor for preclinical heart dysfunction. There was a greater prevalence (61%) of cardiac dysfunction in obesity patients—which was subclinical and without known CVD—diagnosed by GLS. In another case study, including 589 patients with obesity and 100 non-obese controls [37], the value of GLS was significantly lower in obesity compared to non-obese controls ($p < 0.001$). The values of global work index, efficiency, constrictive work, and wasted work in obese patients were also significantly lower compared to non-obese controls ($p < 0.05$), while global wasted work was significantly higher than that in nonobese controls ($p < 0.001$). It seems that higher BMI in patients with obesity may produce a greater risk of subclinical LV systolic dysfunction with coexisting normal LVEF [37].

The STE may also be helpful in the identification of subclinical signs of cardiomyopathy due to diabetes, although significant heterogeneity in the strain values has been reported in the literature [43]. A meta-analysis involving 41 valid studies, which included 6668 subjects with T2DM and 7218 controls, was carried out. The mean and mean difference (MD) in each group were pooled for left ventricular (LV) GLS, LV global circumferential strain (GCS), LV global radial strain (GRS), LV longitudinal systolic strain rate, left atrial reservoir strain (RS), and right ventricular GLS for the assessment. Interestingly, those patients with T2DM had 2 units lower LVGLS overall compared to control subjects (17.5%, vs. 19.5, mean difference = −1.96). The values of other strains were also lesser in patients with T2DM: LVGCS (mean difference = −0.89; LVGRS (mean difference = −5.03; LVSR (mean difference = −0.06; LARS (mean difference = −8.41; and right ventricular GLS (mean difference = −2.41. Higher body mass index was identified as the single greatest contributor to worse LVGLS, LVGCS, and LV strain rate. There was a worsening of right ventricular GLS in patients with higher HBA1c. Myocardial strains were decreased in the whole heart in T2DM patients. The largest decline was found in LA reservoir strain, followed by right ventricular GLS and LVGLS. In patients with T2DM, a greater body mass index was also associated with worse LV strain values. Table 3 illustrates the six stages of chronic heart failure among patients with biological risk factors indicating pre-heart failure via speckle-tracking echocardiography.

Table 3. Six stages of chronic heart failure among patients with biological risk factors indicating Pre-heart failure via speckle-tracking echocardiography.

Stages of Chronic Heart Failure	Obesity	Diabetes	Hypertension	CAD
Pre-heart Failure Stage 1 Dysfunction of the twist	Normal or slightly increased LVT, UNTR	Normal or slightly decreased GLS	Increased LVT	Slightly decreased LVT, UNTR
Pre-heart failure Stage 2 Sub-endocardial dysfunction	Slightly decreased GLS, LVT, UNTR	Slightly decreased GLS, LVT, UNTR	Increased LVT, delayed UNTR Decreased GLS	Decreased LVT, GLS
Pre-heart failure Stage 3 Asymptomatic physio-pathological remodeling+	Slightly increased LVT, preserved GLS	Moderate decreased GLS, LVT, UNTR	slightly increased LVT, delayed UNTR, preserved GLS	Decreased LVT, GLS
Complete heart failure Stage 4, LVH, symptomatic, diastolic dysfunction	Decreased LVT, UNTR, depressed GLS	Decreased GLS, LVT, UNTR	Preserved LVT, delayed or decreased UNTR, decreased GLS	Decreased LVT, decreased and delayed UNTR, decreased GLS
Complete heart failure Stage 5, Severe LVH	Decreased LVT, UNTR, GLS	Decreased LVT, UNTR, GLS	Decreased LVT UNTR, GLS	Decreased LVT, UNTR, GLS
Complete heart failure Stage 6, Severe LVH	Decreased LVT, UNTR, GLS	Decreased LVT, UNTRGLS	Decreased LVT, UNTR, GLS	Decreased LVT, UNTR, GLS

LVT = left ventricular twist, UNTR = untwist rate, CAD = coronary artery disease, GLS = global longitudinal strain, LVH = left ventricular hypertrophy.

In a case control study, 150 subjects with mean age; 62 ± 10 years, 82 males (55%) with metabolic syndrome were included, and they were divided into two groups based on the presence of T2DM [44]. Apart from liver fibrosis, LV diastolic dysfunction was found in 52% of the patients with diabetes and in 36% of non-diabetic patients with metabolic syndrome (p = 0.04). In subjects with diabetes, there was an increased left atrial stiffness (40% vs. 24%, p = 0.03) and decreased left ventricular GLS (47% vs. 16%, p < 0.0001) in nondiabetics via 2D-STE. The link between stiffness of the liver and subclinical cardiac dysfunction detected by 2D-STE in metabolic syndrome patients with T2DM is justified. These parameters derived from left atrial and left ventricular 2D-STE may have greater sensitivity compared to the older measurements and a substantial link to fibrosis of the liver.

In a case study, four groups of subjects were designated as Group I ($n = 56$), with T2DM and moderate HTN [45]; Group II ($n = 52$), with T2DM with normal blood pressure; Group III ($n = 54$), with grade II HTN without diabetes; and Group IV ($n = 30$, control), healthy subjects. The results showed that diastolic dysfunction was significantly ($p < 0.05$) more common in patients with a combination of T2DM and moderate HTN in contrast to patients with diabetes without HTN or those with HTN without diabetes. The combination of T2DM and HTN was associated with an increase in the longitudinal global deformation of the left ventricle compared with patients who had only one of these diseases ($p < 0.05$). A decrease in the global area strain, an early marker of LV systolic dysfunction, was expressed ($p < 0.05$) in patients with T2DM, regardless of the presence of concomitant HTN [45]. This study further shows the utility and necessity of tissue Doppler echocardiography and STE in the diagnosis of subclinical HF. The results reemphasize that high prevalence of subclinical systolic–diastolic LV dysfunction in T2DM may be aggravated in the presence of concomitant HTN in patients without obvious clinical signs of PHF and other CVDs. Although twist dysfunction is the beginning of PHF [48], there is a need to examine GLS in these patients [49].

9. Oxidative Function and Dysfunction in the Pathogenesis of Pre-Heart Failure

Risk factors may cause inflammation due to oxidative dysfunction in the cardiac ultra-structure. However, it is not well known how oxidative function of the myocardium and oxidative dysfunction can cause physio-pathological remodeling, predisposing to HF [2,48,50]. During metabolism, a small amount of reactive oxygen species (ROS) is generated in the cells for cell signaling, which may be spontaneously modulated by the endogenous antioxidant defenses to perform the oxidative function of the myocardium. Protective factors such as Indo-Mediterranean diets and physical training without any harmful health behaviors may be helpful in sustaining the oxidative functions of the heart and delay the development of cardiac cell dysfunction responsible for PHF. Oxidative stress is characterized by an imbalance in the generation of ROS and the availability of endogenous antioxidant enzymes and other defenses [18]. In the myocardial cells, the development and progression of maladaptive myocardial remodeling could be an early stage of HF with decline in GLS. Cardiac electrophysiology could be impaired by targeting contractile machinery and cardiac components via the dysfunction of proteins that are crucial to excitation–contraction coupling (ECC), including sodium channels, L-type calcium channels, potassium channels, ryanodine receptor (RyR), sarcoendoplasmic reticulum calcium ATPase (SERCA pump), and the sodium–calcium exchanges [2,18,51,52].

The biological alterations may begin with physiological cardiac cell remodeling to pathological remodeling due to uncontrolled increase in oxidative stress, inflammation and infiltration of fibrosis, which may increase many folds due to presence of other behavioral and biological risk factors [9–11]. In the presence of the risk factors, the generation of ROS could be enormous compared to the buffering ability of the cellular endogenous antioxidant defenses [2,7,48,50]. The increased production of ceramide due to high sugar and high fat in the diet also induces high advanced glycation end-products (AGE) and triamino-methyl-N-oxide (TMAO), which can predispose oxidative dysfunction due to over-loading of cytosolic calcium in the cardiac cells. These biological alterations may result in damage to protein and DNA, peroxidation of lipids with an increase in troponin T [2], and cell damage due to apoptosis [2,51,52]. Figure 3 reveals oxidative dysfunction in the heart due to Western diets and other risk factors with a decrease in antioxidant defenses, causing mitochondrial dysfunction and leading to electrophysiological dysfunction due to derangement in cation-transporting proteins (including L-type calcium channels, potassium channels, RyR). SERCA and NCX also showed twist and sub-endocardial dysfunction.

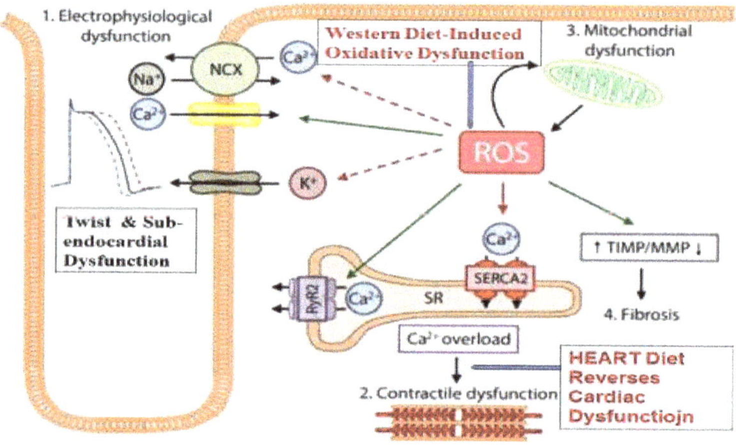

Figure 3. Oxidative dysfunction in the heart due to Western diets and other risk factors leading to decreases in antioxidant defenses causes mitochondrial dysfunction, leading to electrophysiological dysfunction with twist and sub-endocardial dysfunction. Exogenous antioxidants (HEART diet) and lifestyle modification may improve antioxidative function with reduced Ca overloading and reversal of mitochondrial and electrophysiological dysfunction. (Modified from Reference [18], Vander Pol A, Euro J Heart Failure 2019, under the license http://creativecommons.org/licenses/by-nc/4.0/, accessed on 1 January 2024).

Oxidative dysfunction may alter the activity of RyR and SERCA, which are associated with cellular calcium mobilization. In turn, this can lead to a decline in myofilament calcium sensitivity and protein function as well as in energy metabolism, resulting in diastolic dysfunction [51]. Thus, oxidative dysfunction induces loss of cardiac cells due to apoptosis, development of fibrosis, and alteration of physiological remodeling to pathological remodeling of the myocardium. Oxidative dysfunction can predispose a pro-fibrotic function as a physiological adaptation by causing the proliferation of cardiac cells, such as fibroblasts and matrix metallo-proteinases, for extracellular remodeling, initiating the hypertrophy of the heart [2,51,52]. The hypertrophy of the heart may be in multiple stages, depending upon the extent of oxidative dysfunction in the cardiac cells. These stages of PHF may be classified into six stages of HF, depending upon the effects of pathological remodeling on cardiovascular hemodynamics and electrophysiological dysfunction causing dysregulation in the twist and sub-endocardial function. Table 4 shows the clinical and echocardiographic features of six stages of heart failure proposed in patients with risk factors of heart disease.

It is crucial to know the biology of the six stages of HF via an understanding of twist function which may be the earliest change during oxidative dysfunction of the heart and that may require grading of the LV twist [39,53,54]. It is clear from the apex of the left ventricle that the cardiac twist or torsion indicates the mean longitudinal gradient of the net difference in the clockwise and counterclockwise rotation of the apex and base of the LV [55–57]. There is a storage of the cell energy during deformation of the LV twist in the sub-endocardial fiber matrix to complete the recoil. It is possible that it may facilitate release of restoring forces, causing further deformation in the recoil of twist and thus contributing to diastolic relaxation of the LV with early diastolic filling [48,55–57]. Interestingly, despite the normal ejection fraction, the systolic function could also be dysfunctional. Therefore, there may be greater subtle defects with a decline in the left ventricular systolic long axis at earlier stages of PHF. Bath et al. [12] have also emphasized the above points about the earlier diagnosis before the clinical manifestations of CHF. Lastly, HF may be reversed if behavioral risk factors—in particular, nutritional factors—are duly considered during the management of HF [58–61]. The clinical value of GLS is being gradually established to ascertain the atrial function [62] in patients with acute myocardial infarction [63,64].

Table 4. Clinical and echocardiographic features of six stages of heart failure proposed in patients with risk factors of heart disease.

Stages (HT on HF)	Manifestations	2D Echocardiography	2D Speckle-Tracking Echo	3D Speckle-Tracking Echo
Stage A	Mild to moderate Oxidative dysfunction, neuro-humoral dysfunction begins.	Increasing filling pressure with abnormal relaxation.	Dysfunctional untwist rate. LA strain reduced.	Dysfunctional untwist rate, LA strain reduced.
Stage B	Moderate oxidative dysfunction, hyper-rotation. Sub-endocardial dysfunction.	Dysfunction of systole.	Dysfunctional untwist rate and increased diastolic pressure. LA strain decreased.	Dysfunctional untwist rate and increased diastolic pressure. LA strain decreased.
Stage C, PHF	Asymptomatic Physio-Pathological remodeling+.	EF% normal > 53%	Normal GLS −20–−23% ≥−27.0% area strain	Normal GLS −17–21% Normal AS −31–−36%
Stage D, PHF and HFpEF	Pathological remodeling disease without symptoms of HF but elevated Natriuretic peptide, dyspnea on exertion.	EF% ≥ 50% Systolic LV dysfunction.	EF% 40–49% Impaired GLS −16–20% Impaired GCS, GRS Impaired early diastolic SR, right ventricular LS, and global RV longitudinal SR.	Impaired GLS−16–20% Impaired AS −27–31%
Stage E, HFmrEF	Structural heart disease with symptoms of HF.	EF% 40–49% Grade 1, diastolic dysfunction.	Reduced GLS−12–16%, reduced GCS, GRS, treat with ACE, ARB.ARNI.	GLS ≤ −16% AS ≤ −27%
Stage F, HFrEF	Refractory class III HF.	EF% < 40%	All above GLS < −12%, treated with ARNI	GLS < −13% AS < −27%

HT = hypertension, HF = heart failure, AS = aortic stenosis, HFpEF = Heart Failure with Preserved Ejection Fraction, GLS = global longitudinal strain, GCS = global circumferential strain, GRS = global radial strain, LV = left ventricle, SR = strain rate, ARNI = angiotensin receptor neprilysin inhibitor, mrEF = mild reduction in ejection fraction (EF), r = reduction in EF, LA = left atrial (Adapted from Reference [2]).

10. Modalities in Assessing Heart Failure

It seems that conventional echocardiography (CE) with color-encoded Doppler, pulsed, and continuous-wave Doppler are the best modalities for non-invasive imaging in the quantitative assessment of valvular heart diseases and cardiomyopathy [12,14]. However, left ventricular GLS is now considered as a more sensitive measure of LV myocardial systolic function compared to LVEF. Therefore, recent guidelines for diagnosis and treatment of valvular heart disease, CAD and cardiomyopathies emphasize the value of LVEF as one of the main parameters for the clinical decision-making [1,14]. Despite these considerations, it has been found that left ventricular GLS is a classical sensitive marker of LV contractility and inotropic state [1–3]. Therefore, GLS could be a better indicator of prognosis compared to LVEF in patients at increased risk of HF. Recent techniques for imaging, such as strain and computerized magnetic resonance, can quantify structural alterations—including cardiac mass, fibrosis, shape, and functional parameters; EF; blood flow; and other alterations most accurately with improved temporality. It is possible that the same information can have tremendous benefit in the planning of treatment and monitoring of health care in these patients. STE via 2D echo and 3D echo, and computerized magnetic resonance imaging (CMSI) may be beneficial in the identification of ACC Stage A, B, and C patients by detecting abnormal subclinical myocardial dysfunction and deformation at the earliest stages of PHF.

11. Limitations and Disadvantages

STE via the 3D echo is a novel technique, but it has limitations due to lower temporal resolution compared with 2D echo. Apart from this, the acquisition requires excellent windows and quality of image. The normal metrics in STE are quite vendor-sensitive; therefore, the normal values range for GLS, range from −18.5% to −22.0%. Moreover, there are gender-specific differences in the normal values of GLS, global circumferential strain, and central circumferential strain. Another disadvantage is that speckle motion is measured more at the endocardial border and less in the myocardium. In addition, the cardiologist/physician must understand and standardize the methodology and generate data for normal values that can be standardized for clinical use for each modality and across modalities to factor these into guidelines.

12. Prevention Strategies for Pre-Heart Failure and Consequences for Clinical Cardiologists

The pillars in the prevention strategies for pre-heart failure are related to lifestyle changes, the regulation of blood glucose levels, modification of risk factors for the development of CVDs and the treatment of HF combined with an element of psychological intervention to adhere to daily treatment and follow up educational programs by health professionals. Achieving euglycemia reduces the risk of major cardiovascular events, such as myocardial infarction or stroke, and the likelihood of developing pre-heart failure [65]. Other strategies involve early healthy screening tests, smoking cessation, limiting alcohol consumption and salt intake, reducing body weight, and avoiding obesity by eating moderate amounts daily and regular aerobic exercise are the cornerstones in terms of lifestyle change. The susceptible patient must also focus on a nutritious diet, such as the Mediterranean diet, which is low in sodium, fats, and carbohydrates but high in fruits, fibers, vegetables, and whole grains—foods which are rich in antioxidants—to manage blood pressure and weight [66]. Regular physical activity, as recommended by the Physical Activity Guidelines for Americans, plays a vital role, with a suggestion of at least 150 min of moderate aerobic exercise weekly to maintain cardiovascular health [67]. Achieving and maintaining a healthy weight range, ensuring adequate sleep, and reducing or managing stress levels and blood pressure are crucial. It is also important to take medication as prescribed and to avoid medications which may trigger arrhythmia.

Missed opportunities of early diagnosis are incidents in which different actions by the clinical cardiologist involved could have resulted in more desirable events. Thus, early screening and diagnosis can give the cardiologists more time to find a treatment that can keep the symptoms of the patients in check and prevent total HF and other health issues. The cardiologist must undertake a blood test for early markers, an electrocardiogram (ECG) to determine ant derangement in electrical activity of the heart and follow up with an echocardiogram or ultrasound technique to determine any dysfunction of the sound wave and valves of the myocardiogram. It is well established that HF is a complex clinical syndrome presenting as symptoms and signs common to other diagnoses, frequently in patients with multiple co-morbidities. HH itself is not a diagnosis, but it is the common clinical presentation of a variety of cardiac conditions. Correct diagnosis involves a combination of the clinical presentation, the results from general and specific investigations and diagnoses, and the ability of the clinical cardiologist to analyze the information for early intervention and treatment. Misdiagnosis can occur at any level during the development of HF and can occur because of factors related to the patient, the cardiologist and the health systems. As such, delayed diagnosis can lead to early—as well as excess—morbidity and mortality in these patients [40].

13. Conclusions

The continuous progression of the disease, occurring in PHF, begins with cardiac and neuro-humoral dysfunction initiated by behavioral and biological risk factors of HF. It seems that 2D echocardiography is considered the primary non-invasive imaging tool for the diagnosis of patients with CHF and for the assessment of prognosis. Both 2D echo and 3D echo have been used as the cornerstone or gold standard in the assessment of cardiomyopathy and valvular heart disease (VHD) and the method of choice in identifying and evaluating the morphology and severity of aortic, pulmonary, tricuspid, and mitral valve diseases. The combined use of 2D echo and 3D echo and STE has improved image quality and early diagnosis of cardiac dysfunction by identifying GLS and area strains as well as twist and untwist of the heart. Since ejection fraction does not indicate HF in many patients with PHF, and it is inconclusive in cardiac-dysfunction-causing PHF, there is an urgent need to develop and create evidence on the role of cardiac strain in the early diagnosis of HF. It is possible that GLS by STE analysis could be an accurate method in the early diagnosis of the severity of CAD in patients presenting with non-ST segment elevation (NSTSE) myocardial infarction as well as in subjects with behavioral risk factors and biological risk factors, who are likely to have PHF.

Funding: This research received no external funding.

Institutional Review Board Statement: Not applicable.

Informed Consent Statement: Not applicable.

Data Availability Statement: No new data were created or analyzed in this study. Data sharing is not applicable to this article.

Conflicts of Interest: The authors declare no conflict of interest.

References

1. Heidenreich, P.A.; Bozkurt, B.; Aguilar, D.; Allen, L.A.; Byun, J.J.; Colvin, M.M.; Deswal, A.; Drazner, M.H.; Dunlay, S.M.; Evers, L.R.; et al. 2022 AHA/ACC/HFSA Guideline for the Management of Heart Failure: A Report of the American College of Cardiology/American Heart Association Joint Committee on Clinical Practice Guidelines. *J. Am. Coll. Cardiol.* **2022**, *79*, e263–e421. [PubMed]
2. Singh, R.B.; Fedacko, J.; Pella, D.; Fatima, G.; Elkilany, G.; Moshiri, M.; Hristova, K.; Jakabcin, P.; Vaňova, N. High Exogenous Antioxidant, Restorative Treatment (Heart) for Prevention of the Six Stages of Heart Failure: The Heart Diet. *Antioxidants* **2022**, *11*, 1464. [CrossRef] [PubMed]
3. Stages of Heart Failure. Available online: https://my.clevelandclinic.org/health/diseases/17069-heart-failure-understanding-heart-failure#management-and-treatment (accessed on 1 January 2024).
4. Muraru, D.; Niero, A.; Rodriguez-Zanella, H.; Cherata, D.; Badano, L. Three-dimensional speckle-tracking echocardiography: Benefits and limitations of integrating myocardial mechanics with three dimensional imaging. *Cardiovasc. Diagn. Ther.* **2018**, *8*, 101–117. [CrossRef]
5. Congestive Heart Failure: Symptoms, Causes, Treatment, Types, Stages. Available online: https://www.webmd.com/heart-disease/guide-heart-failure (accessed on 2 January 2024).
6. Mann, D.L.; Bristow, M.R. Mechanisms and models in heart failure: The biomechanical model and beyond. *Circulation* **2005**, *111*, 2837–2849. [CrossRef] [PubMed]
7. Halabi, A.; Yang, H.; Wright, L.; Potter, E.; Huynh, Q.; Negishi, K.; Marwick, T.H. Evolution of myocardial dysfunction in asymptomatic patients at risk of heart failure. *JACC Cardiovasc. Imaging* **2021**, *14*, 350–361. [CrossRef]
8. Rahamim, E.; Nachman, D.; Yagel, O.; Yarkoni, M.; Elbaz-Greener, G.; Amir, O.; Asleh, R. Contemporary Pillars of heart failure with reduced ejection fraction medical therapy. *J. Clin. Med.* **2021**, *10*, 4409. [CrossRef]
9. Singh, R.B.; Elkilany, G.; Fedacko, J.; Hristova, K. (Eds.) The Six Stages of Chronic Heart Failure. In *Chronic Heart Failure;* Pathophysiology and Management; Elsevier: Cambridge, MA, USA, 2024.
10. Singh, R.B.; Komatsu, T.; Lee, M.C.; Dewi, M.; Watanabe, S. Effects of behavioral risk factors, with reference to smoking on cardiomyocyte dysfunction. *World Heart J.* **2020**, *11*, 191–196.
11. Elkilany, G.N.; Merril, E.; Ajash, H.; Singh, R.B.; Elkilany, G.Y.; Allah, S.B.; Nanda, N.C.; Singh, J.; Kabbash, I.; Sozzi, F. Prediction of preclinical myocardial dysfunction among obese diabetics with preserved ejection fraction using tissue doppler imaging and speckle tracking echocardiography. *World Heart J.* **2020**, *12*, 31–40.
12. Baath, A.S.; Singh, J.; Elsaady, A.; Singh, R.B.; Elmahal, M.; Lohana, P.; Elkilany, G.E. Advances in cardiac imaging via speckle tracking echocardiography and allied techniques: Early diagnosis of heart failure. *World Heart J.* **2021**, *13*, 391–406.
13. Singh, R.B.; Fedacko, J.; Elkilany, G.; Hristova, K.; Palmiero, P.; Pella, D.; Cornelissen, G.; Isaza, A.; Pella, D. 2020 Guidelines on Pre-Heart Failure in the Light of 2D and 3D Speckle Tracking Echocardiography. A Scientific Statement of the International College of Cardiology. *World Heart J.* **2020**, *12*, 51–70.
14. Singh, R.B.; Sozzi, F.B.; Fedacko, J.; Hristova, K.; Fatima, G.; Pella, D.; Cornelissen, G.; Isaza, A.; Pella, D.; Singh, J.; et al. Pre-heart failure at 2D- and 3D-speckle tracking echocardiography: A comprehensive review. *Echocardiography* **2022**, *39*, 302–309. [CrossRef] [PubMed]
15. Singhal, S.; Singh, R.B.; Haruichi, R.; Takahashi, T. Can high sugar diets induce chronic heart failure? *World Heart J.* **2017**, *9*, 189–192.
16. Tikellis, C.; Thomas, M.C.; Harcourt, B.E.; Coughlan, M.T.; Pete, J.; Bialkowski, K.; Tan, A.; Bierhaus, A.; Cooper, M.E.; Forbes, J.M. Cardiac inflammation associated with a Western diet is mediated via activation of RAGE by AGEs. *Am. J. Physiol. Endocrinol. Metab.* **2008**, *295*, E323–E330. [CrossRef]

17. Ilgenli, T.F.; Akpinar, O. Acute effects of smoking on right ventricular function. *Swiss Med. Wkly.* **2007**, *137*, 91–96.
18. van der Pol, A.; van Gilst, W.H.; Voors, A.A.; van der Meer, P. Treating oxidative stress in heart failure: Past, present and future. *Eur. J. Heart Fail.* **2019**, *21*, 425–435. [CrossRef] [PubMed]
19. Singh, R.B.; Fedacko, J.; Goyal, R.; Rai, R.H.; Nandave, M.; Tonk, R.K.; Gaur, S.S.; Gautam, R.; Chibisov, S. Pathophysiology and significance of troponin t, in heart failure, with reference to behavioural risk factors. *World Heart J.* **2020**, *12*, 15–22.
20. Henein, M.Y.; Vancheri, S.; Longo, G.; Vancheri, F. The Impact of Mental Stress on cardiovascular Health—Part II. *J. Clin. Med.* **2022**, *11*, 4405. [CrossRef] [PubMed]
21. Suzuki, H.; Matsumoto, Y.; Kaneta, T.; Sugimura, K.; Takahashi, J.; Fukumoto, Y.; Takahashi, S.; Shimokawa, H. Evidence for brain. activation in patients with Takotsubo cardiomyopathy. *Circ. J.* **2014**, *78*, 256–258. [CrossRef]
22. Pimple, P.; Lima, B.B.; Hammadah, M. Psychological distress and subsequent cardiovascular events in individuals with coronary artery disease. *J. Am. Heart Assoc.* **2019**, *8*, e011866. [CrossRef]
23. Song, H.; Fang, F.; Arnberg, F.K.; Mataix-Cols, D.; de la Cruz, L.F.; Almqvist, C.; Fall, K.; Lichtenstein, P.; Thorgeirsson, G.; Valdimarsdóttir, U.A. Stress related disorders and risk of cardiovascular disease: Population based, sibling controlled cohort study. *BMJ* **2019**, *365*, l1255, Correction *BMJ* **2019**, *365*, l1850. [CrossRef]
24. Barbiero, S.; Aimo, A.; Castiglione, V.; Giannoni, A.; Vergaro, G.; Passino, C.; Emdin, M. Healthy hearts at hectic pace: From daily life stress to abnormal cardiomyocyte function and arrhythmias. *Eur. J. Prev. Cardiol.* **2018**, *25*, 1419–1430. [CrossRef] [PubMed]
25. Aweimer, A.; Engemann, L.; Amar, S.; Ewers, A.; Afshari, F.; Maiß, C.; Kern, K.; Lücke, T.; Mügge, A.; El-Battrawy, I.; et al. Stress-Mediated Abnormalities in Regional Myocardial Wall Motion in Young Women with a History of Psychological Trauma. *J. Clin. Med.* **2023**, *12*, 6702. [CrossRef] [PubMed]
26. Cristine, J.R.; Tami, A.M. Disruption of circadian rhythms and sleep on critical illness and the impact on cardiovascular events. *Curr. Pharm. Des.* **2015**, *21*, 3505–3511.
27. Khayat, R.; Abraham, W.; Patt, B. Central sleep apnea is a predictor of cardiac readmission in hospitalised patients with systolic heart failure. *J. Card. Fail.* **2012**, *18*, 534–540. [CrossRef]
28. Seyedian, S.M.; Majididi, S.; Asgharinejad, L.; Esmaeili, M.; Ahmadi, F.; Salemzadeh, M. Acute and chronic effect of cigarette on right and left ventricular diastolic function. *Jentashapir J. Health Res.* **2015**, *6*, e25889. [CrossRef]
29. Morimoto, A.; Tatsumi, Y.; Deura, K.; Mizuno, S.; Ohno, Y.; Watanabe, S. Impact of cigarette smoking on impaired insulin secretion and insulin resistance in Japanese men: The Saku Study. *J. Diabetes Investig.* **2013**, *4*, 274–280. [CrossRef] [PubMed]
30. Zhang, Y.B.; Meng, Y.H.; Chang, S.; Zhang, R.Y.; Shi, C. High fructose causes cardiac hypertrophy via mitochondrial signalling pathway. *Am. J. Transl. Res.* **2016**, *8*, 4869–4880. [PubMed]
31. Santos, C.X.; Anilkumar, N.; Zhang, M.; Brewer, A.C.; Shah, A.M. Redox signalling in cardiac myocytes. *Free Radic. Biol. Med.* **2011**, *50*, 777–793. [CrossRef]
32. Butler, T.J.; Ashford, D.; Seymour, A.M. Western diet increases cardiac ceramide content in healthy and hypertrophied hearts. *Nutr. Metab. Cardiovas. Dis.* **2017**, *27*, 991–998. [CrossRef]
33. Singh, R.B.; Hristova, K.; Fedacko, J.; Singhal, S.; Khan, S.; Wilson, D.W.; Takahashi, T.; Sharma, Z. Antioxidant vitamins and oxidative stress in chronic heart failure. *World Heart J.* **2015**, *7*, 257–264.
34. Weir, R.A.P.; Tsorlalis, I.K.; Steedman, T.; Dargie, H.J.; Fraser, R.; McMurray, J.J.V.; Connell, J.M.C. Aldosterone and cortisol predict medium-term left ventricular remodelling following myocardial infarction. *Eur. J. Heart Fail.* **2011**, *13*, 1305–1313. [CrossRef] [PubMed]
35. Singh, R.B.; Elkilany, G.; Visen, A.S.; Visen, S.S. The puzzle of diastole and imbalance of risk factors and protective factors. *World Heart J.* **2021**, *13*, 487–497.
36. Snelder, S.M.; de Groot-de Laat, L.E.; Biter, L.U.; Castro Cabezas, M.; Pouw, N.; Birnie, E.; Boxma-de Klerk, B.M.; Klaassen, R.A.; Zijlstra, F.; van Dalen, B.M. Subclinical cardiac dysfunction in obesity patients is linked to autonomic dysfunction: Findings from the CARDIOBESE study. *ESC Heart Fail.* **2020**, *7*, 3726–3737. [CrossRef] [PubMed]
37. Huang, J.; Li, G.A.; Wang, J.; Jiao, Y.W.; Qian, Z.F.; Fan, L.; Tang, L.M. Evaluation of subclinical left ventricular systolic dysfunction in obese patients by global myocardial work. *Diabetol. Metab. Syndr.* **2023**, *15*, 254. [CrossRef]
38. Elkilany, G.E.; Ghobashi, A.S.; Salama, M.; Singh, J.; Allah, S.B.; Elmahal, M.; Singh, R.; Nanda, N.C. Prevalence and sub clinical detection of concomitant dilated cardiomyopathy in subjects with bicuspid aortic valves. *J. Cardiol. Cardiovasc. Ther.* **2020**, *16*, 4–8.
39. Elkilany, G.N.; Baath Allah, S.; Lohana, P.; Sozzi, F.; Singh, J.; Khorshid, M.; Singh, R.B.; Aiash, H. Sub-clinical detection of left ventricular myocardial dysfunction in valvular heart diseases: A state-of-the-art review in a speckle tracking echocardiography and myocardial performance. *J. Cardiol. Res. Rev. Rep.* **2020**, *1*, 11–13. [CrossRef]
40. Oh, J.K.; Park, J.H. Role of strain echocardiography in patients with hypertension. *Clin. Hypertens.* **2022**, *28*, 6. [CrossRef]
41. Imbalzano, E.; Zito, C.; Carerj, S.; Oreto, G.; Mandraffino, G.; Cusmà-Piccione, M.; di Bella, G.; Saitta, C.; Saitta, A. Left ventricular function in hypertension: New insight by speckle tracking echocardiography. *Echocardiography* **2011**, *28*, 649–657. [CrossRef]
42. Kang, S.J.; Lim, H.S.; Choi, B.J.; Choi, S.Y.; Hwang, G.S.; Yoon, M.H.; Tahk, S.J.; Shin, J.H. Longitudinal strain and torsion assessed by two-dimensional speckle tracking correlate with the serum level of tissue inhibitor of matrix metalloproteinase-1, a marker of myocardial fibrosis, in patients with hypertension. *J. Am. Soc. Echocardiogr.* **2008**, *21*, 907–911. [CrossRef]
43. Ghoreyshi-Hefzabad, S.M.; Jeyaprakash, P.; Vo, H.Q.; Gupta, A.; Ozawa, K.; Pathan, F.; Negishi, K. Subclinical systolic dysfunction detected by 2D speckle tracking echocardiography in adults with diabetes mellitus: Systematic review and meta-analysis of 6668 individuals with diabetes mellitus and 7218 controls. *Int. J. Cardiovasc. Imaging* **2023**, *39*, 977–989. [CrossRef]

44. Apostu, A.; Malita, D.; Arnautu, S.F.; Tomescu, M.C.; Gaiță, D.; Popescu, A.; Mare, R.; Gidea, R.; Arnautu, D.A. Significant Association between Subclinical Left Cardiac Dysfunction and Liver Stiffness in Metabolic Syndrome Patients with Diabetes Mellitus and Non-Alcoholic Fatty Liver Disease. *Medicina (Kaunas)* **2023**, *59*, 328. [CrossRef] [PubMed]
45. Tsvetkov, V.A.; Krutikov, E.S.; Chistyakova, S.I. Subclinical left ventricular dysfunction in patients with type 2 diabetes mellitus. *Probl. Endokrinol.* **2020**, *66*, 56–63. [CrossRef] [PubMed]
46. Simmonds, S.J.; Cuijpers, I.; Heymans, S.; Jones, E.A.V. Cellular and molecular differences between HFpEF and HFrEF: A step ahead in an improved pathological understanding. *Cells* **2020**, *9*, 242. [CrossRef] [PubMed]
47. Shin, S.H.; Suh, Y.J.; Baek, Y.S.; Lee, M.J.; Park, S.D.; Kwon, S.W.; Woo, S.I.; Kim, D.H.; Park, K.S.; Kwan, J. Impact of area strain by 3D speckle tracking on clinical outcome in patients after acute myocardial infarction. *Echocardiography* **2016**, *33*, 1854–1859. [CrossRef] [PubMed]
48. Sengupta, P.P.; Tajik, A.J.; Chandrasekaran, K.; Khandheria, B.K. Twist mechanics of the left ventricle: Principles and application. *JACC Cardiovasc. Imaging* **2008**, *1*, 366–376. [CrossRef]
49. Haugaa, K.H.; Dejgaard, L.A. Global longitudinal strain: Ready for clinical use and guideline implementation. *J. Am. Coll. Cardiol.* **2018**, *71*, 1958–1959. [CrossRef]
50. Santoro, C.; Arpino, G.; Esposito, R.; Lembo, M.; Paciolla, I.; Cardalesi, C.; de Simone, G.; Trimarco, B.; De Placido, S.; Galderisi, M. 2D and 3D strain for detection of subclinical anthracycline cardiotoxicity in breast cancer patients: A balance with feasibility. *Eur. Heart J. Cardiovasc. Imaging* **2017**, *18*, 930–936. [CrossRef]
51. Takimoto, E.; Kass, D.A. Role of oxidative stress in cardiac hypertrophy and remodeling. *Hypertension* **2007**, *49*, 241–248. [CrossRef]
52. Singh, R.B.; Elkilany, G.; Fedacko, J.; Hristova, K.; Palmiero, P.; Singh, J.; Manal, M.A.; Badran, H.M. Evolution of the natural history of myocardial twist and diastolic dysfunction as cardiac dysfunction. In *Chronic Heart Failure, Pathophysiology and Management*; Singh, R.B., Fedacko, J., Elkilany, G., Hristova, K., Eds.; Elsevier: Cambridge, MA, USA, 2024.
53. Marwick, T.H. Methods used for the assessment of LV systolic function: Common currency or tower of Babel? *Heart* **2013**, *99*, 1078–1086. [CrossRef]
54. Baath Allah, S.; Elmahal, M.; Askar, M.H.; Singh, J.; Khorshid, M.H.; Lohana, P.; Fedacko, J.; Elkilany, G.E.N. Myocardial deformation imaging meta-analysis in two cohorts of patients from UAE and Heart Hospital Hamad Medical Corporation: A potential role in assessment of coronary artery disease severity and myocardial viability. *J. Clin. Exp. Cardiolog.* **2020**, *11*, 1–10.
55. Johnson, C.; Kuyt, K.; Oxborough, D.; Stout, M. Practical tips and tricks in measuring strain, strain rate and twist for the left and right ventricles. *Echo. Res. Pract.* **2019**, *6*, 87–98. [CrossRef] [PubMed]
56. Abou, R.; van der Bijl, P.; Bax, J.J.; Delgado, V. Global longitudinal strain: Clinical use and prognostic implications in contemporary practice. *Heart* **2020**, *106*, 1438–1444. [CrossRef] [PubMed]
57. Singh, R.B.; Torshin, V.I.; Chibisov, S.; Goyal, R.K.; Watanabe, S.; Nakagami, H.; Mogi, M.; Nayak, B.N. Can protective factors inculcate molecular adaptations of cardiomyocyte in the prevention of chronic heart failure? *World Heart J.* **2019**, *11*, 149–157.
58. ICC; ICN; ISCN Committee. 2023 ICC/ICN/ISC Guideline for the Management of Heart Failure: A Report of the International College of Cardiology/International College of Nutrition and Indian Society of Chrono Medicine, Joint Committee on Guidelines, for Developing and Newly Industrialized Countries. ICC, ICN, ISCN Committee. *World Heart J.* **2023**, *14*, 127–151.
59. Aggrawal, M.; Bozkurt, B.; Panjrath, G.; Aggrawal, B.; Ostfield, R.J.; Barnard, N.D.; Gaggin, H.; Freemen, A.; Allen, K.; Madan, S.; et al. Lifestyle modifications for preventing and treating heart failure. *J. Amer. Coll. Cardiol.* **2018**, *72*, 2391–2405. [CrossRef]
60. Singh, R.B.; Elkilany, G.; Hristova, K.; Fedacko, J.; Elmarghi, O.; Chakravorty, S.; Shukla, A.K.; Jain, M.; Agarwal, A.; Yaduvanshi, A.; et al. Current controversies persisting in the recent guidelines for identification and classification of chronic heart failure. *World Heart J.* **2022**, *14*, 257–263.
61. Khouri, M.G.; Peshock, R.M.; Ayers, C.R.; de Lemos, J.A.; Drazner, M.H. A 4-tiered classification of left ventricular hypertrophy based on left ventricular geometry: The Dallas heart study. *Circ. Cardiovasc. Imaging* **2010**, *3*, 164–171. [CrossRef]
62. Yuda, S. Current clinical applications of speckle tracking echocardiography for assessment of left atrial function. *J. Echocardiogr.* **2021**, *19*, 129–140. [CrossRef]
63. Lacalzada, J.; de la Rosa, A.; Izquierdo, M.M.; Jiménez, J.J.; Iribarren, J.L.; García-González, M.J.; López, B.M.; Duque, M.A.; Barragán, A.; Hernández, C.; et al. Left ventricular global longitudinal systolic strain predicts adverse remodeling and subsequent cardiac events in patients with acute myocardial infarction treated with primary percutaneous coronary intervention. *Int. J. Cardiovasc. Imaging* **2015**, *31*, 575–584. [CrossRef]
64. Cong, T.; Sun, Y.; Shang, Z.; Wang, K.; Su, D.; Zhong, L.; Zhang, S.; Yang, Y. Prognostic value of speckle tracking echocardiography in patients with ST-elevation myocardial infarction treated with late percutaneous intervention. *Echocardiography* **2015**, *32*, 1384–1391. [CrossRef]
65. Hordern, M.D.; Coombes, J.S.; Cooney, L.M.; Jeffrises, L.; Prins, J.B.; Marwick, T.H. Effects of exercise intervention on myocardial function in type 2 diabetes. *Heart* **2009**, *95*, 1343–1349. [CrossRef] [PubMed]

66. Laffond, A.; Picon, C.R.; Rodriguez- Munoz, P.M.; Vela, R.J.; Vinhaspre-Hernandez, R.R.; Navas-Echavarria, N.; Sánchez-González, J.L. Mediterranean diet for primary and secondary prevention of cardiovascular disease and mortality: An updated systematic review. *Nutrients* **2023**, *15*, 3356. [CrossRef] [PubMed]
67. American Heart Association. American Heart Association Recommendations for Physical Activity in Adults and Kids. Available online: https://www.heart.org/en/healthy-living/fitness/fitness-basics/aha-recs-for-physical-activity-in-adults (accessed on 28 January 2024).

diagnostics

MDPI

Article

Neuroimaging and Genetic Markers of Cerebral Small Vessel Disease and Cognitive Outcomes: A Systematic Review and Meta-Analysis (NEUROGEN-SVD Study)

Chelsea Jin [1,2], Roy G. Beran [1,2,3,4,5,6,7] and Sonu M. M. Bhaskar [1,3,6,7,8,*]

1 Global Health Neurology Lab, Sydney, NSW 2150, Australia
2 South West Sydney Clinical Campuses, UNSW Medicine and Health, University of New South Wales (UNSW), Sydney, NSW 2170, Australia
3 NSW Brain Clot Bank, NSW Health Pathology, Sydney, NSW 2170, Australia
4 School of Medicine, Western Sydney University, Sydney, NSW 2000, Australia
5 Griffith Health, School of Medicine and Dentistry, Griffith University, Southport, QLD 4215, Australia
6 Clinical Sciences Stream, Ingham Institute for Applied Medical Research, Liverpool, NSW 2170, Australia
7 Department of Neurology & Neurophysiology, Liverpool Hospital and South West Sydney Local Health District, Liverpool, NSW 2150, Australia
8 Department of Neurology, Division of Cerebrovascular Medicine and Neurology, National Cerebral and Cardiovascular Center (NCVC), Suita 564-8565, Osaka, Japan
* Correspondence: sonu.bhaskar@globalhealthneurolab.org; Tel.: +81-90-9274-1265

Abstract: Background/Objectives: Cerebral small vessel disease (CSVD) is a leading cause of cognitive decline and dementia. The comparative prognostic value of MRI-based neuroimaging markers and genetic risk factors such as the APOE ε4 allele for cognitive outcomes remains uncertain. The objectives of this study were to estimate the pooled prevalence of cognitive impairment in CSVD, evaluate the associations of key neuroimaging markers (white matter hyperintensities [WMHs], cerebral microbleeds [CMBs], lacunes) and APOE ε4 with cognitive outcomes, and assess their diagnostic performance. **Methods:** This study included a systematic review and meta-analysis in accordance with PRISMA and MOOSE guidelines, searching five databases (2005–2025). Eligible studies included adults with CSVD and MRI-visible markers reporting cognitive outcomes (mild cognitive impairment [MCI], global cognitive impairment [GCI], all-cause dementia [ACD], vascular dementia [VaD], and Alzheimer's disease [AD]). Thirty-nine studies comprising 18,425 participants were included. Pooled prevalence and associations were estimated using random-effects models, and diagnostic accuracy was evaluated. Certainty of evidence was assessed using the GRADE framework. **Results:** The pooled prevalence of GCI in CSVD was 57% (95% CI: 51–62%), while MCI prevalence was 46% (95% CI: 42–51%). WMHs were strongly associated with VaD (OR 10.35, 95% CI: 7.32–14.64), lacunes with ACD (OR 3.18, 95% CI: 1.24–8.20), and CMBs with AD (OR 1.52, 95% CI: 1.04–2.24). APOE ε4 carriage increased the risk of GCI (OR 1.80, 95% CI: 1.41–2.29). Across markers, diagnostic sensitivity was low, specificity was moderate-to-high, and AUROC values were modest. GRADE certainty ranged from low to moderate, with the highest confidence for WMHs and VaD. **Conclusions:** CSVD-related MRI markers and APOE ε4 are significantly associated with both early and late cognitive outcomes, supporting the integrated vascular–neurodegenerative continuum. The limited diagnostic sensitivity and variable certainty of evidence highlight the need for harmonized definitions, lesion quantification, and multimodal imaging–genetic approaches to improve early detection and risk stratification of CSVD-related cognitive impairment.

https://doi.org/10.3390/diagnostics15202585

Keywords: cerebral small vessel disease (CSVD); cognitive impairment; magnetic resonance imaging; APOE ε4; meta-analysis

1. Introduction

Cerebral small vessel disease (CSVD) encompasses a spectrum of microvascular brain pathologies that play a major role in the development of cognitive impairment and dementia [1,2]. Dementia ranks as the seventh leading cause of mortality globally [3] and is projected to increase in prevalence by 66% by 2050 [4]. CSVD is estimated to contribute to nearly 50% of all dementia cases [5], highlighting the urgency for more accurate diagnostic and predictive tools.

Magnetic resonance imaging (MRI) plays a crucial role in detecting the hallmark features of CSVD [6], including white matter hyperintensities (WMHs) [7], cerebral microbleeds (CMBs) [6,8], and lacunes [9]. The relative associations of these neuroimaging markers with different subtypes and stages of cognitive impairment remain inadequately characterized.

In addition to imaging, genetic risk factors, particularly the apolipoprotein E ε4 (APOE ε4) allele [10], have emerged as important contributors to cognitive decline in CSVD populations [11]. Yet, the combined utility of neuroimaging and genetic profiling in understanding and predicting CSVD-related cognitive outcomes is not well established [12]. Moreover, the mechanistic interplay between vascular injury and neurodegenerative processes, potentially mediated by APOE ε4 [13,14], warrants further investigation [15].

The NEUROGEN-SVD (NEUROimaging and GENetic determinants in Small Vessel Disease–related dementia) study was designed to address key gaps in understanding the relationship between CSVD and cognitive decline through a comprehensive meta-analysis and systematic review. Its objectives are threefold: first, to estimate the pooled prevalence of cognitive impairment among individuals with CSVD; second, to evaluate associations between MRI-based CSVD markers and cognitive outcomes; and third, to assess the impact of genetic risk factors, particularly the APOE ε4 allele, on cognitive impairment in CSVD populations.

2. Materials and Methods

2.1. Literature Search and Study Selection

The NEUROGEN-SVD study was conducted in accordance with PRISMA 2020 (Supplemental Table S1) and the Meta-analysis Of Observational Studies in Epidemiology (MOOSE) (Supplemental Table S2) guidelines. A comprehensive search strategy was employed across five databases, namely, PubMed, Embase, Cochrane Library, Scopus, and Web of Science, for studies published between January 2005 and March 2025. The search terms combined keywords and MeSH terms such as "cerebral small vessel disease," "CSVD," "MRI," "APOE ε4," "vascular cognitive impairment," and "dementia." The full search strategy is detailed in the Online Supplemental Information. Study screening and selection were guided by the PRISMA flowchart (Figure 1).

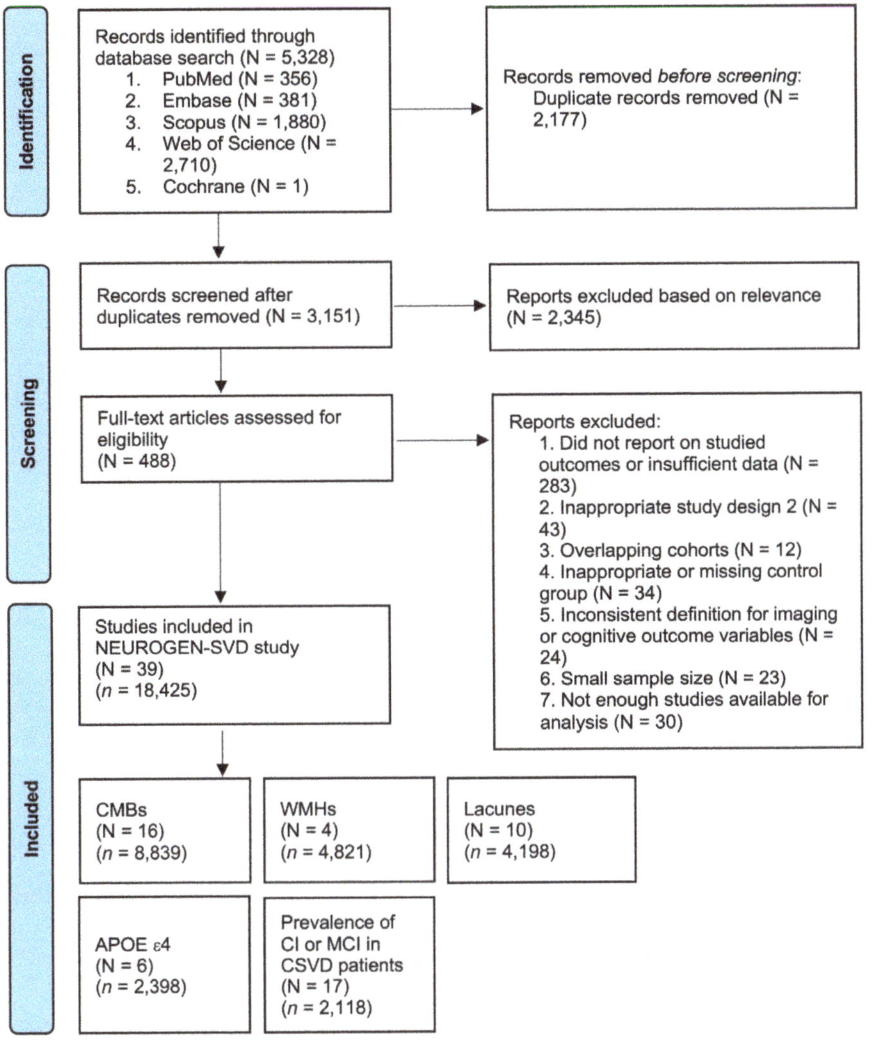

Figure 1. PRISMA flow diagram of study selection process for the NEUROGEN-SVD meta-analysis. The above illustration depicts the study selection flow according to the PRISMA guidelines, leading to the inclusion of studies in the meta-analysis. Abbreviations: NEUROGEN-SVD: NEUROimaging and GENetic determinants in Small Vessel Disease–related dementia; APOE: apolipoprotein E; GCI: global cognitive impairment; CMBs: cerebral microbleeds; CSVD: cerebral small vessel disease; MCI: mild cognitive impairment; WMHs: white matter hyperintensities.

2.2. Inclusion and Exclusion Criteria

Inclusion criteria encompassed (a) adult human subjects (≥18 years); (b) the presence of CSVD as defined by STRIVE criteria; (c) cognitive outcomes defined by clear diagnostic criteria; (d) comparative data between cognitively impaired and unimpaired individuals with respect to neuroimaging markers and/or APOE ε4 genotype; and (e) sample size ≥ 20. Exclusion criteria included non-English publications (unless translated); pediatric/animal studies; case reports; reviews; lack of full text; and studies without extractable data.

2.3. Definitions of CSVD Imaging Markers, Genetic Risk Factors, and CSVD

STRIVE [16] neuroimaging standards were used to define the selected imaging markers for CSVD. The markers with sufficient studies meeting eligibility criteria included in this meta-analysis were white matter hyperintensities (WMHs): the presence of moderate-severe (score of 2–3 on the Fazekas [17] scale); cerebral microbleeds (CMBs): the presence of any CMBs; lacunes: the presence of any lacunes; and genetic risk factor: APOE ε4 carrier status. The criteria used to diagnose CSVD varied across included studies and are detailed in Tables 1–3.

2.4. Definitions of Cognitive Outcomes

The primary outcome was global cognitive impairment (GCI) as defined by each study. Secondary outcomes included mild cognitive impairment (MCI), all-cause dementia (ACD), vascular dementia (VaD), and Alzheimer's disease (AD). Imaging markers were WMH, lacunes, and CMBs; the genetic marker was APOE ε4 carrier status. Where reported, WMH severity (Fazekas) was captured for dose–response analyses. The diagnostic criteria varied and are detailed in Tables 1–3.

2.5. Data Extraction

The titles and abstracts of all articles were initially reviewed using Endnote (Clarivate Analytics, London, UK) to exclude articles that did not meet the eligibility criteria. The remaining articles were comprehensively examined to determine their suitability for inclusion in the meta-analysis, in accordance with the defined eligibility criteria. Data extraction was conducted using a dedicated data extraction sheet, recording the following information from each study:

1. Baseline study demographics: author, country, publication year, cohort size, and study design.
2. Patient demographics: age and sex.
3. CSVD neuroimaging marker: WMHs, CMBs, and lacunes.
4. Genetic risk factor: APOE ε4 allele carrier status.
5. CSVD neuroimaging marker and genetic risk factor characteristics: imaging marker score, CSVD diagnostic criteria, and MRI sequence.
6. Cognitive outcome: MCI, ACD, VaD, AD, and GCI.
7. Cognitive outcome characteristics: cognitive diagnostic criteria.

2.6. Methodological Quality Assessment of Included Studies

The methodological quality of included studies was assessed using the modified Jadad analysis (MJA) [18], which was completed independently by the primary researcher (Supplemental Table S1). The risk of bias due to funding was also evaluated by assessing the declaration of funding sources and conflicts of interest for each study (Supplemental Table S2).

Table 1. Characteristics of prevalence studies included in the NEUROGEN-SVD meta-analysis.

Study ID	Author	Year	Country	Study Design	CSVD Criteria	Cognitive Outcome	GCI Criteria	MCI Criteria	Age (Mean +/− SD)	Female (*n*)	Number of Patients with Cognitive Outcome	Cohort Size	Prevalence
4	Dobrynina et al. [19] (a)	2024	Russia	Cross-sectional	STRIVE	GCI	MOCA	-	59.9 (7.6)	-	111	166	66.9%
4	Dobrynina et al. [19] (b)	2024	Russia	Cross-sectional	STRIVE	MCI	-	DSM-IV	59.9 (7.6)	73	71	166	42.8%
6	Ferro et al. [20]	2017	Netherlands	Cross-sectional	WMHs Fazekas score 2–3, or lacunar infarcts, non-lacunar infarcts, CMBs, intracranial hemorrhage, or WMHs Fazekas 1 and $\geq$2 vascular risk factors	MCI	-	AHA/ASA VCID	-	-	61	131	46.6%
7	Han et al. [21]	2024	China	Prospective	STRIVE	MCI	-	AHA/ASA VCID	66.2 (6.7)	-	36	69	52.2%
12	Ke et al. [22]	2022	China	Cross-sectional	WMHs Fazekas score 2–3 and/or lacunar infarcts, with or without PVS, CMBs, brain atrophy	GCI	AHA/ASA VCID		-	54	81	137	59.1%
14	Lee et al. [23]	2017	Korea	Prospective	Moderate-severe periventricular WMHs, severe deep WMHs	MCI	Study-specific protocol	Study-specific protocol	74.0 (6.9)	45	33	72	45.8%
17	Liao et al. [24]	2024	China	Cross-sectional	2 or more of WMHs Fazekas score 2–3, lacunes, moderate–severe PVS, CMBs	GCI	MMSE	-	65.9 (10.9)	29	39	94	41.5%
18	Liu et al. [25]	2021	China	Cross-sectional	STRIVE	GCI	Study-specific protocol	-	69.0 (7.8)	92	112	199	56.3%
25	Song et al. [26]	2022	China	Cross-sectional	STRIVE	GCI	NINDS-CSN	-	61.1 (5.0)	45	79	156	50.6%
26	Sun et al. [27]	2022	China	Cross-sectional	WMHs Fazekas score 2–3 and/or lacunar infarct with or without PVS, CMBs, brain atrophy	GCI	MMSE, MOCA	-	-	108	135	242	55.8%
27	Tang et al. [28]	2022	China	Cross-sectional prospective	WMHs Fazekas score 2–3 and at least one of CMBs, lacunes, PVS	GCI	MOCA	-	65.5 (7.7)	48	83	133	62.4%

Table 1. *Cont.*

Study ID	Author	Year	Country	Study Design	CSVD Criteria	Cognitive Outcome	GCI Criteria	MCI Criteria	Age (Mean +/− SD)	Female (n)	Number of Patients with Cognitive Outcome	Cohort Size	Prevalence
31	Wang et al. [29]	2023	China	Cross-sectional	WMHs Fazekas score 2–3 OR Fazekas score 1 and vascular risk factors	GCI	MMSE	-	-	14	31	51	60.8%
32	Wei et al. [30]	2019	China	Cross-sectional	WMHs	GCI	MOCA, CDR	-	62.7 (9.1)	53	78	113	69.0%
34	Xing et al. [31]	2021	China	Cross-sectional	WMHs Fazekas score 2–3	MCI	-	DSM-IV	66.4 (6.9)	37	44	77	57.1%
35	Xu et al. [32]	2024	China	Cross-sectional	STRIVE	MCI	-	Study-specific protocol	61.7 (9.2)	76	87	185	47.0%
37	Zhu et al. [33]	2024	China	Cross-sectional	WMHs, lacunes, CMBs, PVS	GCI	MOCA		67.0 (7.0)	71	100	227	44.1%
38	Zhu et al. [34]	2021	China	Cross-sectional	WMHs Fazekas score 3–6 (sum of PVWMHs and DWMHs)	MCI	-	Study-specific protocol	65.5 (6.2)	24	23	66	34.8%

Abbreviations: AHA/ASA VCID: American Heart Association/American Stroke Association Vascular Cognitive Impairment and Dementia; CDR: Clinical Dementia Rating; GCI: global cognitive impairment; CMBs: cerebral microbleeds; CSVD: cerebral small vessel disease; DSM-IV: Diagnostic and Statistical Manual of Mental Disorders, Fourth Edition; DWMHs: deep white matter hyperintensities; MCI: mild cognitive impairment; MMSE: Mini-Mental State Examination; MOCA: Montreal Cognitive Assessment; PVS: perivascular spaces; PVWMHs: periventricular white matter hyperintensities; STRIVE: Standards for Reporting Vascular Changes on Neuroimaging; WMHs: white matter hyperintensities. Note: (a), (b) indicate separate sub-analyses or cohorts from the same study reporting distinct cognitive outcomes.

Table 2. Characteristics of studies assessing CSVD neuroimaging markers.

Study ID	Author	Year	Country	Study Design	MRI Sequence	Cohort Size	Cognitive Outcome	MCI Criteria	GCI Criteria	VaD Criteria	AD Criteria	ACD Criteria	Age (Mean +/− SD)	Female (n)	Imaging Marker (n)	Imaging Marker (%)
								CMBs								
2	Chen et al. [35]	2018	China	Prospective	SWI	82	ACD	-	-	-	-	CDR	68.2 (10.1)	20	14	17.1%
3	Ding et al. [36] (a)	2017	Iceland	Prospective	T2 * GRE	2601	ACD	-	-	-	-	DSM-IV	74.6 (4.8)	1532	20	0.8%
3	Ding et al. [36] (b)	2017	Iceland	Prospective	T2 * GRE	2601	VaD	-	-	ADDTC	-	-	74.6 (4.8)	1532	6	0.2%
3	Ding et al. [36] (c)	2017	Iceland	Prospective	T2 * GRE	2601	AD	-	-	-	NINCDS-ADRDA	-	74.6 (4.8)	1532	12	0.5%

Table 2. *Cont.*

Study ID	Author	Year	Country	Study Design	MRI Sequence	Cohort Size	Cognitive Outcome	MCI Criteria	GCI Criteria	VaD Criteria	AD Criteria	ACD Criteria	Age (Mean +/− SD)	Female (*n*)	Imaging Marker (*n*)	Imaging Marker (%)
5	Fan et al. [37]	2021	China	Cross-sectional	SWI	293	GCI	-	MOCA	-	-	-		174	39	13.3%
8	Hilal et al. [38] (a)	2015	Singapore	Cross-sectional	-	572	GCI	-	Study-specific protocol	-	-	-	70.5 (6.8)	313	152	26.6%
8	Hilal et al. [38] (b)	2015	Singapore	Cross-sectional	-	347	MCI	Study-specific protocol	-	-	-	-	68.6 (5.8)	157	62	17.9%
8	Hilal et al. [38] (c)	2015	Singapore	Cross-sectional	-	204	ACD	-	-	-	-	DSM-IV	68.8 (6.5)	98	12	5.9%
11	Jacob et al. [39] (a)	2023	Netherlands	Prospective	T2 *	498	ACD	-	-	-	-	DSM-IV	65.7 (8.8)	217	22	4.4%
11	Jacob et al. [39] (b)	2023	Netherlands	Prospective	T2 *	424	VaD	-	-	NINDS-AIREN	-	-	64.5 (8.6)	217	14	3.3%
11	Jacob et al. [39] (c)	2023	Netherlands	Prospective	T2 *	428	AD	-	-	-	NIA-AA	-	64.6 (8.7)	217	4	0.9%
15	Li et al. [40]	2021	China	Retrospective	SWI	270	MCI	NIA-AA	-	-	-	-	66.5 (8.7)	118	48	17.8%
17	Liao et al. [24]	2024	China	Cross-sectional	-	94	GCI	-	MMSE	-	-	-	65.9 (10.9)	29	25	26.6%
16	Li et al. [41] (a)	2020	China	Prospective	T2 * GRE	792	MCI	Study-specific protocol	-	-	-	-	72.6 (7.1)	366	124	15.7%
16	Li et al. [41] (b)	2020	China	Prospective	T2 * GRE	792	AD	-	-	-	NINCDS-ADRDA	-	72.6 (7.1)	366	39	4.9%
18	Liu et al. [25]	2021	China	Cross-sectional	-	199	GCI	-	Other	-	-	-	69.0 (7.8)	92	34	17.1%
21	Paradise et al. [42]	2019	Australia	Prospective	SWI	267	ACD	-	-	-	-	DSM-IV/DSM-V	-	-	8	3.0%
23	Romero et al. [43] (a)	2018	US	Prospective	T2 * GRE	1296	ACD	-	-	-	-	DSM-IV	72 (8)	585	17	1.3%
23	Romero et al. [43] (b)	2018	US	Prospective	T2 * GRE	1296	VaD	-	-	NINDS-AIREN	-	-	72 (8)	586	4	0.3%
23	Romero et al. [43] (c)	2018	US	Prospective	T2 * GRE	1296	AD	-	-	-	NINCDS-ADRDA	-	72 (8)	587	13	1.0%
24	Shaikh et al. [44] (a)	2022	US	Cross-sectional	-	94	MCI	NIA-AA	-	-	-	-	69.6 (7.5)	52	7	7.4%

Table 2. *Cont.*

Study ID	Author	Year	Country	Study Design	MRI Sequence	Cohort Size	Cognitive Outcome	MCI Criteria	GCI Criteria	VaD Criteria	AD Criteria	ACD Criteria	Age (Mean +/− SD)	Female (*n*)	Imaging Marker (*n*)	Imaging Marker (%)
24	Shaikh et al. [44] (b)	2022	US	Cross-sectional	-	89	AD	-	-	-	NIA-AA	-	68.6 (6.9)	52	1	1.1%
28	Uetani et al. [45] (a)	2013	Japan	Cross-sectional	SWI	347	GCI	International Working Group on MCI	-	NINDS-AIREN	NINCDS-ADRDA	-	74.3 (8.8)	217	153	44.1%
28	Uetani et al. [45] (b)	2013	Japan	Cross-sectional	SWI	83	MCI	International Working Group on MCI	-	-	-	-	74.1 (9.5)	55	21	25.3%
28	Uetani et al. [45] (c)	2013	Japan	Cross-sectional	SWI	296	ACD	-	-	NINDS-AIREN	NINCDS-ADRDA	-	-	188	132	44.6%
28	Uetani et al. [45] (d)	2013	Japan	Cross-sectional	SWI	60	VaD	-	-	NINDS-AIREN	-	-	73.3 (10.0)	40	24	40.0%
28	Uetani et al. [45] (e)	2013	Japan	Cross-sectional	SWI	194	AD	-	-	NINDS-AIREN	NINCDS-ADRDA	-	74.3 (9.5)	134	77	39.7%
33	Wrigley et al. [46] (a)	2024	Australia	Retrospective	SWI	219	MCI	Study-specific protocol	-	-	-	-	-	112	37	16.9%
33	Wrigley et al. [46] (b)	2024	Australia	Retrospective	SWI	234	ACD	-	-	-	-	Study-specific protocol	-	115	50	21.4%
35	Xu et al. [32]	2024	China	Retrospective		185	MCI	Study-specific protocol	-	-	-	-	61.7 (9.2)	76	18	9.7%
39	Zonneveld et al. [47] (a)	2014	Netherlands	Cross-sectional	SWI	311	MCI	Petersen's criteria	-	-	-	-	64.2 (10.2)	125	40	12.9%
39	Zonneveld et al. [47] (b)	2014	Netherlands	Cross-sectional	SWI	509	ACD	-	-	-	-	Other	-	-	107	21.0%
39	Zonneveld et al. [47] (c)	2014	Netherlands	Cross-sectional	SWI	180	VaD	-	-	NINDS-AIREN	-	-	61.5 (10.0)	80	9	5.0%
39	Zonneveld et al. [47] (d)	2014	Netherlands	Cross-sectional	SWI	417	AD	-	-	-	NINCDS-ADRDA		65.2 (10.0)	204	79	18.9%

Table 2. *Cont.*

Study ID	Author	Year	Country	Study Design	MRI Sequence	Cohort Size	Cognitive Outcome	MCI Criteria	GCI Criteria	VaD Criteria	AD Criteria	ACD Criteria	Age (Mean +/− SD)	Female (n)	Imaging Marker (n)	Imaging Marker (%)
Moderate–severe WMHs																
4	Dobrynina et al. [19]	2024	Russia	Cross-sectional	-	126	MCI	DSM-V	-	-	-	-	59.9 (7.6)	73	64	50.8%
22	Rennie et al. [48] (a)	2024	Sweden	Cross-sectional	-	2994	MCI	ICD-10	-	-	-	-	70.2 (9.2)	1592	388	13.0%
22	Rennie et al. [48] (b)	2024	Sweden	Cross-sectional	-	2213	AD	-	-	-	ICD-10	-	69.0 (8.6)	1299	188	8.5%
22	Rennie et al. [48] (c)	2024	Sweden	Cross-sectional	-	1653	VaD	-	-	ICD-10	-	-	68.1 (8.4)	942	92	5.6%
28	Uetani et al. [45] (a)	2013	Japan	Cross-sectional	-	83	MCI	International Working Group on MCI	-	-	-	-	74.1 (9.5)	55	20	24.1%
28	Uetani et al. [45] (b)	2013	Japan	Cross-sectional	-	194	AD	-	-	-	NINCDS-ADRDA	-	74.3 (9.5)	134	64	33.0%
28	Uetani et al. [45] (c)	2013	Japan	Cross-sectional	-	60	VaD	-	-	NINDS-AIREN	-	-	73.3 (10.0)	40	24	40.0%
39	Zonneveld et al. [47] (a)	2014	Netherlands	Cross-sectional	-	311	MCI	Petersen's criteria	-	-	-	-	64.2 (10.2)	125	44	14.1%
39	Zonneveld et al. [47] (b)	2014	Netherlands	Cross-sectional	-	417	AD	-	-	-	NINCDS-ADRDA	-	65.2 (10.0)	204	81	19.4%
39	Zonneveld et al. [47] (c)	2014	Netherlands	Cross-sectional	-	180	VaD	-	-	NINDS-AIREN	-	-	61.5 (10.0)	80	10	5.6%
Lacunes																
5	Fan et al. [37]	2021	China	Retrospective	T1, T2, FLAIR, DWI	293	GCI	-	MOCA	-	-	-	-	174	130	44.4%
8	Hilal et al. [38]	2015	Singapore	Cross-sectional	T2, FLAIR	204	ACD	-	-	-	-	DSM-IV	68.8 (6.5)	98	13	2.3%
8	Hilal et al. [38]	2015	Singapore	Cross-sectional	T2, FLAIR	572	GCI	-	Study-specific protocol	-	-	DSM-IV	70.5 (6.8)	313	99	17.3%

Table 2. *Cont.*

Study ID	Author	Year	Country	Study Design	MRI Sequence	Cohort Size	Cognitive Outcome	MCI Criteria	GCI Criteria	VaD Criteria	AD Criteria	ACD Criteria	Age (Mean +/− SD)	Female (*n*)	Imaging Marker (*n*)	Imaging Marker (%)
9	Hilal et al. [49]	2021	Singapore	Cross-sectional	T1, T2, FLAIR	253	MCI	Study-specific protocol	-	-	-	-	70.2 (6.1)	129	36	14.2%
11	Jacob et al. [39]	2023	Netherlands	Prospective	T1, FLAIR	498	ACD	-	-	-	-	DSM-V	65.7 (8.8)	217	42	8.4%
15	Li et al. [40]	2021	China	Retrospective	-	270	MCI	NIA-AA	MMSE	-	-	-	66.5 (8.7)	-	118	15.2%
17	Liao et al. [24]	2024	China	Cross-sectional	-	94	GCI	-	MMSE	-	-	-	65.9 (10.9)	-	29	37.2%
18	Liu et al. [25]	2021	China	Cross-sectional	-	199	GCI	-	Study-specific protocol	-	-	-	69.0 (7.8)	-	85	42.7%
28	Uetani et al. [45] (a)	2013	Japan	Cross-sectional	T2WI GRE FLAIR	296	ACD	-	-	NINDS-AIREN	NINCDS-ADRDA	-	-	188	58	19.6%
28	Uetani et al. [45] (b)	2013	Japan	Cross-sectional	T2WI GRE FLAIR	83	MCI	International Working Group on MCI	-	-	-	-	74.1 (9.5)	-	13	15.7%
28	Uetani et al. [45] (c)	2013	Japan	Cross-sectional	T2WI GRE FLAIR	347	GCI	International Working Group on MCI	-	NINDS-AIREN	NINCDS-ADRDA	-	74.3 (8.8)	-	217	20.5%
29	Wang et al. [50]	2022	China	Cross-sectional	-	442	ACD	-	-	-	-	DSM-IV	71.6 (11.3)	-	205	10.9%
30	Wang et al. [51]	2024	China	Cross-sectional	FLAIR	1230	MCI	Petersen's criteria	-	-	-	-	69.4 (4.3)	-	720	8.0%

Abbreviations: ACD: all-cause dementia; AD: Alzheimer's Disease; ADDTC: Alzheimer's Disease Diagnostic and Treatment Centers; CDR: Clinical Dementia Rating; GCI: global cognitive impairment; CMB: cerebral microbleed; CSVD: cerebral small vessel disease; DSM-IV: Diagnostic and Statistical Manual of Mental Disorders, Fourth Edition; DSM-V: Diagnostic and Statistical Manual of Mental Disorders, Fifth Edition; DWI: Diffusion-weighted Imaging; FLAIR: Fluid-Attenuated Inversion Recovery; GRE: Gradient-recalled Echo; ICD-10: International Classification of Diseases, 10th Revision; MCI: mild cognitive impairment; MMSE: Mini-Mental State Examination; MOCA: Montreal Cognitive Assessment; MRI: Magnetic Resonance Imaging; NIA-AA: National Institute on Aging and Alzheimer's Association; NINCDS-ADRDA: National Institute of Neurological and Communicate Disorders and Stroke/Alzheimer's Disease and Related Disorders Association; NINDS-AIREN: National Institute of Neurological Disorders and Stroke/Association Internationale pour la Recherche et l'Enseignement en Neurosciences; PVS: perivascular spaces; STRIVE: Standards for Reporting Vascular Changes on Neuroimaging; SWI: Susceptibility-weighted Imaging; T2WI: T2-weighted Imaging; VaD: vascular dementia; WMH: white matter hyperintensities. Notes: The symbol "*" in MRI sequence (e.g., T2* GRE) indicates gradient-echo T2-weighted imaging sequences used for susceptibility detection (commonly employed for visualizing cerebral microbleeds); (a), (b), and (c) indicate separate sub-analyses or cohorts from the same study reporting distinct cognitive outcomes.

Table 3. Characteristics of studies assessing APOE ε4 carrier status.

Study ID	Author	Year	Country	Study Design	Cohort Size	Cognitive Outcome	GCI Criteria	Age (Mean +/− SD)	Female (*n*)	APOE ε4 Carrier (*n*)	APOE ε4 Carrier (%)
1	Brickman et al. [52]	2015	USA	Cross-sectional	694	GCI	DSM-IV	80.4 (5.7)	462	14	2.0%
10	Hong et al. [53]	2011	Korea	Cross-sectional	216	GCI	DSM-IV	68.4 (9.3)	161	17	7.9%
13	Kim et al. [54]	2013	Korea	Cross-sectional	364	GCI	DSM-IV, Petersen's criteria	68.3 (8.5)	212	35	9.6%
19	Nicoll et al. [13]	2010	UK	Prospective	310	GCI	Neuropathological	-	188	56	18.1%
20	Paradela et al. [14]	2023	Brazil	Prospective	648	GCI	CDR	74.7 (12.0)	339	76	11.7%
36	Yu et al. [55]	2023	China	Cross-sectional	166	GCI	CDR	76.5 (7.7)	101	19	11.4%

Abbreviations: APOE: apolipoprotein E; CDR: Clinical Dementia Rating; GCI: global cognitive impairment; DSM-IV: Diagnostic and Statistical Manual of Mental Disorders,
Fourth Edition.

2.7. Certainty of Evidence Assessment (NEUROGEN-SVD)

As part of the NEUROGEN-SVD study, the certainty of evidence for each cognitive outcome was evaluated using the Grading of Recommendations, Assessment, Development and Evaluations (GRADE) framework. Outcomes were assessed for risk of bias (study design limitations and diagnostic variability), inconsistency (heterogeneity in effect estimates), indirectness (applicability of study populations and outcome definitions), imprecision (width of confidence intervals and sample size), and publication bias (funnel plots, Egger's, and Deek's tests). A NEUROGEN-SVD Summary of Findings (SoF) table was constructed, presenting pooled effect estimates, absolute effects, and certainty ratings for the associations between CSVD neuroimaging/genetic markers and cognitive outcomes.

2.8. Statistical Analyses

All statistical analyses were conducted using STATA version 13.0 (StataCorp, College Station, TX, USA). Descriptive statistics were used to summarize baseline characteristics and key outcomes across included studies. Where necessary, means and standard deviations (SDs) were imputed from medians and interquartile ranges (IQRs) using the method described by Wan et al. [56].

The pooled prevalence of GCI among patients with CSVD was calculated using the metaprop command, implementing a random-effects model to account for inter-study variability. Exact binomial methods (cimethod(exact) and ftt) were used to derive 95% confidence intervals (CIs) for prevalence estimates. To estimate the associations between CSVD markers (neuroimaging and genetic) and cognitive outcomes, a random-effects meta-analysis using the DerSimonian-Laird (DL) method was applied via the metan package. Exploratory subgroup analyses were performed when at least three or more studies reported comparable data for a given exposure–outcome pair (e.g., imaging marker type and dementia subtype). However, stratified subgroup analyses by diagnostic framework, MRI modality, or cohort setting, as well as meta-regression, were not feasible because most studies did not report compatible covariates. Results were reported as pooled odds ratios (ORs) with corresponding 95% CIs and visualized using forest plots, which also detailed study weights and between-study heterogeneity.

The midas package in STATA was used to evaluate diagnostic performance. Summary receiver operating characteristic (SROC) curves were generated, and pooled estimates of sensitivity, specificity, and area under the ROC curve (AUROC) were reported. Statistical heterogeneity was assessed using the I^2 statistic (with the following thresholds: <30% = low, 30–50% = moderate, 50–75% = substantial, >75% = high) and Cochran's Q test. Between-study variance was estimated using Tau^2. Sensitivity analyses were performed using the metaninf function to assess the influence of individual studies on pooled effect sizes. Funnel plots were generated to visually assess asymmetry, and Egger's test was applied using the metabias and metafunnel packages to formally test for publication bias. Additional assessments included Deek's funnel-plot asymmetry test and Fagan's nomogram (implemented through the midas package) to evaluate potential bias in diagnostic accuracy estimates. All statistical tests were two-tailed, and $p < 0.05$ was considered statistically significant.

3. Results

3.1. Description of Included Studies

The systematic search across five major databases identified 5328 records. Following the removal of duplicates, 3151 unique records were retained for screening. Abstracts were reviewed for relevance, resulting in the exclusion of 2345 studies that did not address CSVD, cognitive outcomes, or genetic risk factors. Full-text review of the remaining

488 studies led to the exclusion of 449 for the following reasons: absence of relevant outcomes or insufficient data (n = 283); inappropriate study design (n = 43); overlapping cohorts (n = 12); unsuitable control groups (n = 34); unclear or inconsistent definitions of imaging or cognitive variables (n = 24); small sample size below eligibility thresholds (n = 23); and imaging or genetic markers with insufficient numbers of studies for pooled analysis (n = 30).

Thirty-nine (39) studies met all inclusion criteria and were included in the final meta-analysis, encompassing 18,425 participants. Where data from the same database were available across multiple reports, priority was given to the largest or most recent cohort to avoid duplication. When data originated from the same database but were reported on distinct subgroups or different imaging/genetic markers, overlapping participants were excluded when calculating the total sample size.

Of the included studies, 16 reported on CMBs, 4 focused on moderate-to-severe WMHs, 10 on lacunes, 6 on APOE ε4 allele carrier status, and 17 provided prevalence estimates of cognitive outcomes in CSVD populations. These studies covered a wide geographic span, including Asian, European, and North American cohorts, enhancing generalizability. Clinical characteristics, diagnostic definitions, and cognitive outcome measures are detailed in Tables 1–3.

Pooled prevalence estimates of cognitive impairment outcomes are summarized in Table 4 and displayed as a forest plot in Figure 2. Associations between individual CSVD imaging markers and APOE ε4 status with cognitive outcomes are provided in Tables 5 and 6, organized by imaging/genetic risk factor (Table 5) and by cognitive outcome (Table 6). Corresponding odds ratios, study weights, and heterogeneity estimates are presented in Figures 3–6.

Table 4. Pooled prevalence of cognitive outcomes in CSVD patients: summary effects and heterogeneity.

Subgroup	Studies (N)	Participants (n)	Summary Effects					Heterogeneity			
			Crude Prevalence	Pooled Prevalence	95% CI	z	p	Q	I^2 (%)	p (Q)	τ^2
Overall	16	2118	53.5%	53%	0.49–0.58	33.88	$p < 0.001$	62.89	76.15	$p < 0.001$	0.02
GCI	10	1518	55.9%	57%	0.51–0.62	28.48	$p < 0.001$	42.15	78.65	$p < 0.001$	0.02
MCI	7	766	46.3%	46%	0.42–0.51	31.02	$p < 0.001$	8.91	32.67	$p = 0.18$	0.00

Abbreviations: ACI: all-cause dementia; GCI: global cognitive impairment; MCI: mild cognitive impairment; 95% CI: 95% confidence interval; CSVD: cerebral small vessel disease. Notes: (1) "Crude prevalence" is the unweighted proportion; "Pooled prevalence" is the random-effects estimate; (2) Random-effects meta-analysis used the DerSimonian–Laird (DL) method; (3) Heterogeneity: Q (Cochran's Q, χ^2 with df = N − 1); I^2 (% between-study variability); τ^2 (between-study variance, DerSimonian–Laird); (4) z and p are two-sided tests of the pooled effect; (5) One study with overlapping cohorts across MCI and GCI was excluded from the overall pooled estimate.

Table 5. Associations between CSVD imaging/genetic markers and cognitive outcomes: results by marker/genetic factor.

Imaging/ Genetic Marker	Cognitive Subgroup	Studies (N)	Participants (n)	Summary Effects DL		Heterogeneity				
				Pooled OR (95% CI)	p, z	Cochran's Q	p (Q)	H	I^2 (%)	τ^2
CMBs	MCI	8	2301	1.93 [1.48; 2.51]	$p < 0.001$, z = 4.813	11.11	0.134	1.260	37.0	0.051
	GCI	5	1505	1.70 [0.92; 3.16]	$p = 0.091$, z = 1.689	13.45	0.009	1.834	70.3	0.310
	ACD	9	5987	1.92 [1.41; 2.60]	$p < 0.001$, z = 4.20	15.75	0.046	1.403	49.2	0.101
	VaD	5	4399	4.70 [2.10; 10.52]	$p < 0.001$, z = 3.77	11.49	0.022	1.695	65.2	0.535
	AD	7	5369	1.52 [1.04; 2.24]	$p = 0.033$, z = 2.14	12.40	0.054	1.437	80.8	2.275

Table 5. *Cont.*

Imaging/ Genetic Marker	Cognitive Subgroup	Studies (N)	Participants (*n*)	Summary Effects DL		Heterogeneity				
				Pooled OR (95% CI)	*p*, z	Cochran's Q	*p* (Q)	H	I^2 (%)	τ^2
WMHs	MCI	4	3513	2.42 [1.57; 3.74]	$p < 0.001$, z = 3.980	5.86	0.118	1.398	48.8	0.094
	VaD	3	1815	10.35 [7.32; 14.64]	$p < 0.001$, z = 13.229	1.90	0.388	0.974	0.0	0.000
	AD	3	2621	2.78 [1.27; 6.09]	$p = 0.011$, z = 2.558	12.98	0.002	2.548	84.6	0.394
Lacunes	MCI	4	1836	2.70 [1.25; 5.84]	$p = 0.011$, z = 2.528	14.40	0.002	2.191	79.2	0.398
	GCI	5	1505	2.41 [1.33; 4.40]	$p = 0.004$, z = 2.88	10.07	0.039	1.586	60.3	0.246
	ACD	4	1440	3.18 [1.24; 8.20]	$p = 0.017$, z = 2.40	20.51	0.000	2.615	85.4	0.664
APOE ε4	GCI	6	2398	1.80 [1.41, 2.29]	$p < 0.001$, z = 4.729	6.13	0.294	1.107	68.3	0.017

Abbreviations: ACD: all-cause dementia; AD: Alzheimer's disease; APOE: apolipoprotein E; CMBs: cerebral microbleeds; GCI: global cognitive impairment; MCI: mild cognitive impairment; OR: odds ratios; VaD: vascular dementia; WMHs: white matter hyperintensities; 95% CI: 95% confidence interval; df: degrees of freedom. Notes: (1) Model: random-effects DerSimonian-Laird (DL); (2) Effect size is pooled OR (95% CI); z and *p* test log(OR) = 0 (two-sided); (3) Heterogeneity: Q (Cochran's Q, χ^2 with df = N − 1); H ($\sqrt{[Q/df]}$); I^2 (%); τ^2 (DL); (4) For N ≤ 3, interpret I^2/H cautiously due to low df; (5) When a study reports multiple cognitive outcomes, each is listed separately and analysed as an independent comparison.

Table 6. Associations between CSVD imaging/genetic markers and cognitive outcomes: results by cognitive outcome.

Cognitive Outcome	Marker	Studies (N)	Participants (*n*)	Summary Effects DL		Heterogeneity				
				Pooled OR (95% CI)	*p*, z	Cochran's Q	*p* (Q)	H	I^2	τ^2
MCI	CMB	8	2301	1.93 [1.48; 2.51]	$p < 0.001$, z = 4.813	11.11	0.134	1.260	37.0	0.051
	WMHs	4	3513	2.42 [1.57; 3.74]	$p < 0.001$, z = 3.980	5.86	0.118	1.398	48.8	0.094
	Lacunes	4	1836	2.70 [1.25; 5.84]	$p = 0.011$, z = 2.528	14.40	0.002	2.191	79.2	0.398
GCI	CMBs	5	1505	1.70 [0.92; 3.16]	$p = 0.091$, z = 1.689	13.45	0.009	1.834	70.3	0.310
	Lacunes	5	1505	2.41 [1.33; 4.40]	$p = 0.004$, z = 2.88	10.07	0.039	1.586	60.3	0.246
	APOE ε4 allele	6	2398	1.80 [1.41, 2.29]	$p < 0.001$, z = 4.729	6.13	0.294	1.107	68.3	0.017
ACD	CMBs	9	5987	1.92 [1.41; 2.60]	$p < 0.001$, z = 4.20	15.75	0.046	1.403	49.2	0.101
	Lacunes	4	1440	3.18 [1.24; 8.20]	$p = 0.017$, z = 2.40	20.51	0.000	2.615	85.4	0.664
VaD	CMBs	5	4399	4.70 [2.10; 10.52]	$p < 0.001$, z = 3.77	11.49	0.022	1.695	65.2	0.535
	WMHs	3	1815	10.35 [7.32; 14.64]	$p < 0.001$, z = 13.229	1.90	0.388	0.974	0.0	0.000
AD	CMBs	7	5369	1.52 [1.04; 2.24]	$p = 0.033$, z = 2.14	12.40	0.054	1.437	80.8	2.275
	WMHs	3	2621	2.78 [1.27; 6.09]	$p = 0.011$, z = 2.558	12.98	0.002	2.548	84.6	0.394

Abbreviations: ACD: all-cause dementia; AD: Alzheimer's disease; APOE: apolipoprotein E; CMBs: cerebral microbleeds; GCI: global cognitive impairment; MCI: mild cognitive impairment; OR: odds ratios; VaD: vascular dementia; WMHs: white matter hyperintensities; 95% CI: 95% confidence interval; df: degrees of freedom. Notes: (1) Model: random-effects, DerSimonian and Laird (DL); (2) Pooled OR (95% CI) per marker within each outcome; z and *p* test log(OR) = 0 (two-sided); (3) Heterogeneity: Q (Cochran's Q, χ^2 with df = N − 1); H ($\sqrt{[Q/df]}$); I^2 (%); τ^2 (between-study variance); (4) Studies reporting multiple markers/outcomes appear as separate rows and are analysed as independent comparisons.

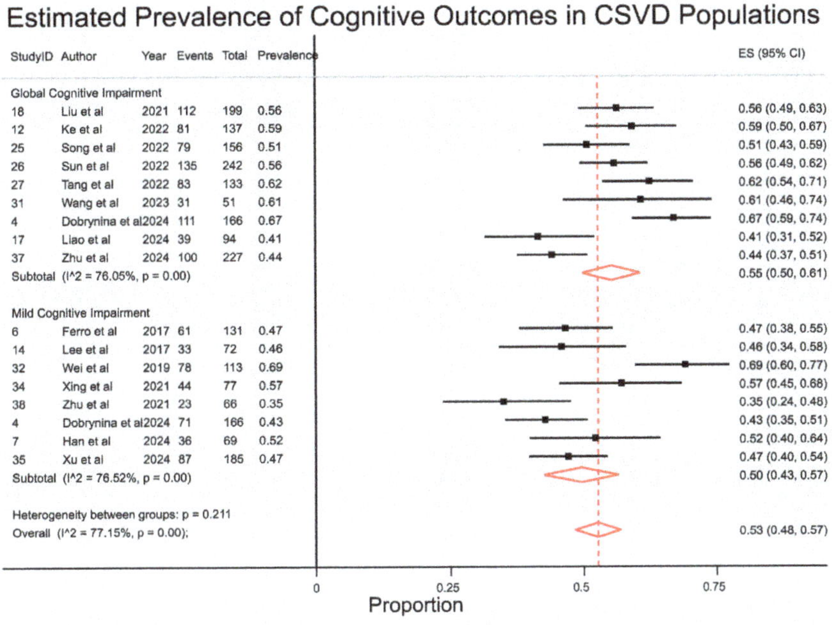

Figure 2. Forest plot of pooled prevalence of cognitive outcomes in CSVD patients. Abbreviations: CSVD: cerebral small vessel disease [19–34].

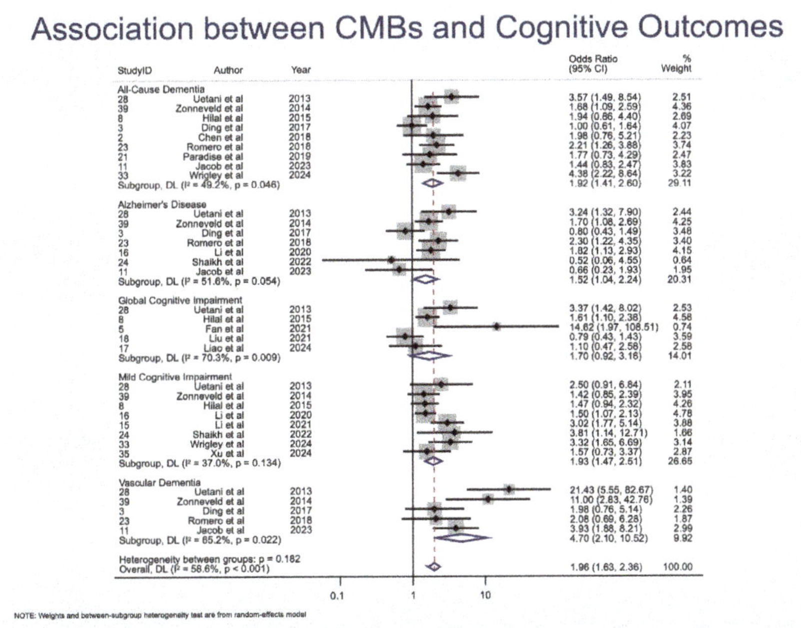

Figure 3. Forest plot of associations between cerebral microbleeds (CMBs) and cognitive outcomes. Abbreviations: CMBs: cerebral microbleeds [24,25,32,35–47].

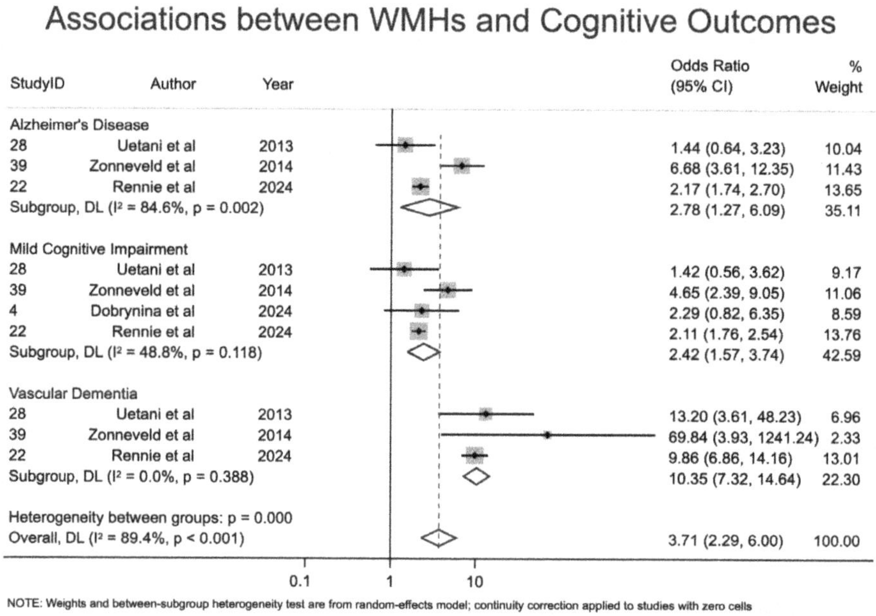

Figure 4. Forest plot of associations between moderate-to-severe white matter hyperintensities (WMHs) and cognitive outcomes. Abbreviations: WMHs: white matter hyperintensities [19,45,47,48].

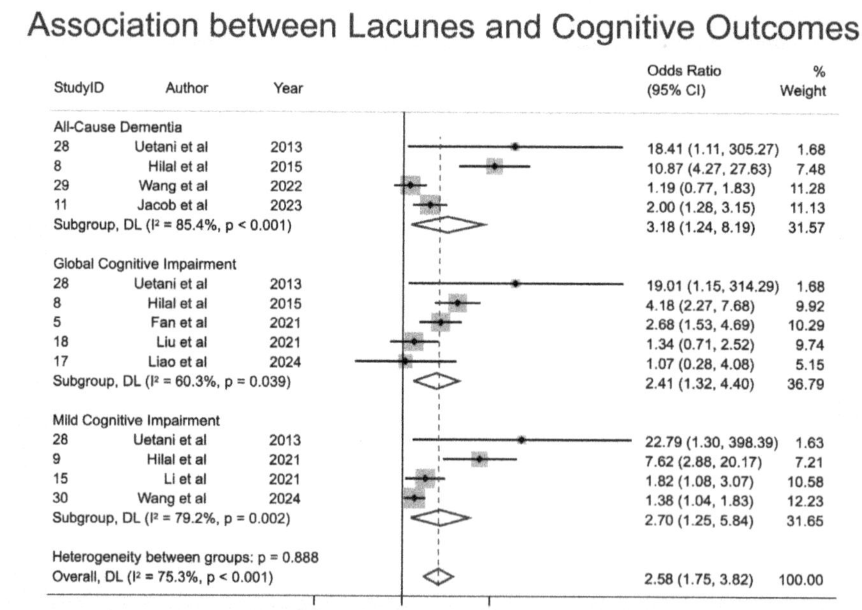

Figure 5. Forest plot of associations between lacunes and cognitive outcomes [24,25,37–40,45,49–51].

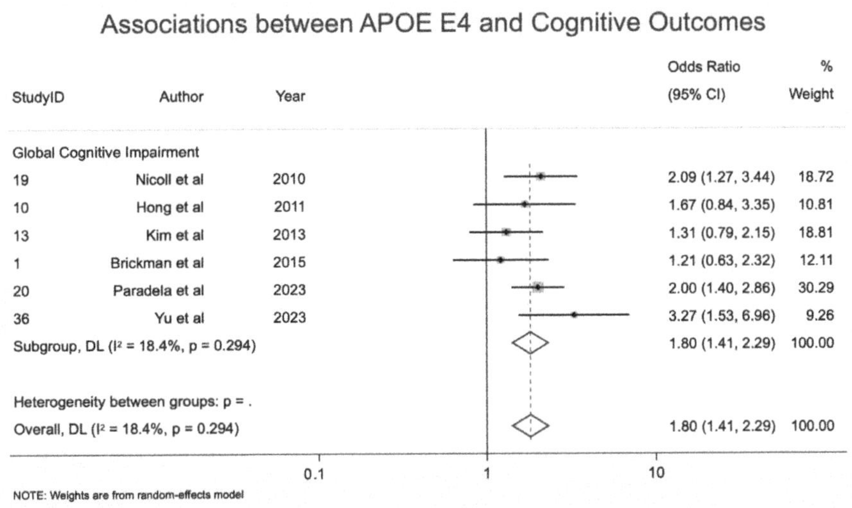

Figure 6. Forest plot of associations between APOE ε4 carrier status and cognitive outcomes [13,14,52–55]. Abbreviations: APOE: apolipoprotein E.

Although not the primary focus of this analysis, the diagnostic performance of imaging and genetic markers was also examined. These results, including pooled sensitivity, specificity, and AUROC estimates, are presented in Supplemental Table S5 and visualized through SROC curves in Supplemental Figures S1 and S2.

The methodological quality of included studies was appraised using a modified Jadad scale (Supplemental Table S3), while potential funding-related bias was assessed in Supplemental Table S4. Sensitivity analyses were conducted to evaluate the influence of individual studies on pooled results, with findings displayed in Supplemental Figure S3. To assess publication bias, funnel plots, Egger's regression, and Deek's funnel-plot asymmetry test were applied, with results presented in Supplemental Tables S6 and S7 and Supplemental Figures S4–S6. Fagan's nomogram analysis, included for exploratory purposes, is shown in Supplemental Figure S7.

3.2. *Prevalence of Cognitive Outcomes in CSVD Patients*

Sixteen studies [19–34], comprising a total of 2118 participants, were included in the pooled prevalence analysis of cognitive outcomes in patients with CSVD (Figure 2). The overall meta-analysis demonstrated that cognitive impairment was present in more than half of individuals with CSVD, with a pooled prevalence of 53% (95% CI: 49–58%; z = 33.88; $p < 0.001$) (Table 6; Figure 2). Between-study heterogeneity was substantial ($I^2 = 76.15\%$, $p < 0.001$), reflecting differences in study populations, diagnostic thresholds, and neuropsychological batteries. Figure 2 provides a stratified visualization of prevalence estimates by cognitive outcome subtype, specifically distinguishing between clinically defined GCI and MCI.

3.2.1. Prevalence of GCI in CSVD Patients

Ten studies [19,22,24–30,33] (n = 1518) specifically assessed the prevalence of GCI among CSVD patients. The pooled prevalence was 57% (95% CI: 51–62%; z = 28.48; $p < 0.001$), indicating that nearly three out of five patients with radiological evidence of CSVD exhibit a measurable global cognitive decline (Table 6; Figure 2). Heterogeneity across these studies was high ($I^2 = 78.65\%$, $p < 0.001$), suggesting variability in diagnostic methods and cohort characteristics. Some studies drew participants from memory clinics and stroke registries, whereas others used community-based samples, which may partly explain the observed inconsistency. Despite this heterogeneity, the direction of effect was consistent, underscoring GCI as a common and clinically significant outcome of CSVD.

3.2.2. Prevalence of MCI in CSVD Patients

Seven studies [19–21,23,31,32,34] (n = 766) examined the prevalence of MCI among CSVD patients. The pooled prevalence was 46% (95% CI: 42–51%; z = 31.02; $p < 0.001$), with only moderate heterogeneity ($I^2 = 32.67\%$, $p = 0.18$) (Table 6; Figure 2). This finding suggests that almost half of CSVD patients exhibit early-stage cognitive decline, often preceding overt dementia. The lower heterogeneity compared to GCI analyses may reflect more consistent use of standardized MCI criteria, such as Petersen's framework or modified NINDS-CSN definitions, across contributing studies.

3.2.3. Geographical Subgroup Analysis

When stratified by region, notable differences emerged. Asian cohorts (predominantly from China, Korea, and Singapore) reported a pooled prevalence of GCI around 60–65%, often higher than Western cohorts (Europe and North America), where prevalence estimates clustered between 45 and 55%. For MCI, Asian studies reported slightly higher prevalence (48–50%) compared to Western studies (42–45%), though heterogeneity was lower overall. This suggests that cultural and methodological differences may partly explain variability in GCI prevalence, whereas MCI prevalence estimates appear more stable across regions.

3.3. Association Between Neuroimaging Markers of CSVD and the APOE ε4 Allele with Cognitive Outcomes

3.3.1. Analysis by Imaging Marker

Cerebral Microbleeds (CMBs)

Sixteen studies [24,25,32,35–47] (n = 8612) examined the association between CMBs and cognitive outcomes. The presence of CMBs was significantly associated with MCI (OR 1.93, 95% CI: 1.48–2.51, $p < 0.001$), all-cause dementia (ACD; OR 1.92, 95% CI: 1.41–2.60, $p < 0.001$), vascular dementia (VaD; OR 4.70, 95% CI: 2.10–10.52, $p < 0.001$), and AD (OR 1.52, 95% CI: 1.04–2.24, $p = 0.033$). No statistically significant association was observed with overall GCI (OR 1.70, 95% CI: 0.92–3.16, $p = 0.091$) (Table 5; Figure 3). Heterogeneity ranged from moderate for MCI ($I^2 = 37.0\%$) and ACD ($I^2 = 49.2\%$) to substantial for VaD ($I^2 = 65.2\%$) and GCI ($I^2 = 70.3\%$), and was highest for AD ($I^2 = 80.8\%$).

White Matter Hyperintensities (WMHs)

Four studies [19,45,47,48] (n = 4821) investigated moderate-to-severe WMHs. Strong associations were found with MCI (OR 2.42, 95% CI: 1.57–3.74, $p < 0.001$), VaD (OR 10.35, 95% CI: 7.32–14.64, $p < 0.001$), and AD (OR 2.78, 95% CI: 1.27–6.09, $p = 0.011$) (Table 5; Figure 4). Heterogeneity was low for VaD ($I^2 = 0.0\%$), moderate for MCI ($I^2 = 48.8\%$), and high for AD ($I^2 = 84.6\%$). These findings suggest WMHs are a particularly robust imaging marker for vascular dementia and exert significant influence on both early and late cognitive outcomes.

Lacunes

Ten studies [24,25,37–40,45,49–51] (n = 4198) assessed lacunes, showing significant associations with MCI (OR 2.70, 95% CI: 1.25–5.84, $p = 0.011$), GCI (OR 2.41, 95% CI: 1.33–4.40, $p = 0.004$), and ACD (OR 3.18, 95% CI: 1.24–8.20, $p = 0.017$) (Table 5; Figure 5). Heterogeneity was substantial for GCI ($I^2 = 60.3\%$) and high for both MCI ($I^2 = 79.2\%$) and

ACD ($I^2 = 85.4\%$). This likely reflects differences in imaging protocols, diagnostic thresholds, and classification challenges in distinguishing lacunes from enlarged perivascular spaces.

APOE ε4 Allele

Six studies [13,14,52–55] ($n = 2398$) examined APOE ε4 carrier status. Carriers had significantly higher odds of GCI (OR 1.80, 95% CI: 1.41–2.29, $p < 0.001$) (Table 5; Figure 6). Heterogeneity was substantial ($I^2 = 68.3\%$, $p = 0.017$), likely reflecting population differences in allele frequency and interactions with vascular risk factors.

3.3.2. Analysis by Cognitive Outcome

MCI: Twelve studies [19,32,38,40,41,44–49,51] ($n = 6904$) indicated that lacunes (OR 2.70, 95% CI: 1.25–5.84), WMHs (OR 2.42, 95% CI: 1.57–3.74), and CMBs (OR 1.93, 95% CI: 1.48–2.51) were significantly associated with MCI (Table 6).

GCI: Eleven studies [13,14,24,25,37,38,45,52–55] ($n = 3903$) showed that lacunes (OR 2.41, 95% CI: 1.33–4.40) and APOE ε4 allele status (OR 1.80, 95% CI: 1.41–2.29) were the strongest predictors of GCI. No significant association was observed with CMBs (OR 1.70, 95% CI: 0.92–3.16, $p = 0.091$) (Table 6).

ACD: Ten studies [35,36,38,39,42,43,45–47,50] ($n = 6429$) demonstrated significant associations with lacunes (OR 3.18, 95% CI: 1.24–8.20) and CMBs (OR 1.92, 95% CI: 1.41–2.60) (Table 6).

VaD: Six studies [36,39,43,45,47,48] ($n = 6214$) found very strong associations with WMHs (OR 10.35, 95% CI: 7.32–14.64) and CMBs (OR 4.70, 95% CI: 2.10–10.52) (Table 6).

AD: Eight studies [36,39,41,43–45,47,48] ($n = 8030$) reported significant associations with WMHs (OR 2.78, 95% CI: 1.27–6.09) and CMBs (OR 1.52, 95% CI: 1.04–2.24) (Table 6).

Publication bias was formally assessed using funnel plots, Egger's regression, and Deek's test. Visual inspection of funnel plots suggested small-study effects in some analyses, particularly for lacunes and APOE ε4. Egger's test was significant in a subset of comparisons, while Deek's test indicated asymmetry in diagnostic accuracy analyses for CMBs and lacunes but not for WMHs or APOE ε4 (Supplemental Tables S6 and S7; Supplemental Figures S4–S6).

These findings suggest that some degree of publication bias may be present, particularly for markers with fewer contributing studies, but overall results remained directionally consistent across sensitivity analyses (Supplemental Figure S3).

3.3.3. Diagnostic Performance of Neuroimaging Markers of CSVD and the APOE ε4 Allele for Cognitive Outcomes

Although the primary aim of this meta-analysis was to evaluate associations between CSVD markers and cognitive outcomes, the diagnostic performance of these imaging and genetic markers was also assessed. This exploratory analysis provides insight into their clinical utility for identifying patients at risk of cognitive decline. Data were available for all markers and cognitive outcome subtypes, with the exception of WMHs in relation to VaD and AD, where too few studies precluded pooled estimates.

Diagnostic performance was modest. Across markers, sensitivities were consistently low, ranging from 22% (95% CI: 0.13–0.35) for CMBs predicting AD to 50% (95% CI: 0.25–0.75) for CMBs predicting VaD. Specificities were generally moderate-to-high, spanning 66% (95% CI: 0.38–0.86) for WMHs with MCI to 90% (95% CI: 0.61–0.98) for lacunes with ACD. This pattern indicates that these markers are more reliable for ruling out cognitive impairment in unaffected individuals than for accurately identifying those at risk.

The APOE ε4 allele displayed similar limitations. Its sensitivity for predicting GCI was only 35% (95% CI: 0.29–0.41), whereas specificity was moderate at 76% (95% CI: 0.73–0.79). These values suggest that while APOE ε4 positivity raises the likelihood of cognitive decline, a large proportion of affected patients will not be captured by genetic screening alone.

Discriminatory ability, as assessed by AUROC, was also modest across most markers. AUROC values ranged from poor (0.44; 95% CI: 0.40–0.49 for lacunes with MCI) to moderate (0.81; 95% CI: 0.78–0.85 for CMBs with VaD). These results imply that while some combinations of markers and outcomes (such as CMBs with VaD) demonstrate reasonable discriminative potential, most fall short of thresholds typically considered clinically actionable. The full set of pooled sensitivities, specificities, likelihood ratios, and AUROC values is provided in Supplemental Table S5, with SROC curves shown in Supplemental Figures S1 and S2.

3.3.4. Certainty of Evidence (NEUROGEN-SVD)

The certainty of evidence across NEUROGEN-SVD outcomes ranged from low to moderate. Evidence supporting the association between WMHs and VaD was rated moderate certainty, reflecting a very strong effect size (OR 10.35) and low heterogeneity, despite variability in diagnostic criteria. Associations between WMHs and AD, as well as lacunes with GCI, were rated low certainty due to the heterogeneity and indirectness of outcome definitions. The APOE ε4–GCI association was also rated low certainty, reflecting inconsistency across populations. The NEUROGEN-SVD Summary of Findings (Table 7) provides pooled effect sizes, absolute risk differences, and certainty ratings for each outcome.

Table 7. NEUROGEN-SVD GRADE summary of findings: certainty of evidence for CSVD imaging and genetic markers in cognitive outcomes.

Outcome	No. of Studies (Participants)	Study Design	Relative Effect (95% CI)	Assumed Risk (Control)	Risk with Marker	Absolute Effect (per 1000)	Certainty of Evidence	Reasons
CMBs → MCI	8 (~2301)	Observational (meta-analysis, random-effects)	OR 1.93 (1.48–2.51)	300 per 1000	450 per 1000	150 more per 1000	⊕⊕○○ Low	−1 risk of bias, −1 inconsistency, +1 moderate effect
CMBs → ACD	9 (~5987)	Observational (meta-analysis, random-effects)	OR 1.92 (1.41–2.60)	350 per 1000	500 per 1000	150 more per 1000	⊕⊕⊕○ Moderate	−1 risk of bias, +1 consistent effect
CMBs → VaD	5 (~4399)	Observational (meta-analysis, random-effects)	OR 4.70 (2.10–10.52)	200 per 1000	560 per 1000	360 more per 1000	⊕⊕○○ Low to Moderate	−1 heterogeneity, +1 strong effect
CMBs → AD	7 (~5369)	Observational (meta-analysis, random-effects)	OR 1.52 (1.04–2.24)	250 per 1000	360 per 1000	110 more per 1000	⊕⊕○○ Low	−1 imprecision, −1 heterogeneity
WMHs → MCI	4 (~3513)	Observational (meta-analysis, random-effects)	OR 2.42 (1.57–3.74)	300 per 1000	520 per 1000	220 more per 1000	⊕⊕○○ Low to Moderate	−1 risk of bias, −1 inconsistency, +1 effect size
WMHs → VaD	4 (~1815)	Observational (meta-analysis, random-effects)	OR 10.35 (7.32–14.64)	200 per 1000	740 per 1000	540 more per 1000	⊕⊕⊕○ Moderate	−1 diagnostic variability, +1 very strong effect, +1 low heterogeneity
WMHs → AD	3 (~2621)	Observational (meta-analysis, random-effects)	OR 2.78 (1.27–6.09)	250 per 1000	480 per 1000	230 more per 1000	⊕⊕○○ Low	−1 inconsistency, −1 indirectness, −1 imprecision
Lacunes → MCI	4 (~1836)	Observational (meta-analysis, random-effects)	OR 2.70 (1.25–5.84)	300 per 1000	560 per 1000	260 more per 1000	⊕⊕○○ Low	−1 heterogeneity, −1 indirectness
Lacunes → GCI	5 (~1505)	Observational (meta-analysis, random-effects)	OR 2.41 (1.33–4.40)	400 per 1000	610 per 1000	210 more per 1000	⊕⊕○○ Low	−1 marker misclassification, −1 heterogeneity

Table 7. *Cont.*

Outcome	No. of Studies (Participants)	Study Design	Relative Effect (95% CI)	Assumed Risk (Control)	Risk with Marker	Absolute Effect (per 1000)	Certainty of Evidence	Reasons
Lacunes → ACD	4 (~1440)	Observational (meta-analysis, random-effects)	OR 3.18 (1.24–8.20)	350 per 1000	640 per 1000	290 more per 1000	⊕⊕○○ Low	−1 inconsistency, −1 imprecision
APOE ε4 → GCI	6 (~2398)	Observational (meta-analysis, random-effects)	OR 1.80 (1.41–2.29)	400 per 1000	570 per 1000	170 more per 1000	⊕⊕○○ Low	−1 inconsistency, −1 indirectness

Assumed risk (control) represents the baseline probability of the outcome in patients without the CSVD marker or APOE ε4 allele, expressed per 1000 individuals. This value was derived from the median or pooled control-group risk across included studies. Risk with the marker was calculated by converting the pooled odds ratio (OR) into an absolute risk using the following formula:

$$R_1 = \frac{\text{OR} \times \text{R}_0}{1 - \text{R}_0 + (\text{OR} \times \text{R}_0)} \quad (1)$$

where R_0 is the assumed baseline risk. Absolute effect is the difference between the risk with marker and the assumed risk, expressed as the number of additional (or fewer) events per 1000 individuals. Confidence intervals for absolute effects were derived by applying the same conversion using the lower and upper bounds of the pooled OR. Abbreviations: NEUROGEN-SVD = Neuroimaging and Genetic Markers in Small Vessel Disease study; GCI = global cognitive impairment; MCI = mild cognitive impairment; ACD = all-cause dementia; VaD = vascular dementia; AD = Alzheimer's disease; WMHs = white matter hyperintensities; CMBs = cerebral microbleeds; OR = odds ratio. GRADE Working Group grades of evidence: ⊕⊕⊕⊕ High: Very confident that the true effect lies close to the estimate; ⊕⊕⊕○ Moderate: Moderately confident; true effect likely close but may differ; ⊕⊕○○ Low: Limited confidence; true effect may differ substantially; ⊕○○○ Very low: Very little confidence; true effect likely substantially different.

4. Discussion

This NEUROGEN-SVD study offers one of the most comprehensive syntheses to date on the relationship between CSVD, genetic susceptibility, and cognitive outcomes. Drawing on data from nearly 18,500 participants across 39 studies, the meta-analysis demonstrates that CSVD is a major driver of cognitive decline, extending beyond VaD to encompass the full spectrum of impairment, from MCI to ACD and AD. These findings reinforce CSVD as a central determinant of late-life cognitive trajectories and highlight its dual role in both vascular and neurodegenerative pathways.

While prior meta-analyses have largely focused on the prevalence of VaD, with estimates ranging between 36 and 67% depending on diagnostic criteria [57,58], few have examined GCI and MCI more broadly within CSVD populations. The current meta-analysis advances this field by estimating the pooled prevalence of GCI at 57% (95% CI: 51–62%), which is notably higher than a previous estimate of 44.1% [33]. It also reports an MCI prevalence of 46%, underscoring that CSVD is not only a driver of late-stage dementia but is also prominently involved in early cognitive decline. The relatively consistent prevalence of MCI across regions reinforces its potential as a stable and clinically meaningful early marker. The higher heterogeneity in GCI prevalence likely reflects the broad and variable definitions of GCI, which encompass a spectrum from MCI through to overt dementia.

These findings support an evolving model of CSVD-related GCI as a continuum. Rather than being defined solely by advanced pathology, CSVD contributes to subtle, microstructural tissue damage that disrupts connectivity long before overt lesion burden is visible [59]. This highlights opportunities for early intervention through aggressive management of vascular risk factors, lifestyle modification, and cognitive rehabilitation before cognitive deterioration becomes irreversible.

Previous meta-analyses have tended to examine individual imaging markers or their associations with specific cognitive domains [60] or dementia [61–63] outcomes. Based on current understanding, this is the first meta-analysis to systematically compare multiple CSVD markers (WMHs, CMBs, and lacunes) and genetic risk (APOE ε4) across a range of cognitive outcomes, from MCI through ACD, VaD, AD, and GCI. These results show that associations were generally stronger for later stages of impairment (ACD, VaD, and AD) than for MCI. This gradient is consistent with the progressive nature of CSVD, in which the cumulative burden of lesions disrupts white matter tracts and cortical–subcortical connectivity, gradually leading to widespread network dysfunction and cognitive impairment [59,64]. Weaker associations observed with MCI likely reflect that lesion burden in early stages may not have crossed the threshold needed to cause a clinically detectable decline.

Among individual markers, WMHs demonstrated the strongest and most consistent associations with cognitive outcomes, particularly VaD, where odds were increased more than tenfold. These findings align with prior evidence and reinforce the view of WMHs as markers of diffuse ischemia from chronic microvascular compromise, directly linked to sustained cerebral hypoperfusion [2,65]. Lacunes also showed significant associations across all outcomes, though with greater heterogeneity, likely reflecting variations in imaging protocols and challenges in distinguishing them from perivascular spaces [16]. CMBs demonstrated weaker associations overall, though they became significant in dementia outcomes, suggesting their cognitive impact becomes apparent only once a threshold burden is reached [66].

The APOE ε4 allele was significantly associated with GCI, extending its relevance beyond AD risk alone. Emerging evidence suggests that APOE ε4 may exacerbate vascular injury through mechanisms such as neuroinflammation [67,68] and blood–brain barrier (BBB) dysfunction, independent of amyloid-β [69]. These observations support the Integrated Vascular–Neurodegenerative Continuum Hypothesis [2], which posits that vascular and neurodegenerative processes converge through overlapping pathways, lowering the threshold for GCI. This concept is reinforced by the observed associations between CSVD markers and AD, as well as by the influence of APOE ε4 across multiple cognitive subtypes. White matter lesions in AD are themselves heterogeneous: while many reflect chronic hypoperfusion [70] or amyloid angiopathy [71], others—particularly posterior periventricular lesions [72]—may arise from primary neurodegenerative mechanisms such as tau-mediated axonal degeneration [73–75]. Recognizing these dual etiologies is essential for interpreting the burden of WMHs in mixed dementias. Related biomarkers, including soluble low-density lipoprotein receptor-related protein-1 (sLRP-1), have been implicated in both vascular [76] and amyloidogenic cascades [77]; however, current data do not support sLRP-1 as a CSVD-specific marker distinct from AD or other comorbid conditions.

In addition to associations, the diagnostic performance of these markers was explored. Across neuroimaging and genetic measures, sensitivities were uniformly low, specificities were moderate-to-high, and AUROC values were modest. This indicates that while these markers have diagnostic weight when present, they lack sufficient sensitivity to serve as standalone screening tools. The findings underscore that cognitive decline in CSVD is unlikely to arise from isolated vascular or neurodegenerative processes but rather from their overlapping contributions [2]. This has two major implications. First, early systematic screening for GCI in CSVD populations is warranted, particularly targeting MCI, which appears as a consistent and robust early marker. Second, multimodal diagnostic strategies, integrating imaging markers, genetic risk, vascular burden, and clinical data, are needed to enhance diagnostic accuracy and risk stratification. Emerging BBB-permeability imaging, including arterial-spin labeling (ASL) [78], dynamic contrast-enhanced MRI [79], and novel PET tracers [80], provides complementary mechanistic insight into microvascular injury. Incorporating BBB metrics with WMHs and APOE status may refine early CSVD detection [81]. For Asian populations, where prevalence is higher, aggressive management of hypertension and diabetes may have particularly strong preventive effects.

High heterogeneity was observed across several outcomes, especially for associations involving GCI and AD with CMBs and lacunes. This likely reflects variation in imaging protocols, definitions of markers, diagnostic criteria, and study populations, which spanned hospital-based cohorts to community samples. Differences in adjustment for confounders and inter-rater reliability further contributed to variability. Funnel plots and formal bias tests suggested some small-study effects, particularly for lacunes and APOE ε4, though the overall direction of associations was consistent in sensitivity analyses. These observations highlight the pressing need for standardized imaging protocols, operational definitions of CSVD markers, and harmonized neuropsychological assessments to improve comparability and reduce heterogeneity. Insights from monogenic CSVD syndromes such as CADASIL (NOTCH3 mutations) [82,83] and COL4A1/2-related angiopathies [84] illuminate pure vascular pathways leading to cognitive impairment. Observations from these disorders, where lesion burden predicts cognitive decline independent of classical risk factors [85,86], reinforce the vascular mechanisms highlighted in our meta-analysis.

While this meta-analysis demonstrates robust associations between CSVD markers and cognitive outcomes, it is equally important to assess the certainty of these findings. Using the GRADE framework within NEUROGEN-SVD, the strengths and limitations of the available evidence were systematically evaluated. Certainty of evidence ranged from low to moderate, with the strongest confidence observed for WMHs as predictors of VaD, and lower confidence for associations with GCI and AD due to heterogeneity, diagnostic variability, and potential publication bias (Table 7). The NEUROGEN-SVD GRADE assessment therefore underscores both the strengths and gaps in the current evidence base. The most consistent and reliable signal was the strong association between WMHs and VaD, supported by moderate-certainty evidence, reinforcing WMHs as a key imaging marker of vascular cognitive outcomes. Associations with GCI and AD were downgraded to low certainty, reflecting variation in diagnostic criteria, imaging protocols, and study designs. These findings emphasize two key points: first, the directional consistency of associations across multiple markers suggests a genuine biological link between CSVD pathology and cognitive decline; second, the overall low-to-moderate certainty highlights the urgent need for harmonized definitions, standardized imaging protocols, and longitudinal multimodal studies. Incorporating lesion severity quantification, advanced neuroimaging, and genetic–vascular interaction models within the NEUROGEN-SVD framework will be critical for improving evidence certainty and advancing precision medicine strategies for CSVD-related cognitive impairment. Given the predominantly cross-sectional evidence, future multicenter cohorts with serial imaging and harmonized cognitive batteries are essential to track the transition from MCI to dementia, validate temporal associations, and test whether early multimodal risk scores predict trajectory.

Limitations

This meta-analysis has several limitations. First, pooled estimates could only be derived for WMHs, CMBs, and lacunes; data on other CSVD markers, such as cortical superficial siderosis and perivascular spaces, were insufficient. Second, most studies reported presence/absence rather than severity or burden of lesions, limiting the ability to assess dose–response relationships. Third, the analysis of APOE ε4 was restricted to GCI, as too few studies reported across other cognitive outcomes. Fourth, most included studies were cross-sectional, restricting insights into progression from MCI to dementia. Fifth, it was impossible to evaluate the combined predictive value of imaging and genetic markers simultaneously. Sixth, although limited exploratory subgroup analyses were performed, detailed stratified subgroup analyses and meta-regression could not be undertaken due to insufficient stratified reporting. No data imputation was attempted. Finally, substantial heterogeneity across studies likely influenced the precision and reliability of pooled estimates.

5. Conclusions

The NEUROGEN-SVD study provides one of the most comprehensive evidence syntheses to date linking CSVD imaging and genetic markers with cognitive outcomes. This meta-analysis confirms that WMHs, lacunes, CMBs, and the APOE ε4 allele are significantly associated with both early (MCI and GCI) and late (ACD, VaD, and AD) stages of cognitive impairment, reinforcing the concept of a vascular neurodegenerative continuum. Certainty of evidence ranged from low to moderate, with the strongest confidence in WMHs as a predictor of vascular dementia. These findings highlight the urgent need for harmonized diagnostic frameworks, standardized imaging protocols, and longitudinal multimodal studies within NEUROGEN-SVD and related initiatives. Strengthening the certainty of evidence will be critical to advancing precision strategies for early detection, risk stratification, and prevention of CSVD-related cognitive decline.

Supplementary Materials: The following supporting information can be downloaded at: https://www.mdpi.com/article/10.3390/diagnostics15202585/s1, Table S1: PRISMA-2020 Checklist; Table S2: MOOSE Checklist; Table S3: Modified Jadad Analysis for Methodological Quality; Table S4: Funding Bias Scores for Studies; Table S5: Summary Data and Performance Estimates for Meta-analysis on the Diagnostic Accuracy of CSVD Neuroimaging Markers and APOE ε4 Allele Carrier Status and Cognitive Outcomes; Table S6: Outputs from Egger's Regression Test; Table S7: Outputs from Deek's Funnel-Plot Asymmetry Test; Figure S1: Summary Receiver Operating Characteristic (SROC) Curves for Imaging Markers of CSVD and APOE ε4 Allele Carrier Status (1); Figure S2: Summary Receiver Operating Characteristic (SROC) Curves for Imaging Markers of CSVD and APOE ε4 Allele Carrier Status (2); Figure S3: Sensitivity Analysis on the Association between Imaging and Genetic Markers of CSVD and Cognitive Outcomes; Figure S4: Graphs of Egger's Regression Test; Figure S5: Graphs of Funnel Plots; Figure S6: Graphs of Deek's Funnel-Plot Asymmetry Test; Figure S7: Graphs of Fagan's Normogram. References [87,88] are cited in the Supplementary Materials.

Author Contributions: S.M.M.B. conceived the NEUROGEN-SVD study; supervised the project; and contributed to study planning, drafting, and critical revision. C.J. and S.M.M.B. conducted the literature review, data collection, statistical synthesis, and drafting of the manuscript. R.G.B. contributed to study design discussions, interpretation, validation, and critical revisions. All authors have read and agreed to the published version of the manuscript.

Funding: This work forms part of the Neuroimaging and Genetic Markers in Small Vessel Disease (NEUROGEN-SVD) study. Financial support for a separate project was received through the Grant-in-Aid for Scientific Research (KAKENHI) (PI: S.M.M.B.), funded by the Japan Society for the Promotion of Science (JSPS), Japanese Ministry of Education, Culture, Sports, Science and Technology (MEXT), Japan (grant ID: 23KF0126). S.M.M.B. was awarded the JSPS International Fellowship, supported by MEXT and the Australian Academy of Science, for the period 2023–2025 (Grant ID: P23712). The funding bodies had no role in the study design, data collection, analysis, interpretation, or manuscript preparation.

Institutional Review Board Statement: Not applicable.

Informed Consent Statement: Not applicable.

Data Availability Statement: The original contributions presented in this study are included in the article/Supplementary Material. Further inquiries can be directed to the corresponding author.

Acknowledgments: We acknowledge the Japan Society for the Promotion of Science (JSPS) International Fellowship (grant ID: P23712) and the Grant-in-Aid for Scientific Research (KAKENHI, grant ID: 23KF0126) for financial support.

Conflicts of Interest: S.M.M.B. reports leadership or fiduciary roles (paid or unpaid) with the following organizations: Visiting Director, National Cerebral and Cardiovascular Center (Osaka, Japan, 2023–2025); District Chair for Diversity, Equity, and Inclusion, Rotary District 9675 (Sydney, Australia); Chair, Founding Member, and Manager, Global Health and Migration Hub Community, Global Health Hub Germany (Berlin, Germany); editorial board memberships at *PLOS One*, *BMC Neurology*, *Frontiers in Neurology*, *Frontiers in Stroke*, *Frontiers in Public Health*, *Journal of Aging Research*, *Neurology International*, *Diagnostics*, and *BMC Medical Research Methodology*; Member, College of Reviewers, Canadian Institutes of Health Research (CIHR), Government of Canada; Director of Research, World Headache Society (Bengaluru, India); Member, Scientific Review Committee, Cardiff University Biobank (Cardiff, UK); Chair, Rotary Reconciliation Action Plan (RAP), Rotary District 9675 (NSW, Australia); Healthcare and Medical Adviser, Japan Connect (Osaka, Japan); and Expert Adviser/Reviewer, Cariplo Foundation (Milan, Italy). These roles are unrelated to the submitted work. C.J. and R.G.B. declare no conflicts of interest.

References

1. Pantoni, L. Cerebral small vessel disease: From pathogenesis and clinical characteristics to therapeutic challenges. *Lancet Neurol.* **2010**, *9*, 689–701. [CrossRef]
2. Jin, C.; Bhaskar, S. Unifying Vascular Injury and Neurodegeneration: A Mechanistic Continuum in Cerebral Small Vessel Disease and Dementia. *Eur. J. Neurosci.* **2025**, *62*, e70246. [CrossRef]
3. World Health Organisation. *Dementia*; WHO: Geneva, Switzerland, 2025; Volume 2025.
4. GBD 2019 Dementia Forecasting Collaborators. Estimation of the global prevalence of dementia in 2019 and forecasted prevalence in 2050: An analysis for the Global Burden of Disease Study 2019. *Lancet Public Health* **2022**, *7*, e105–e125. [CrossRef]
5. Wardlaw, J.M.; Smith, C.; Dichgans, M. Small vessel disease: Mechanisms and clinical implications. *Lancet Neurol.* **2019**, *18*, 684–696. [CrossRef]
6. Tan, R.; Spring, K.J.; Killingsworth, M.; Bhaskar, S. Comparative Diagnostic and Prognostic Performance of SWI and T2-Weighted MRI in Cerebral Microbleed Detection Following Acute Ischemic Stroke: A Meta-Analysis and SPOT-CMB Study. *Medicina* **2025**, *61*, 1566. [CrossRef]
7. Rastogi, A.; Weissert, R.; Bhaskar, S.M.M. Emerging role of white matter lesions in cerebrovascular disease. *Eur. J. Neurosci.* **2021**, *54*, 5531–5559. [CrossRef]
8. Tipirneni, S.; Stanwell, P.; Weissert, R.; Bhaskar, S.M.M. Prevalence and Impact of Cerebral Microbleeds on Clinical and Safety Outcomes in Acute Ischaemic Stroke Patients Receiving Reperfusion Therapy: A Systematic Review and Meta-Analysis. *Biomedicines* **2023**, *11*, 2865. [CrossRef]
9. Wardlaw, J.M.; Smith, E.E.; Biessels, G.J.; Cordonnier, C.; Fazekas, F.; Frayne, R.; Lindley, R.I.; O'Brien, J.T.; Barkhof, F.; Benavente, O.R.; et al. Neuroimaging standards for research into small vessel disease and its contribution to ageing and neurodegeneration. *Lancet Neurol.* **2013**, *12*, 822–838. [CrossRef] [PubMed]
10. Di Battista, A.M.; Heinsinger, N.M.; Rebeck, G.W. Alzheimer's Disease Genetic Risk Factor APOE-ε4 Also Affects Normal Brain Function. *Curr. Alzheimer Res.* **2016**, *13*, 1200–1207. [CrossRef]
11. Fongang, B.; Sargurupremraj, M.; Jian, X.; Mishra, A.; Damotte, V.; de Rojas, I.; Skrobot, O.; Bis, J.C.; Fan, K.-H. A genome-wide association meta-analysis of all-cause and vascular dementia. *Alzheimer's Dement.* **2024**, *20*, 5973–5995.
12. Song, X.; Jin, C.; Luan, M.; Zheng, X. Roles of neuroimaging markers and biomarkers in cerebral small vessel disease and their associations with cognitive function. *Front. Neurol.* **2025**, *16*, 1483842. [CrossRef]
13. Nicoll, J.A.; Savva, G.M.; Stewart, J.; Matthews, F.E.; Brayne, C.; Ince, P.; Medical Research Council Cognitive Function and Ageing Study. Association between APOE genotype, neuropathology and dementia in the older population of England and Wales. *Neuropathol. Appl. Neurobiol.* **2011**, *37*, 285–294. [CrossRef] [PubMed]
14. Paradela, R.S.; Justo, A.F.O.; Paes, V.R.; Leite, R.E.P.; Pasqualucci, C.A.; Grinberg, L.T.; Naslavsky, M.S.; Zatz, M.; Nitrini, R.; Jacob-Filho, W.; et al. Association between APOE-ε4 allele and cognitive function is mediated by Alzheimer's disease pathology: A population-based autopsy study in an admixed sample. *Acta Neuropathol. Commun.* **2023**, *11*, 205. [CrossRef]
15. Mahley, R.W. Apolipoprotein E: From cardiovascular disease to neurodegenerative disorders. *J. Mol. Med.* **2016**, *94*, 739–746. [CrossRef]
16. Duering, M.; Biessels, G.J.; Brodtmann, A.; Chen, C.; Cordonnier, C.; de Leeuw, F.E.; Debette, S.; Frayne, R.; Jouvent, E.; Rost, N.S.; et al. Neuroimaging standards for research into small vessel disease-advances since 2013. *Lancet Neurol.* **2023**, *22*, 602–618. [CrossRef]
17. Fazekas, F.; Chawluk, J.B.; Alavi, A.; Hurtig, H.I.; Zimmerman, R.A. MR signal abnormalities at 1.5 T in Alzheimer's dementia and normal aging. *AJR Am. J. Roentgenol.* **1987**, *149*, 351–356. [CrossRef]

18. Jadad, A.R.; Moore, R.A.; Carroll, D.; Jenkinson, C.; Reynolds, D.J.; Gavaghan, D.J.; McQuay, H.J. Assessing the quality of reports of randomized clinical trials: Is blinding necessary? *Control. Clin. Trials* **1996**, *17*, 1–12. [CrossRef]
19. Dobrynina, L.A.; Kremneva, E.I.; Shamtieva, K.V.; Geints, A.A.; Filatov, A.S.; Gadzhieva, Z.S.; Gnedovskaya, E.V.; Krotenkova, M.V.; Maximov, I.I. Cognitive Impairment in Cerebral Small Vessel Disease Is Associated with Corpus Callosum Microstructure Changes Based on Diffusion MRI. *Diagnostics* **2024**, *14*, 1838. [CrossRef] [PubMed]
20. Ferro, D.A.; van Veluw, S.J.; Koek, H.L.; Exalto, L.G.; Biessels, G.J.; Utrecht Vascular Cognitive Impairment (VCI) Study Group. Cortical Cerebral Microinfarcts on 3 Tesla MRI in Patients with Vascular Cognitive Impairment. *J. Alzheimer's Dis.* **2017**, *60*, 1443–1450. [CrossRef]
21. Han, X.; Wang, Y.; Chen, Y.; Qiu, Y.; Gu, X.; Dai, Y.; Xu, Q.; Sun, Y.; Zhou, Y. Predicting white-matter hyperintensity progression and cognitive decline in patients with cerebral small-vessel disease: A magnetic resonance-based habitat analysis. *Quant. Imaging Med. Surg.* **2024**, *14*, 6621–6634. [CrossRef] [PubMed]
22. Ke, Z.; Mo, Y.; Li, J.; Yang, D.; Huang, L.; Yang, Z.; Qin, R.; Mao, C.; Lv, W.; Huang, Y.; et al. Glymphatic Dysfunction Mediates the Influence of White Matter Hyperintensities on Episodic Memory in Cerebral Small Vessel Disease. *Brain Sci.* **2022**, *12*, 1611. [CrossRef]
23. Lee, J.; Seo, S.W.; Yang, J.J.; Jang, Y.K.; Lee, J.S.; Kim, Y.J.; Chin, J.; Lee, J.M.; Kim, S.T.; Lee, K.H.; et al. Longitudinal cortical thinning and cognitive decline in patients with early- versus late-stage subcortical vascular mild cognitive impairment. *Eur. J. Neurol.* **2018**, *25*, 326–333. [CrossRef]
24. Liao, M.; Wang, M.; Li, H.; Li, J.; Yi, M.; Lan, L.; Ouyang, F.; Shi, L.; Fan, Y. Discontinuity of deep medullary veins in SWI is associated with deep white matter hyperintensity volume and cognitive impairment in cerebral small vessel disease. *J. Affect. Disord.* **2024**, *350*, 600–607. [CrossRef] [PubMed]
25. Liu, J.; Zhao, W.; Gui, Q.; Zhang, Y.; Guo, Z.; Liu, W. Addition of Abeta(42) to Total Cerebral Small Vessel Disease Score Improves the Prediction for Cognitive Impairment in Cerebral Small Vessel Disease Patients. *Neuropsychiatr. Dis. Treat.* **2021**, *17*, 195–201. [CrossRef]
26. Song, H.; Ruan, Z.; Gao, L.; Lv, D.; Sun, D.; Li, Z.; Zhang, R.; Zhou, X.; Xu, H.; Zhang, J. Structural network efficiency mediates the association between glymphatic function and cognition in mild VCI: A DTI-ALPS study. *Front. Aging Neurosci.* **2022**, *14*, 974114. [CrossRef]
27. Sun, W.; Huang, L.; Cheng, Y.; Qin, R.; Xu, H.; Shao, P.; Ma, J.; Yao, Z.; Shi, L.; Xu, Y. Medial Temporal Atrophy Contributes to Cognitive Impairment in Cerebral Small Vessel Disease. *Front. Neurol.* **2022**, *13*, 858171. [CrossRef] [PubMed]
28. Tang, J.; Zhang, M.; Liu, N.; Xue, Y.; Ren, X.; Huang, Q.; Shi, L.; Fu, J. The Association Between Glymphatic System Dysfunction and Cognitive Impairment in Cerebral Small Vessel Disease. *Front. Aging Neurosci.* **2022**, *14*, 916633. [CrossRef]
29. Wang, X.; Shi, Y.; Chen, Y.; Gao, Y.; Wang, T.; Li, Z.; Wang, Y. Blood-Brain Barrier Breakdown is a Sensitive Biomarker of Cognitive and Language Impairment in Patients with White Matter Hyperintensities. *Neurol. Ther.* **2023**, *12*, 1745–1758. [CrossRef] [PubMed]
30. Wei, N.; Deng, Y.; Yao, L.; Jia, W.; Wang, J.; Shi, Q.; Chen, H.; Pan, Y.; Yan, H.; Zhang, Y.; et al. A neuroimaging marker based on diffusion tensor imaging and cognitive impairment due to cerebral white matter lesions. *Front. Neurol.* **2019**, *10*, 81. [CrossRef]
31. Xing, Y.; Yang, J.; Zhou, A.; Wang, F.; Wei, C.; Tang, Y.; Jia, J. White Matter Fractional Anisotropy Is a Superior Predictor for Cognitive Impairment Than Brain Volumes in Older Adults with Confluent White Matter Hyperintensities. *Front. Psychiatry* **2021**, *12*, 633811. [CrossRef]
32. Xu, J.; Zhong, H.; Wu, L. Magnetic resonance diffusion tensor imaging for detecting the cerebral microstructure changes in patients with CSVD-induced mild cognitive impairment. *J. Neurophysiol.* **2024**, *132*, 1937–1942. [CrossRef] [PubMed]
33. Zhu, F.; Yao, J.; Feng, M.; Sun, Z. Establishment and evaluation of a clinical prediction model for cognitive impairment in patients with cerebral small vessel disease. *BMC Neurosci.* **2024**, *25*, 35. [CrossRef]
34. Zhu, W.; Huang, H.; Yang, S.; Luo, X.; Zhu, W.; Xu, S.; Meng, Q.; Zuo, C.; Liu, Y.; Wang, W.; et al. Cortical and Subcortical Grey Matter Abnormalities in White Matter Hyperintensities and Subsequent Cognitive Impairment. *Neurosci. Bull.* **2021**, *37*, 789–803. [CrossRef]
35. Chen, Y.K.; Xiao, W.M.; Li, W.; Ni, Z.X.; Liu, Y.L.; Xu, L.; Qu, J.F.; Ng, C.; Xiang, Y.T. Microbleeds in fronto-subcortical circuits are predictive of dementia conversion in patients with vascular cognitive impairment but no dementia. *Neural Regen. Res.* **2018**, *13*, 1913–1918. [CrossRef]
36. Ding, J.; Sigurethsson, S.; Jonsson, P.V.; Eiriksdottir, G.; Meirelles, O.; Kjartansson, O.; Lopez, O.L.; van Buchem, M.A.; Gudnason, V.; Launer, L.J. Space and location of cerebral microbleeds, cognitive decline, and dementia in the community. *Neurology* **2017**, *88*, 2089–2097. [CrossRef]
37. Fan, Y.; Xu, Y.; Shen, M.; Guo, H.; Zhang, Z. Total Cerebral Small Vessel Disease Burden on MRI Correlates with Cognitive Impairment in Outpatients with Amnestic Disorders. *Front. Neurol.* **2021**, *12*, 747115. [CrossRef]
38. Hilal, S.; Xin, X.; Ang, S.L.; Tan, C.S.; Venketasubramanian, N.; Niessen, W.J.; Vrooman, H.; Wong, T.Y.; Chen, C.; Ikram, M.K. Risk Factors and Consequences of Cortical Thickness in an Asian Population. *Medicine* **2015**, *94*, e852. [CrossRef]

39. Jacob, M.A.; Cai, M.; van de Donk, V.; Bergkamp, M.; Marques, J.; Norris, D.G.; Kessels, R.P.C.; Claassen, J.; Duering, M.; Tuladhar, A.M.; et al. Cerebral Small Vessel Disease Progression and the Risk of Dementia: A 14-Year Follow-Up Study. *Am. J. Psychiatry* **2023**, *180*, 508–518. [CrossRef]
40. Li, X.; Yuan, J.; Qin, W.; Yang, L.; Yang, S.; Li, Y.; Hu, W. Higher Total Cerebral Small Vessel Disease Burden Was Associated with Mild Cognitive Impairment and Overall Cognitive Dysfunction: A Propensity Score-Matched Case-Control Study. *Front. Aging Neurosci.* **2021**, *13*, 695732. [CrossRef] [PubMed]
41. Li, L.; Wu, D.H.; Li, H.Q.; Tan, L.; Xu, W.; Dong, Q.; Tan, L.; Yu, J.T.; Alzheimer's Disease Neuroimaging Initiative. Association of Cerebral Microbleeds with Cognitive Decline: A Longitudinal Study. *J. Alzheimer's Dis.* **2020**, *75*, 571–579. [CrossRef] [PubMed]
42. Paradise, M.; Seruga, A.; Crawford, J.D.; Chaganti, J.; Thalamuthu, A.; Kochan, N.A.; Brodaty, H.; Wen, W.; Sachdev, P.S. The relationship of cerebral microbleeds to cognition and incident dementia in non-demented older individuals. *Brain Imaging Behav.* **2019**, *13*, 750–761. [CrossRef]
43. Romero, J.R.; Beiser, A.; Himali, J.J.; Shoamanesh, A.; DeCarli, C.; Seshadri, S. Cerebral microbleeds and risk of incident dementia: The Framingham Heart Study. *Neurobiol. Aging* **2017**, *54*, 94–99. [CrossRef]
44. Shaikh, I.; Beaulieu, C.; Gee, M.; McCreary, C.R.; Beaudin, A.E.; Valdes-Cabrera, D.; Smith, E.E.; Camicioli, R. Diffusion tensor tractography of the fornix in cerebral amyloid angiopathy, mild cognitive impairment and Alzheimer's disease. *Neuroimage Clin.* **2022**, *34*, 103002. [CrossRef]
45. Uetani, H.; Hirai, T.; Hashimoto, M.; Ikeda, M.; Kitajima, M.; Sakamoto, F.; Utsunomiya, D.; Oda, S.; Sugiyama, S.; Matsubara, J.; et al. Prevalence and topography of small hypointense foci suggesting microbleeds on 3T susceptibility-weighted imaging in various types of dementia. *AJNR Am. J. Neuroradiol.* **2013**, *34*, 984–989. [CrossRef]
46. Wrigley, S.; Cody, R.; Amadoru, S.; Huynh, A.; Galante, O.; Mandrawa, C.; Yassi, N.; Yates, P. Prevalence and associations of cerebral microbleeds in an Australian Memory clinic cohort. *Intern. Med. J.* **2025**, *55*, 608–616. [CrossRef] [PubMed]
47. Zonneveld, H.I.; Goos, J.D.; Wattjes, M.P.; Prins, N.D.; Scheltens, P.; van der Flier, W.M.; Kuijer, J.P.; Muller, M.; Barkhof, F. Prevalence of cortical superficial siderosis in a memory clinic population. *Neurology* **2014**, *82*, 698–704. [CrossRef] [PubMed]
48. Rennie, A.; Ekman, U.; Shams, S.; Ryden, L.; Samuelsson, J.; Zettergren, A.; Kern, S.; Oppedal, K.; Blanc, F.; Hort, J.; et al. Cerebrovascular and Alzheimer's disease biomarkers in dementia with Lewy bodies and other dementias. *Brain Commun.* **2024**, *6*, fcae290. [CrossRef] [PubMed]
49. Hilal, S.; Liu, S.; Wong, T.Y.; Vrooman, H.; Cheng, C.Y.; Venketasubramanian, N.; Chen, C.L.H.; Zhou, J.H. White matter network damage mediates association between cerebrovascular disease and cognition. *J. Cereb. Blood Flow Metab.* **2021**, *41*, 1858–1872. [CrossRef]
50. Wang, J.; Zhou, Y.; He, Y.; Li, Q.; Zhang, W.; Luo, Z.; Xue, R.; Lou, M. Impact of different white matter hyperintensities patterns on cognition: A cross-sectional and longitudinal study. *Neuroimage Clin.* **2022**, *34*, 102978. [CrossRef]
51. Wang, J.; Han, X.; Li, Y.; Fa, W.; Zhao, M.; Li, C.; Mao, M.; Hou, T.; Wang, Y.; Cong, L.; et al. Strategic Lacunes Associated with Mild Cognitive Impairment in Rural Chinese Older Adults: A Population-Based Study. *Stroke* **2024**, *55*, 1288–1298. [CrossRef]
52. Brickman, A.M.; Schupf, N.; Manly, J.J.; Stern, Y.; Luchsinger, J.A.; Provenzano, F.A.; Narkhede, A.; Razlighi, Q.; Collins-Praino, L.; Artero, S.; et al. APOE epsilon4 and risk for Alzheimer's disease: Do regionally distributed white matter hyperintensities play a role? *Alzheimer's Dement.* **2014**, *10*, 619–629. [CrossRef] [PubMed]
53. Hong, Y.J.; Yoon, B.; Shim, Y.S.; Cho, A.H.; Shin, H.E.; Kim, Y.I.; Kim, S.Y.; Yang, D.W. APOE epsilon4 allele status in korean dementia patients with severe white matter hyperintensities. *J. Alzheimer's Dis.* **2011**, *24*, 519–524. [CrossRef]
54. Kim, H.J.; Ye, B.S.; Yoon, C.W.; Cho, H.; Noh, Y.; Kim, G.H.; Choi, Y.S.; Kim, J.H.; Jeon, S.; Lee, J.M.; et al. Effects of APOE ε4 on brain amyloid, lacunar infarcts, and white matter lesions: Astudy among patients with subcortical vascular cognitive impairment. *Neurobiol. Aging* **2013**, *34*, 2482–2487. [CrossRef]
55. Yu, M.C.; Chuang, Y.F.; Wu, S.C.; Ho, C.F.; Liu, Y.C.; Chou, C.J. White matter hyperintensities in cholinergic pathways are associated with dementia severity in e4 carriers but not in non-carriers. *Front. Neurol.* **2023**, *14*, 1100322. [CrossRef]
56. Wan, X.; Wang, W.; Liu, J.; Tong, T. Estimating the sample mean and standard deviation from the sample size, median, range and/or interquartile range. *BMC Med. Res. Methodol.* **2014**, *14*, 135. [CrossRef]
57. Elahi, F.M.; Wang, M.A.-O.; Meschia, J.F. Cerebral Small Vessel Disease-Related Dementia: More Questions Than Answers. *Stroke* **2023**, *54*, 648–660. [CrossRef] [PubMed]
58. Peng, D.; Geriatric Neurology Group; Chinese Society of Geriatrics; Clinical Practice Guideline for Cognitive Impairment of Cerebral Small Vessel Disease Writing Group. Clinical practice guideline for cognitive impairment of cerebral small vessel disease. *Aging Med.* **2019**, *2*, 64–73. [CrossRef]
59. Ter Telgte, A.; Duering, M. Cerebral Small Vessel Disease: Advancing Knowledge with Neuroimaging. *Stroke* **2024**, *55*, 1686–1688. [CrossRef]
60. Jansma, A.; de Bresser, J.; Schoones, J.W.; van Heemst, D.; Akintola, A.A. Sporadic cerebral small vessel disease and cognitive decline in healthy older adults: A systematic review and meta-analysis. *J. Cereb. Blood Flow Metab.* **2024**, *44*, 660–679. [CrossRef] [PubMed]

61. Debette, S.; Schilling, S.; Duperron, M.G.; Larsson, S.C.; Markus, H.S. Clinical Significance of Magnetic Resonance Imaging Markers of Vascular Brain Injury: A Systematic Review and Meta-analysis. *JAMA Neurol.* **2019**, *76*, 81–94. [CrossRef]
62. Rensma, S.P.; van Sloten, T.T.; Launer, L.J.; Stehouwer, C.D.A. Cerebral small vessel disease and risk of incident stroke, dementia and depression, and all-cause mortality: A systematic review and meta-analysis. *Neurosci. Biobehav. Rev.* **2018**, *90*, 164–173. [CrossRef] [PubMed]
63. Bos, D.; Wolters, F.J.; Darweesh, S.K.L.; Vernooij, M.W.; de Wolf, F.; Ikram, M.A.; Hofman, A. Cerebral small vessel disease and the risk of dementia: A systematic review and meta-analysis of population-based evidence. *Alzheimer's Dement.* **2018**, *14*, 1482–1492. [CrossRef]
64. Dichgans, M.; Leys, D. Vascular Cognitive Impairment. *Circ. Res.* **2017**, *120*, 573–591. [CrossRef]
65. Hu, H.Y.; Ou, Y.N.; Shen, X.N.; Qu, Y.; Ma, Y.H.; Wang, Z.T.; Dong, Q.; Tan, L.; Yu, J.T. White matter hyperintensities and risks of cognitive impairment and dementia: A systematic review and meta-analysis of 36 prospective studies. *Neurosci. Biobehav. Rev.* **2021**, *120*, 16–27. [CrossRef]
66. Van Veluw, S.J.; Biessels, G.J.; Klijn, C.J.; Rozemuller, A.J. Heterogeneous histopathology of cortical microbleeds in cerebral amyloid angiopathy. *Neurology* **2016**, *86*, 867–871. [CrossRef]
67. Arnaud, L.; Benech, P.; Greetham, L.; Stephan, D.; Jimenez, A.; Jullien, N.; García-González, L.; Tsvetkov, P.O.; Devred, F.; Sancho-Martinez, I.; et al. APOE4 drives inflammation in human astrocytes via TAGLN3 repression and NF-κB activation. *Cell Rep.* **2022**, *40*, 111200. [CrossRef]
68. Ahmad, S.; Yang, W.; Orellana, A.; Frölich, L.; de Rojas, I.; Cano, A.; Boada, M.; Hernández, I.; Hausner, L.; Harms, A.C.; et al. Association of oxidative stress and inflammatory metabolites with Alzheimer's disease cerebrospinal fluid biomarkers in mild cognitive impairment. *Alzheimer's Res. Ther.* **2024**, *16*, 171. [CrossRef]
69. Montagne, A.; Nikolakopoulou, A.M.; Huuskonen, M.T.; Sagare, A.P.; Lawson, E.J.; Lazic, D.; Rege, S.V.; Grond, A.; Zuniga, E.; Barnes, S.R.; et al. APOE4 accelerates advanced-stage vascular and neurodegenerative disorder in old Alzheimer's mice via cyclophilin A independently of amyloid-β. *Nat. Aging* **2021**, *1*, 506–520. [CrossRef]
70. Love, S.; Miners, J.S. White matter hypoperfusion and damage in dementia: Post-mortem assessment. *Brain Pathol.* **2015**, *25*, 99–107. [CrossRef] [PubMed]
71. Graff-Radford, J.; Arenaza-Urquijo, E.M.; Knopman, D.S.; Schwarz, C.G.; Brown, R.D., Jr.; Rabinstein, A.A.; Gunter, J.L.; Senjem, M.L.; Przybelski, S.A.; Lesnick, T.; et al. White matter hyperintensities: Relationship to amyloid and tau burden. *Brain* **2019**, *142*, 2483–2491. [CrossRef] [PubMed]
72. Garnier-Crussard, A.; Bougacha, S.; Wirth, M.; Dautricourt, S.; Sherif, S.; Landeau, B.; Gonneaud, J.; De Flores, R.; de la Sayette, V.; Vivien, D.; et al. White matter hyperintensity topography in Alzheimer's disease and links to cognition. *Alzheimer's Dement.* **2022**, *18*, 422–433. [CrossRef]
73. Zare, A.; Salehi, S.; Bader, J.M.; Wiessler, A.L.; Prokesch, M.; Albrecht, V.; Villmann, C.; Mann, M.; Briese, M.; Sendtner, M. Axonal Tau Reduction Ameliorates Tau and Amyloid Pathology in a Mouse Model of Alzheimer's Disease. *Transl. Neurodegener.* **2025**, *14*, 39. [CrossRef]
74. Pescoller, J.; Dewenter, A.; Dehsarvi, A.; Steward, A.; Frontzkowski, L.; Zhu, Z.; Roemer-Cassiano, S.N.; Palleis, C.; Hirsch, F.; Wagner, F.; et al. Cortical tau deposition promotes atrophy in connected white matter regions in Alzheimer's disease. *Brain* **2025**, awaf339. [CrossRef] [PubMed]
75. Nasrabady, S.E.; Rizvi, B.; Goldman, J.E.; Brickman, A.M. White matter changes in Alzheimer's disease: A focus on myelin and oligodendrocytes. *Acta Neuropathol. Commun.* **2018**, *6*, 22. [CrossRef]
76. He, Z.; Wang, G.; Wu, J.; Tang, Z.; Luo, M. The molecular mechanism of LRP1 in physiological vascular homeostasis and signal transduction pathways. *Biomed. Pharmacother.* **2021**, *139*, 111667. [CrossRef]
77. Yang, W.; Wei, Z.; Wang, T. Unraveling the Role of LRP1 in Alzheimer's Disease: A Focus on Aβ Clearance and the Liver-Brain Axis. *J. Mol. Neurosci.* **2025**, *75*, 43. [CrossRef]
78. Uchida, Y.; Kano, Y.; Kan, H.; Sakurai, K.; Morita, H.; Akagawa, Y.; Matsukawa, N.; Oishi, K. Blood-brain barrier water exchange and paramagnetic susceptibility alterations during anti-amyloid therapy: Preliminary MRI findings. *J. Prev. Alzheimer's Dis.* **2025**, *12*, 100256. [CrossRef]
79. Van den Brink, H.; Doubal, F.N.; Duering, M. Advanced MRI in cerebral small vessel disease. *Int. J. Stroke* **2023**, *18*, 28–35. [CrossRef] [PubMed]
80. Cuadrado-Godia, E.; Dwivedi, P.; Sharma, S.; Ois Santiago, A.; Roquer Gonzalez, J.; Balcells, M.; Laird, J.; Turk, M.; Suri, H.S.; Nicolaides, A.; et al. Cerebral Small Vessel Disease: A Review Focusing on Pathophysiology, Biomarkers, and Machine Learning Strategies. *J. Stroke* **2018**, *20*, 302–320. [CrossRef] [PubMed]
81. Nyúl-Tóth, Á.; Patai, R.; Csiszar, A.; Ungvari, A.; Gulej, R.; Mukli, P.; Yabluchanskiy, A.; Benyo, Z.; Sotonyi, P.; Prodan, C.I.; et al. Linking peripheral atherosclerosis to blood-brain barrier disruption: Elucidating its role as a manifestation of cerebral small vessel disease in vascular cognitive impairment. *Geroscience* **2024**, *46*, 6511–6536. [CrossRef]

82. Heidari, P.; Taghizadeh, M.; Vakili, O. Signaling pathways and molecular mechanisms involved in the onset and progression of cerebral autosomal dominant arteriopathy with subcortical infarcts and leukoencephalopathy (CADASIL); a focus on Notch3 signaling. *J. Headache Pain* **2025**, *26*, 96. [CrossRef]
83. Uchida, Y.; Kan, H.; Sakurai, K.; Arai, N.; Inui, S.; Kobayashi, S.; Kato, D.; Ueki, Y.; Matsukawa, N. Iron leakage owing to blood-brain barrier disruption in small vessel disease CADASIL. *Neurology* **2020**, *95*, e1188–e1198. [CrossRef]
84. Guey, S.; Hervé, D. Main features of COL4A1-COL4A2 related cerebral microangiopathies. *Cereb. Circ. Cogn. Behav.* **2022**, *3*, 100140. [CrossRef]
85. Jolly, A.A.; Nannoni, S.; Edwards, H.; Morris, R.G.; Markus, H.S. Prevalence and Predictors of Vascular Cognitive Impairment in Patients with CADASIL. *Neurology* **2022**, *99*, e453–e461. [CrossRef]
86. Xiao, X.; Guo, L.; Liao, X.; Zhou, Y.; Zhang, W.; Zhou, L.; Wang, X.; Liu, X.; Liu, H.; Xu, T.; et al. The role of vascular dementia associated genes in patients with Alzheimer's disease: A large case-control study in the Chinese population. *CNS Neurosci. Ther.* **2021**, *27*, 1531–1539. [CrossRef] [PubMed]
87. Page, M.J.; McKenzie, J.E.; Bossuyt, P.M.; Boutron, I.; Hoffmann, T.C.; Mulrow, C.D.; Shamseer, L.; Tetzlaff, J.M.; Akl, E.A.; Brennan, S.E.; et al. The PRISMA 2020 statement: An updated guideline for reporting systematic reviews. *BMJ* **2021**, *372*, n71. [CrossRef] [PubMed]
88. Stroup, D.F.; Berlin, J.A.; Morton, S.C.; Olkin, I.; Williamson, G.D.; Rennie, D.; Moher, D.; Becker, B.J.; Sipe, T.A.; Thacker, S.B.; et al. Meta-analysis of Observational Studies in Epidemiology. A Proposal for Reporting. *JAMA* **2000**, *283*, 2008–2012. [CrossRef] [PubMed]

MDPI AG
Grosspeteranlage 5
4052 Basel
Switzerland
Tel.: +41 61 683 77 34

Diagnostics Editorial Office
E-mail: diagnostics@mdpi.com
www.mdpi.com/journal/diagnostics

www.ingramcontent.com/pod-product-compliance
Lightning Source LLC
Chambersburg PA
CBHW040818250926
51850CB00007BA/1204
* 9 7 8 3 7 2 5 8 5 8 7 7 4 *